OPTICAL PROPERTIES OF NANO STRUCTURES

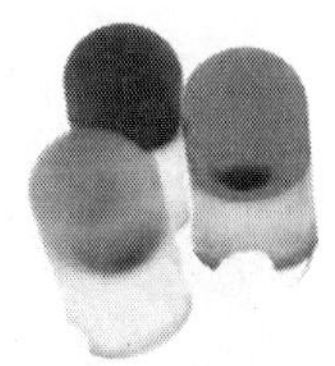

OPTICAL PROPERTIES OF NANO STRUCTURES

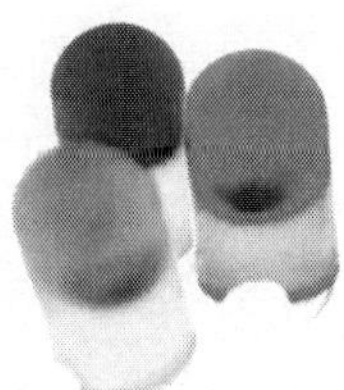

YING FU
MIN QIU

PAN STANFORD PUBLISHING

Published by

Pan Stanford Publishing Pte. Ltd.
Penthouse Level, Suntec Tower 3
8 Temasek Boulevard
Singapore 038988

Email: editorial@panstanford.com
Web: www.panstanford.com

British Library Cataloguing-in-Publication Data
A catalogue record for this book is available from the British Library.

OPTICAL PROPERTIES OF NANOSTRUCTURES

ISBN 978-981-4303-26-2 (Hardcover)
ISBN 978-981-4303-27-9 (eBook)

Printed in the USA

Contents

List of Figures

List of Tables

Preface

The history of technology development is epitomized in Moore's law. Industrial deep-submicron and laboratorial nanometer process technologies have already been fabricating electronic and optical components containing only a few active electrons, and the geometrical sizes of these components are comparable with the characteristic wavelength of the electrons. However, the advanced multimedia infrastructure and service in the future demand further developments in the chip's capability.

Photonic integrated circuits (PICs) are currently orders of magnitude larger in physical dimensions than their microelectronic counterparts. Field-effect-type transistors have reached lengths on the order of 50 nm, while in contrast, passive optical devices, also those based on photonic crystals, have sizes on the order of one wavelength. The sizes of active devices are even larger, essentially depending on the matrix element of the interaction. In order to pursue the steady increase in integration density in photonics such that it rivals the microelectronic footprint size, nanostructure-based high index of refraction and metallic behaviour (negative epsilon) are two mostly studied fundamental issues to shrink optical component sizes and to tackle the sub-wavelength limit.

Nanotechnology has been named as one of the most important areas of forthcoming technology because they promise to form the basis of future generations of electronic and optoelectronic devices. From the point view of technical physics, all these developments greatly reduce the geometric sizes of devices, and thus the number of active electrons in the system. Quantum mechanical considerations about electronic states, electron transports and various scattering processes including light-matter interaction, are thus crucial. However, theoretical study is extremely difficult. Our first numerical simulation work about a three-dimensional energy band structure calculation in 1995 took more than 6 months to complete for one bias-configuration of a nanoscale metal-oxide-semiconductor field-effect transistor (MOSFET). With today's computation workstations the CPU time is reduced to be less than 24 hours.

The main purpose of the book is to discuss electrons and photons in and through nanostructures by the first-principles quantum mechanical theories and fundamental concepts (a unified coverage of nanostructured electronic and optical components) behind nano-electronics and optoelectronics, the material basis, physical phenomena, device physics, as well as designs and applications. The combination of viewpoints presented within the book can help to foster further research and cross-disciplinary interaction needed to surmount the barriers facing future generations of technology design.

In the first part of the book we discuss the concepts of excitons and excitonic polaritons, their nonlinear optical properties in nanostructures and their applications within integrated electronics and optoelectronics. Various theoretical aspects of excitons and excitonic polaritons are introduced followed by a resume of their experimental and application-specific development in nanostructures at the electronic and photonic engineering levels. A number of technical applications are highlighted.

In the second part, we discuss the most important discoveries in the field of surface plasmon polariton (SPP) which has been an important candidate to circumvent the refraction limit. The technique will not only allow us to use light in subwavelength regime, it will also give us more control by manipulating the nanostructures and their assembly. Moreover, it will make it possible to use two-dimensional plasmonics together with conventional electronic devices, as SPP needs metals to be generated. The theoretical aspect about the plasmonics is reviewed. Further developments to include plasmon formation in nanoparticle and thin film, coupling between plasmon and electromagnetic field at the most fundamental level of starting from Maxwell and Schrödinger equations and at the established microwave engineering level are discussed for application-specific plasmonics devices.

Chapter 1

Electrons in Nanostructures

Semiconductors are materials that have moderately good conductivity, which is higher than that of insulators and lower than that of metals. The conductivity of sufficiently pure semiconductors decays by orders of magnitude when they are cooled down from room temperature to liquid helium temperature (at absolute zero temperature, the conductivity almost vanishes). A semiconductor in a very pure state resembles an insulator, whereas in a highly polluted state, it acts like a metal. Furthermore, irradiation with light can transform the semiconductor from insulator-like behavior to metal-like. The optical absorption spectra of semiconductors normally exhibit a threshold. Below the threshold frequency, light can pass through with practically no losses, whereas above it, the light is strongly absorbed.

All these macroscopic properties of a semiconductor can be traced back to a common microscopic origin: its energy band structure and the electron distribution in the energy bands. The energy band structure of a semiconductor consists of energy bands separated by bandgaps. At absolute zero temperature, a pure semiconductor is characterized by having only completely occupied and completely empty energy bands. It is this common microscopic feature that underlies the totality of macroscopic material properties that uniquely define a semiconductor.

The first reference to a characteristic semiconductor property dates back to Faraday, who in 1833 observed an increase of the electric conductivity of silver sulfide with increasing temperature. The term "semiconductor" was introduced in 1911 by Königsberg and Weiss subsequent to a similar term used in a similar context employed by Ebert in 1789 and Bromme in 1851.

In 1874, Braun discovered that contacts between certain metal sulfides and metal tips exhibited different electrical resistance upon reversal of the polarity of the applied voltage. Such point contact structures were used as rectifiers in radio receivers at the beginning of the 20th century. Similar rectifying action was also found for selenium and copper oxide. Copper oxide was used in 1926 by Grondahle as a rectifier, followed by rectifiers using selenium. The first practical application of copper oxide in photocells was accomplished in 1932 by Lang.

The decisive events for the entire future development of semiconductor physics were the invention of the germanium-based bipolar transistor in 1949 and the realization of the field-effect transistor, with the help of silicon at the end of the 1950s. With the introduction of silicon, the development of semiconductor microelectronics began. Later, a similar role was played by compounds involving elements of III–V groups in the periodic table, such as GaAs for the development of semiconductor optoelectronics.

Today's advanced information technology is mainly attributed to the electronic representation and processing of information in a low-cost, high-speed, very compact, and highly reliable fashion. The quest for and accomplishment of continual miniaturization and integration of solid-state electronics have been the key to the success of the computer industry and computer applications.

As the number of transistors integrated in a circuit continues to increase, discrete device dimensions have begun to reach the nanometer regime (10^{-9} m). Such a down-scaling progress of individual device components has been tremendous over the past 30 years: a 1.0 μm gate length metal-oxide-semiconductor field-effect transistor (MOSFET) was reported in 1974 by Dennard *et al.*;[1] 0.1 μm gate length by Sai-Halasz *et al.*[2] in 1987, 70 nm gate length by Hashimoto *et al.*[3] in 1992, 40 nm gate length by Ono *et al.*[4] in 1995, and 30 nm gate length in 1998 by Kawaura *et al.*[5] Vertical p-MOS transistors also have been extensively investigated, and transistors with channel lengths of 130 and 100 nm based on Si as well as GaAs have been fabricated.[6–11]

However, the advanced multimedia infrastructure and service in the future demand further reduction in the chip size. Chip density, represented by memory technology, has followed Moore's law and roughly doubled every other year over the past three decades. The trend remains strong and definite. For example, a 0.15 μm process technology was implemented in the first 4 Gb dynamic random access memory (DRAM) in 1997 and the feature size of DRAM transistors is projected to be 0.18 μm (1 Gb) in 2001, 0.13 μm (4 Gb) in 2004, 0.10 μm (16 Gb) in 2007, and 0.07 μm (64 Gb) in 2010.[12, 13] Nano scale is just a common word in today's daily life such as TSMC's 40 nm process in Radeon HD 5770 Video Card.

When the size of a system becomes comparable to the electron wavelength, quantum effects become dominant.[14] This occurs when transistors are down-scaled and their characteristic dimensions reach the nanometer regime, leading to various new phenomena, for example, electron interference,[15] additive parallel conductance in the absence of magnetic field,[16–18] conductance oscillation[19] and abrupt period changes of conductance oscillation with applied magnetic field,[20, 21] as well as novel electronic and optoelectronic devices such as resonant tunneling diodes[22] and quantum well infrared photodetectors,[23] based on quantum mechanisms.

For nanoelectronics to become a reality, it is essential that new devices and circuits be fabricated with nanometer precision, and that devices and circuits can be designed accurately. Nanofabrication technology[24, 25] has made impressive advances during the last decade by producing artificial semiconductor structures using molecular-beam epitaxy (MBE), metal-organic chemical vapor deposition

(MOCVD), and chemical-beam epitaxy. Accurately controlled feature sizes as small as monolayers of atoms in the growth direction for dissimilar semiconductor materials, or heterostructure systems, have been achieved. Nanoscale lithography and patterning by electron-beam lithography have also been highly developed in the direction perpendicular to the growth direction. Soft lithography[26] can be used to make devices smaller than 100 nm and can pattern curved surfaces and functional materials other than photoresists. It can also fabricate three-dimensional structures and chemically modify surfaces.

The advances in nanofabrication technology have brought quantum effect device concepts to reality and presented a great challenge for device physicists in the theoretical analysis of nanoelectronic devices.[27] In this chapter, we present the physical models about electrons in semiconductor nanostructures.

1.1 ATOMS AND SOLIDS

In 1913, Niels Bohr presented a model of the hydrogen atom, which has one electron. Bohr stated two postulates.

1. *The electron moves only in certain circular orbits, called stationary states.*

Figure. 1.1 shows an electron of mass m_0 and charge $-e$, moving at speed v in a stable circular orbit of radius r, around a nucleus of charge $+e$. The centripetal force is provided by the Coulomb attraction between the electron and the nucleus. From Newton's second law we have

$$\frac{m_0 v^2}{r} = \frac{e^2}{4\pi\epsilon_0 r^2} \tag{1.1}$$

and then the total mechanical energy of the electron is

$$E = \frac{1}{2}mv^2 - \frac{e}{4\pi\epsilon_0 r} = -\frac{e^2}{8\pi\epsilon_0 r} \tag{1.2}$$

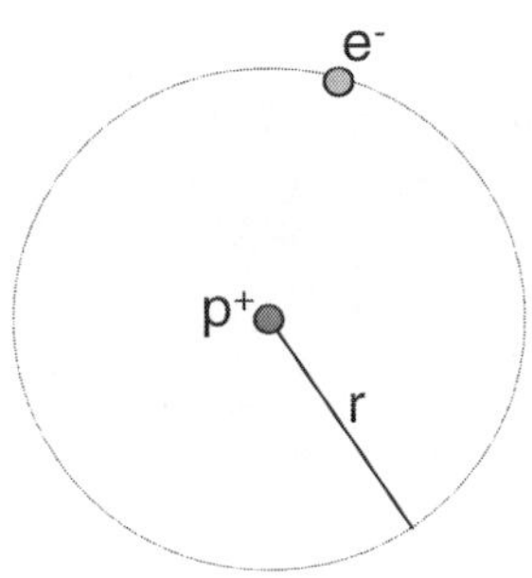

Fig. 1.1 In the Bohr model of the hydrogen atom the negatively charged electron e^- is in a circular orbit around the positively charged proton p^+.

2. *Radiation occurs only when an electron goes from one allowed orbit to another of lower energy. The radiation frequency is* $\hbar\omega = E_m - E_n$*, where* E_m *and* E_n *are the energies of the two states.*

To restrict the allowed values of the orbital radius, we need the "third" postulate:

3. *The angular momentum of the electron is restricted to integer multiples of* $\hbar$*:*

$$m_0 v r = n\hbar \tag{1.3}$$

When $v = n\hbar/m_0 r$ from this equation is equated to $v = \sqrt{e^2/4\pi\epsilon_0 m_0 r}$ of Eq. 1.1, the radius of the nth orbit is

$$r_n = \frac{n^2\hbar^2}{4\pi\epsilon_0 m_0 e^2} \tag{1.4}$$

and the total energy of the nth orbit is

$$E_n = -\frac{m_0 e^4}{32\pi^2\epsilon_0^2\hbar^2}\,\frac{1}{n^2} \tag{1.5}$$

Bohr's theory may be applied to other single-electron systems such as He^+ and Li^{++}, provided the nuclear charge is replaced by Ze, where Z is the atomic number. The energy of the nth level is

$$E_n = -\frac{13.6Z^2}{n^2} \tag{1.6}$$

in terms of electronvolts (eV).

The energy level diagram for hydrogen ($Z = 1$) is shown in Fig. 1.2. Each state is characterized by the integer n, which is called the principal quantum number. When the atom is unexcited, the electron is in the ground state with $n = 1$. The electron may be raised to a higher level by a collision with another electron or by absorbing a photon. Note that the photon energy must correspond exactly to the energy difference between the two states involved in the transition. The electron may return from an excited state to the ground state in one step or via intermediate levels. A single-frequency photon is emitted in the first case, but there are two or more in the second case. Balmer's series corresponds to transitions from higher levels to $n = 2$. Transitions to level $n = 1$ form the Lyman series; those to $n = 3$ form the Paschen series.

All these details about the hydrogen atom are described rigorously by the Schrödinger equation in its time-independent form

$$\left(-\frac{\hbar^2\nabla^2}{2m_0} - \frac{Ze^2}{4\pi\epsilon_0 r}\right)\psi_{n\ell m} = E_n\psi_{n\ell m} \tag{1.7}$$

The energies depend only on the principal quantum number n, which varies from 1 to ∞. The magnitude of the orbital angular momentum, L, of a state is determined by the orbit quantum number ℓ,

$$L = \sqrt{\ell(\ell+1)}\hbar \tag{1.8}$$

where the maximum value of ℓ is restricted by the value of n: $\ell = 0, 1, 2, \ldots, (n-1)$. In order to specify the direction of the angular momentum vector, we need to set

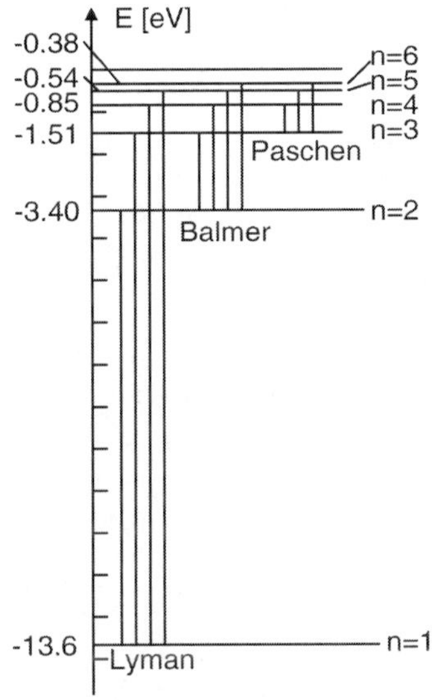

Fig. 1.2 The energy level diagram for hydrogen atom. Light is emitted or absorbed when an electron makes a transition between two states.

up a preferred axis, say, the z axis. The component of the orbit angular momentum along this axis is also quantized

$$L_z = m\hbar \tag{1.9}$$

where the values of the orbital magnetic quantum number m are restricted to $m = 0, \pm 1, \pm 2, \ldots, \pm \ell$.

All states with a given value of n are said to form a shell, whereas states with a given value of ℓ form a subshell. The designations are listed in Table 1.1. The first four letters for the subshells are historical (sharp, principal, diffuse, and fundamental).

In addition, the electron has an intrinsic property called spin that manifests itself according to the following rules. The magnitude of the spin angular momentum, $\boldsymbol{S}$,

Table 1.1 Shell structure of atomic states of electrons.

n	Shell	ℓ	Subshell
1	K	0	s (sharp)
2	L	1	p (principal)
3	M	2	d (diffuse)
4	N	3	f (fundamental)
5	O	4	g
6	P	5	h
.	.	.	.

of the electron is determined by its spin quantum number, $s = 1/2$:

$$S = \sqrt{s(s+1)}\hbar = \frac{\sqrt{3}}{2}\hbar \tag{1.10}$$

In a magnetic field, the z component can assumes only two values

$$S_z = m_s\hbar \tag{1.11}$$

where the spin magnetic quantum number, $m_s = \pm 1/2$. The introduction of spin doubles the number of states allowed for each value of n.

Now, four quantum numbers, n, ℓ, m_ℓ, and m_s are used to classify the states of electrons in all atoms, although the energy associated with a given set of values depends on the atom. The question naturally arises as to why all electrons in an atom do not fall to the ground state. A study of the classification of spectral lines led Pauli in 1925 to make an important statement, now called the Pauli exclusion principle:

No two electrons in an atom can have the same four quantum numbers n, ℓ, m_ℓ, and m_s.

With the aid of the exclusion principle one can see how electrons fill shells (n) and subshells (ℓ). For each value of ℓ there are $(2\ell + 1)$ values of m_ℓ and each subshell can accommodate $2(2\ell + 1)$ electrons. The ground-state electron configurations are indicated in the periodic table by the number of electrons in a subshell as a superscript. For example, $2p^3$ means that there are three electrons in subshell $\ell = 1$.

In isolated atoms the energy levels are sharply defined. Now suppose that two atoms are brought close to each other so that their electron wave functions overlap. As a result of the interaction between the electrons, it turns out that each single state of the isolated atom splits into two states with different energies. As Fig. 1.3 shows, the degree of splitting increases as the interatomic separation decreases. Similarly, if five atoms are placed in close proximity, then each original energy level splits into five new levels. The same process occurs in a solid, where there are roughly 10^{28} atoms/m^3: The energy levels associated with each state of the isolated atom spread into essentially continuous energy bands separated from each other by energy gaps.

Before further examining the various properties of semiconductors it is extremely useful to examine the atomic structure of some of the elements that make up the various semiconductors as listed in Table 1.2.

A very important conclusion can be drawn about the elements making up the semiconductors: The outmost valence electrons are made up of electrons in either the s- or p-type orbitals. While this conclusion is strictly true for elements in the atomic form, it turns out that even in the crystalline semiconductors the electrons in the valence and conduction bands retain this s- or p-type character. The core electrons are usually not of interest, except of some special characterization-type experiments. Note that II–VI such as CdSe are normally complicated and their lattice structures are hexagonal.

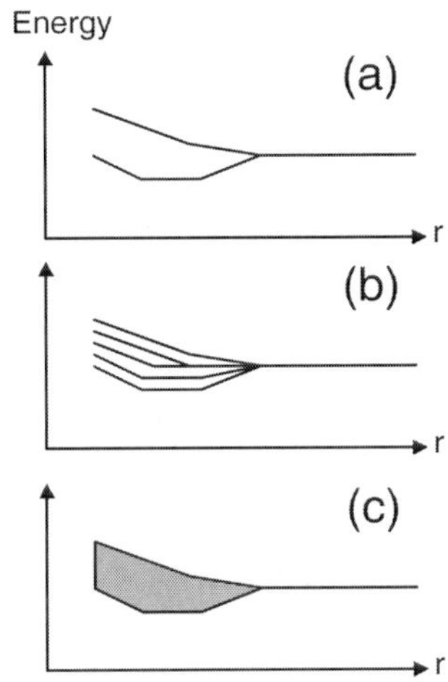

Fig. 1.3 (a) As two atoms are brought closer together, a single atomic level splits into two states with different energies. (b) A single atomic level splits into five when five atoms are in close proximity. (c) In a crystal each atomic level splits into an essentially continuous band of energies.

Here we have assumed that solids are composed of ion cores (i.e., nuclei and those electrons so strongly bound as to be negligibly perturbed from their atomic configuration by their environment in the solid) and valence electrons (i.e., those electrons whose configuration in the solid may differ significantly from that in the isolated atom). However, it is to be remembered that the distinction between core and valence electrons is one of convenience.

In general it is found that when atoms exchange or share valence electrons so that the complement of quantum states is completed, they have a lower electrostatic energy for their combined electron patterns than when they are separate. For example, silicon has four valence electrons grouped into two closely spaced energy levels ($3s$ and $3p$, see Table 1.2), it can combine with other silicon atoms by sharing four valence electrons with four surrounding silicon atoms in an endless array. The atoms around any one atom are centred at the corners of a regular tetrahedron: the tetrahedral bond (Fig. 1.4). This creates the *diamond crystal structure*.

Table 1.2 Atomic structures of typical elements making up common semiconductors.

	IV semiconductors		III-V semiconductors
Si	$1s^2s^22p^6\mathbf{3s^23p^2}$	Ga	$1s^22s^22p^63s^2sp^63d^{10}\mathbf{4s^24p^1}$
Ge	$1s^22s^22p^63s^23p^63d^{10}\mathbf{4s^24p^2}$	As	$1s^22s^22p^63s^2sp^63d^{10}\mathbf{4s^24p^3}$
	II-VI semiconductors		
Cd	$1s^22s^22p^63s^23p^63d^{10}4s^24p^64d^{10}\mathbf{5s^2}$		
Se	$1s^22s^22p^63s^23p^63d^{10}\mathbf{4s^24p^4}$		

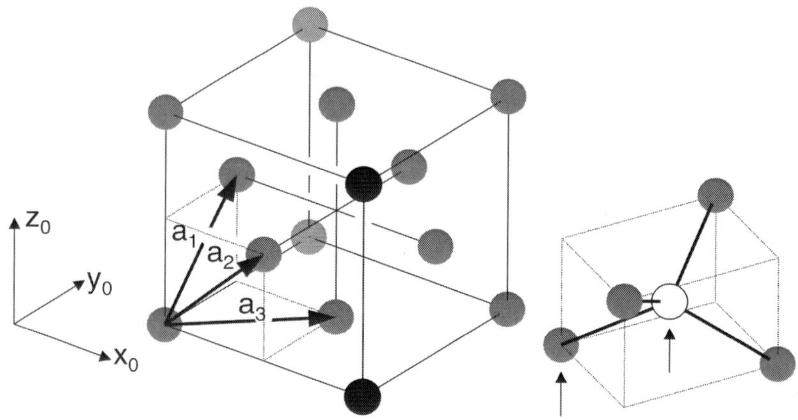

Fig. 1.4 Face-centered cubic lattice and the zincblende crystal structure.

1.2 CRYSTALLINE NATURE OF SOLIDS

The intrinsic property of a crystal is that the environment around a given atom or group of atoms is exactly the same as the environment around another atom or similar group of atoms. To understand and define the crystal structure, two important concepts are introduced, i.e., the lattice and the basis.

The *lattice* represents a set of points in the space that form a periodic structure. Each point sees exactly the same environment. A building block of atoms, called the *basis*, is then attached to each lattice point, yielding a crystal structure.

An important property of a lattice is the ability to define three vectors, $\boldsymbol{a}_1$, $\boldsymbol{a}_2$, and $\boldsymbol{a}_3$, such that any lattice point $\boldsymbol{R}'$ can be obtained from any other lattice point $\boldsymbol{R}$ by a translation

$$\boldsymbol{R}' = \boldsymbol{R} + m_1\boldsymbol{a}_1 + m_2\boldsymbol{a}_2 + m_3\boldsymbol{a}_3 \tag{1.12}$$

where m_1, m_2, and m_3 are three integers. The translation vectors, $\boldsymbol{a}_1$, $\boldsymbol{a}_2$, and $\boldsymbol{a}_3$ are called primitive if the volume of the cell formed by them is the smallest possible.

There are 14 types of lattices in the three-dimensional space. We shall focus on the cubic lattice, which is the structure taken by almost all semiconductors, at least of common interest. There are three kinds of cubic lattices: simple cubic, body-centred cubic, and face-centred cubic. The simple cubic lattice is generated by the primitive vectors of $\boldsymbol{a}_1 = a\boldsymbol{x}_0$, $\boldsymbol{a}_2 = a\boldsymbol{y}_0$, and $\boldsymbol{a}_3 = a\boldsymbol{z}_0$, where $\boldsymbol{x}_0$, $\boldsymbol{y}_0$, and $\boldsymbol{z}_0$ are the three unit vectors of a normal rectangular Cartesian coordinate, and a is the lattice constant.

The *face-centred cubic Bravais lattice* (fcc) (see Fig. 1.4): To construct the fcc lattice we add to the simple cubic lattice an additional point in the center of each square face. The fcc Bravais lattice is of great importance, since an enormous variety of solids crystalize in this form with an atom (or ion) at each lattice site.

A symmetric set of primitive vectors for the fcc lattice is

$$\begin{aligned} \boldsymbol{a}_1 &= \frac{a}{2}(\boldsymbol{y}_0 + \boldsymbol{z}_0) \\ \boldsymbol{a}_2 &= \frac{a}{2}(\boldsymbol{z}_0 + \boldsymbol{x}_0) \\ \boldsymbol{a}_3 &= \frac{a}{2}(\boldsymbol{x}_0 + \boldsymbol{y}_0) \end{aligned} \tag{1.13}$$

Essentially all semiconductors of interest for electronics and optoelectronics have the fcc structure. However, they have two atoms per basis. The coordinates of the two basis atoms are (000) and $(a/4)(111)$ (indicated in Fig. 1.4 by two vertical arrows). If the two atoms of the basis are identical, the structure is called the diamond structure. Semiconductors such as silicon, germanium, and carbon fall into this category. If the two atoms are different, for example, GaAs, AlAs, and CdS, the structure is called zincblende.

Semiconductors with the diamond structure are often called elemental semiconductors, while the zincblende semiconductors are usually called compound semiconductors. The compound semiconductors are also denoted by the positions of the atoms in the periodic table; for example, GaAs, AlAs, and InP are called III–V semiconductors while CdS, CdSe, and CdTe are called II–VI semiconductors.

Many of the properties of crystals and many of the theoretical techniques used to describe crystals derive from the periodicity of crystalline structures. This suggests the use of Fourier analysis as an analytical tool. In the analysis of periodic time varying fields (for example, the acoustic signal analysis and radio signal analysis) we often do much of the analytical work in the frequency domain rather than in the time domain. In analogy with the time–frequency duality, there is a corresponding *real space–reciprocal space* or *wave vector space* duality for crystal problem discussions. Many concepts are best understood in terms of functions of the *wave vector*. We prefer to describe a wave with wavelength λ as a plane wave with wave vector $\boldsymbol{k}$ of magnitude $2\pi/\lambda$ and propagation direction perpendicular to the wave front. The space of the wave vectors is called the reciprocal space, the analogue of the frequency domain for the time problem.

A simple transformation is carried out to map the real space lattice into the reciprocal space ($\boldsymbol{k}$-space)

$$\begin{aligned} \boldsymbol{b}_1 &= 2\pi\frac{\boldsymbol{a}_2 \times \boldsymbol{a}_3}{\boldsymbol{a}_1 \cdot \boldsymbol{a}_2 \times \boldsymbol{a}_3} \\ \boldsymbol{b}_2 &= 2\pi\frac{\boldsymbol{a}_3 \times \boldsymbol{a}_1}{\boldsymbol{a}_1 \cdot \boldsymbol{a}_2 \times \boldsymbol{a}_3} \\ \boldsymbol{b}_3 &= 2\pi\frac{\boldsymbol{a}_1 \times \boldsymbol{a}_2}{\boldsymbol{a}_1 \cdot \boldsymbol{a}_2 \times \boldsymbol{a}_3} \end{aligned} \tag{1.14}$$

The simple cubic Bravais lattice, with cubic primitive cell of side a,

$$\boldsymbol{a}_1 = a\boldsymbol{x}_0 \quad , \quad \boldsymbol{a}_2 = a\boldsymbol{y}_0 \quad , \quad \boldsymbol{a}_3 = a\boldsymbol{z}_0 \tag{1.15}$$

has as its reciprocal a simple cubic lattice with cubic primitive cell of side $2\pi/a$:

$$\boldsymbol{b}_1 = \frac{2\pi}{a}\boldsymbol{x}_0 \;,\; \boldsymbol{b}_2 = \frac{2\pi}{a}\boldsymbol{y}_0 \;,\; \boldsymbol{b}_3 = \frac{2\pi}{a}\boldsymbol{z}_0 \tag{1.16}$$

For fcc Bravais lattice with primitive vectors Eq. 1.13,

$$\begin{aligned} \boldsymbol{b}_1 &= \frac{2\pi}{a}(\boldsymbol{y}_0 + \boldsymbol{z}_0 - \boldsymbol{x}_0) \\ \boldsymbol{b}_2 &= \frac{2\pi}{a}(\boldsymbol{z}_0 + \boldsymbol{x}_0 - \boldsymbol{y}_0) \\ \boldsymbol{b}_3 &= \frac{2\pi}{a}(\boldsymbol{x}_0 + \boldsymbol{y}_0 - \boldsymbol{z}_0) \end{aligned} \tag{1.17}$$

A general vector

$$\boldsymbol{G} = m_1\boldsymbol{b}_1 + m_2\boldsymbol{b}_2 + m_3\boldsymbol{b}_3 \tag{1.18}$$

is called a reciprocal lattice vector, where the m_1, m_2, and m_3 are three integers (either positive or negative). It is worth noting the special relation

$$e^{i\boldsymbol{G}\cdot\boldsymbol{R}} = 1 \tag{1.19}$$

where $\boldsymbol{R}$ is a lattice vector in Eq. 1.12 but often called the direct lattice vector to distinguish it from the reciprocal lattice vector.

Because of the above equation, two wave vectors $\boldsymbol{k}$ and $\boldsymbol{k}'$ satisfying

$$\boldsymbol{k}' = \boldsymbol{k} + \boldsymbol{G} \tag{1.20}$$

are said to be equivalent. This implies that $\boldsymbol{k}$ can be restricted to points that lie within or on a geometric figure, called the Brillouin zone, which has the property that no two interior points are equivalent. In the present case, we consider only the first Brillouin zone, which is the region in the reciprocal space that is closer to the center of the reciprocal space than to any other reciprocal lattice point. The first Brillouin zones for the simple cubic and fcc lattices are shown in Fig. 1.5.

Most importantly we consider $XR - R\Gamma - \Gamma X - XM$ in the $\boldsymbol{k}$-space (in unit of $2\pi/a$)

$$\begin{aligned} \Gamma &= (0,0,0)\;,\; X = \left(\frac{1}{2},0,0\right) \\ R &= \left(\frac{1}{2},\frac{1}{2},\frac{1}{2}\right)\;,\; M = \left(\frac{1}{2},0,\frac{1}{2}\right) \end{aligned} \tag{1.21}$$

for the simple cubic lattice and $XU - UL - L\Gamma - \Gamma X - XW - WK$ in the $\boldsymbol{k}$-space (in unit of $2\pi/a$),

$$\begin{aligned} \Gamma &= (0,0,0)\;,\; X = (1,0,0) \\ L &= \left(\frac{1}{2},\frac{1}{2},\frac{1}{2}\right)\;,\; W = \left(1,0,\frac{1}{2}\right) \\ U &= \left(\frac{1}{4},\frac{1}{4},1\right)\;,\; K = \left(\frac{3}{4},0,\frac{3}{4}\right) \end{aligned} \tag{1.22}$$

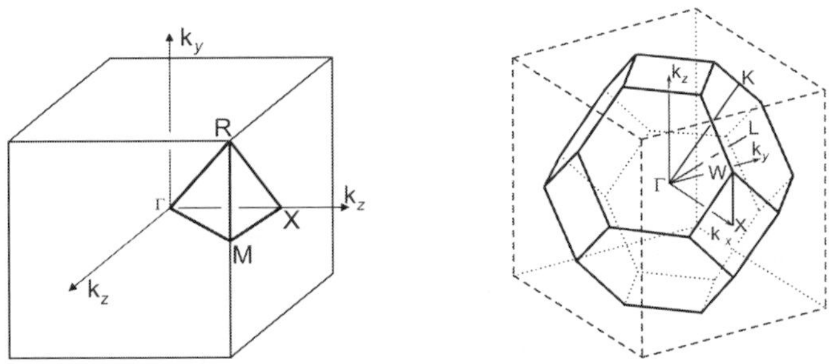

Fig. 1.5 First Brillouin zones for simple cubic lattice (left) and fcc (right) lattice. Points and lines of symmetry are indicated.

for fcc lattice. The diamond structure is an fcc lattice with a basis, as discussed earlier.

So far we have discussed crystal structures that are present in natural semiconductors. These structures are the lowest free energy configuration of the solid state of the atoms. Since the electrical and optical properties of the semiconductors are completely determined by the crystal structures, artificial structures, for example, heteromaterials (among them the well-known superlattices have been fabricating even since mid-1970s inspired by the pioneering work of Esaki and Tsu at IBM) grown by heteroepitaxial crystal growth techniques such as MBE and MOCVD have made a tremendous impact on the semiconductor physics, the semiconductor technology, and the semiconductor electronic and optoelectronic device industry.

Since the new heteroepitaxial techniques allow one to grow heterostructures with atomic control, one can change the periodicity of the crystal in the growth direction. This leads to the concept of superlattices where two (or even more) semiconductors A and B are grown alternately with thickness d_A and d_B, respectively. The periodicity of the superlattice in the growth direction is then $d_A + d_B$. An AlGaAs/GaAs quantum well grown by MBE is illustrated in Fig. 1.6. Superlattices that have been grown can be placed in three general categories: (i) lattice matched, (ii) lattice strained, and (iii) lattice strained with intermediate substrate.

1.3 ELECTRONS IN SOLID

We start with the real-space Schrödinger equation for an electron in a periodic lattice structure

$$\left[\frac{-\hbar^2\nabla^2}{2m_0} + V(\boldsymbol{r})\right]\psi(\boldsymbol{r}) = E\psi(\boldsymbol{r}) \tag{1.23}$$

where the first term represents the kinetic energy and $V(\boldsymbol{r})$ is the periodic potential energy

$$V(\boldsymbol{r} + \boldsymbol{R}) = V(\boldsymbol{r}) \tag{1.24}$$

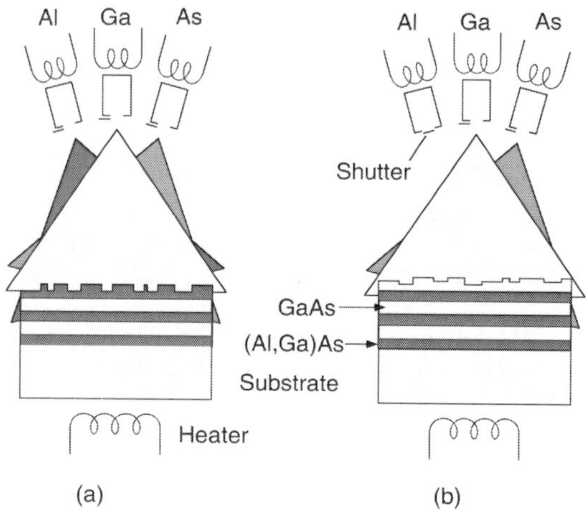

Fig. 1.6 Schematic diagram illustrating the growth of an AlGaAs/GaAs multiple quantum well sample by molecular beam epitaxy. Deposition of (a) (Al,Ga)As, (b) GaAs. The system is under high vacuum.

where $\boldsymbol{R}$ is any lattice vector. m_0 is the free electron mass.

The Bloch theorem states that the solutions of the Schrödinger equation of Eq. 1.23 with periodic condition of Eq. 1.24 have the following properties

$$\begin{aligned} \psi_{n\boldsymbol{k}}(\boldsymbol{r}) &= \frac{1}{\sqrt{N}} u_{n\boldsymbol{k}}(\boldsymbol{r}) \mathrm{e}^{i\boldsymbol{k}\cdot\boldsymbol{r}} \\ u_{n\boldsymbol{k}}(\boldsymbol{r}) &= u_{n\boldsymbol{k}}(\boldsymbol{r}+\boldsymbol{R}) \\ \int_{\text{cell}} & u_{n\boldsymbol{k}}^*(\boldsymbol{r}) u_{n\boldsymbol{k}}(\boldsymbol{r}) \mathrm{d}\boldsymbol{r} = 1 \end{aligned} \tag{1.25}$$

and $E = E_n(\boldsymbol{k})$ is the energy dispersion relationship. Here $N = N_x N_y N_z$, N_x, N_y, and N_z are the numbers of unit cells in the crystal along the x-, y-, and z-direction, n is the energy band index, and $\hbar\boldsymbol{k}$ is the quasi-momentum of the Bloch electron. Here we have imposed periodic boundary conditions on the wave function

$$\psi(\boldsymbol{r}) = \psi(\boldsymbol{r} + N_x \boldsymbol{a}_x) = \psi(\boldsymbol{r} + N_y \boldsymbol{a}_y) = \psi(\boldsymbol{r} + N_z \boldsymbol{a}_z) \tag{1.26}$$

Applying an external force $\boldsymbol{F}$, for example, due to an external electromagnetic field $(\boldsymbol{E}, \boldsymbol{B})$,

$$\begin{aligned} \hbar\dot{\boldsymbol{k}} &= \boldsymbol{F} = -e\left(\boldsymbol{E} + \frac{1}{c}\boldsymbol{v}_n \times \boldsymbol{B}\right) \\ \boldsymbol{v}_n(\boldsymbol{k}) &= \frac{1}{\hbar}\frac{\partial E_n(\boldsymbol{k})}{\partial \boldsymbol{k}} \end{aligned} \tag{1.27}$$

Here $-e$ is the electron charge and $\boldsymbol{v}_n$ is the electron group velocity.

As the atoms of the elements making up the semiconductors are brought together to form the crystal, the valence electronic states are perturbed by the presence of neighboring atoms. While the original atomic functions describing the valence electrons are, of course, no longer eigenstates of the problem, they can be used as a good approximate set of basis states to describe the "crystalline" electrons. This motivates the tight-binding method. For most semiconductor materials of interest, the atomic functions required to describe the outermost electrons are the s, p_x, p_y, and p_z types (see Table 1.2). Moreover, since there are two atoms per basis in a compound semiconductor (here we discuss pure semiconductors), we then require 8 functions to describe the central cell part of the Bloch functions in the form of

$$\psi(\boldsymbol{k}, \boldsymbol{r}) = \sum_{\boldsymbol{R}_i} \sum_{m=1}^{5} \sum_{j=1}^{2} C_{mj}(\boldsymbol{k}) \psi_{mj}(\boldsymbol{r} - \boldsymbol{r}_j - \boldsymbol{R}_i) e^{i\boldsymbol{k} \cdot \boldsymbol{R}_i} \tag{1.28}$$

where the sum over $\boldsymbol{R}_i$ runs over all unit cells, m is the index of the different atomic functions ψ_{mj} used in the basis, and j denotes the atoms in each unit cell.

Once the expansion set for the crystal states has been chosen, the coefficients C_{mj} remain to be determined. To this end, the Schrödinger equation is in the form of a secular determinant

$$|\langle \psi_{mj} | H - E | \psi(\boldsymbol{k}, \boldsymbol{r}) \rangle| = 0 \tag{1.29}$$

where H is the Hamiltonian of the system under investigation.

In theory, one can calculate the matrix elements in the secular determinant, Eq. 1.29, by determining the crystal potential. This, however, is very difficult because of the complexity of the problem. Slater and Koster were the first to advocate the use of the tight-binding method as an empirical technique. In their formalism, the matrix elements of the secular determinant are treated as disposable constants. Energy levels in the band structure can be obtained and fitted with the measurement data by adjusting the disposable constants.

For semiconductors of device application interest, we concentrate on the conduction and valence bands of cubic semiconductors with both diamond (silicon and germanium) and zincblende symmetries (III–V group), see Fig. 1.7. The conduction band consists of three sets of band minima located at the Γ_{15}-point at $\boldsymbol{k} = 0$, the L-points at $\boldsymbol{k} = (\pi/a, \pi/a, \pi/a)$, and along the Δ lines from $(0,0,0)$ to $(\pi/a, 0, 0)$, from $(0,0,0)$ to $(0, \pi/a, 0)$, and from $(0,0,0)$ to $(0,0,\pi/a)$, where a is the lattice constant. The valence band tops are located at Γ_{15}, where two bands normally degenerate, i.e., the heavy-hole and light-hole bands; the third one is the spin-split-off band due to the spin-orbital interaction.

1.4 EIGHT-BAND $k \cdot p$ MODEL

For materials with narrow energy bandgaps the coupling between the conduction-band electron states and valence-band hole states must be included. The well-established eight-band $\boldsymbol{k} \cdot \boldsymbol{p}$ model for III–V semiconductor materials was derived rigorously for bulk materials,[28–32] and has been extensively applied to many low dimensional systems; for example, see Ref.,[33] and the review of Ref.[34] The energy

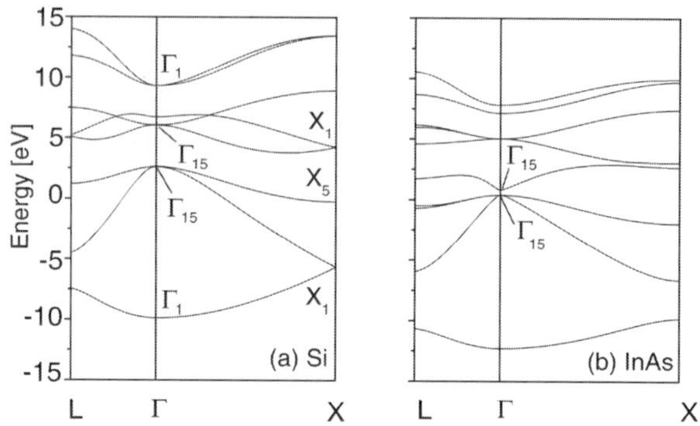

Fig. 1.7 Energy band structures of (a) Si and (b) InAs.

bandgap bowing effect is included via the bowing parameters of the relevant relaxed bulk materials that have been determined experimentally.[34]

In many epitaxial materials such as InAs quantum wells (QWs) or quantum dots (QDs) in GaAs substrate, there is a difference between lattice constants of epitaxial and substrate materials that causes lattice strains. Strain over the deposited materials results in different growth modes such as Volmer-Weber (VW), Frank-van der Merwe (FV), and Stranski-Krastanov (SK) growth.[35] Self-assembled QDs formed by the strain-induced SK mode have been an intensively investigated semiconductor quantum structure.[36–39] Since the spatial coordinate is transformed from the strain crystal $\boldsymbol{x}_\varepsilon$ to the unstrained one $\boldsymbol{x}$ by $\boldsymbol{x}_\varepsilon = (1+\varepsilon)\boldsymbol{x}$, the crystal momentum is transformed as $\boldsymbol{k}_\varepsilon = (1-\varepsilon)\boldsymbol{k}$, i.e.,

$$k_{\alpha\epsilon} = k_\alpha - \sum_{\beta=x,y,z} \epsilon_{\alpha\beta} k_\beta \tag{1.30}$$

where $\alpha = x, y, z$, and k_x, k_y, and k_z are wave vector components in x, y, and z, directions. Here ε denotes the strain in the system (see more discussions below).

By following the notation and phase conventions used by Enders *et al.*,[32] the Hamiltonian of the eight-band $\boldsymbol{k}\cdot\boldsymbol{p}$ model is written as

$$H = H_0 + H_{\text{so}} + D + H_1 + H_2 \tag{1.31}$$

in the basis of atomic-like states $|S\uparrow, X\uparrow, Y\uparrow, Z\uparrow, S\downarrow, X\downarrow, Y\downarrow, Z\downarrow\rangle$, where a s-type state, $|S\rangle$, representing the conduction band edge state, is introduced. $\uparrow$ and $\downarrow$ denote spin up and down.

H_0 represents the energies of the basis states, H_{so} the spin-orbit interaction, D the deformation potential, and H_1 and H_2 the $\boldsymbol{k}\cdot\boldsymbol{p}$ interactions of first and second order, respectively:

$$H_0 = \begin{bmatrix} E_g & 0 & 0 & 0 & 0 & 0 & 0 & 0 \\ 0 & -\Delta'_{\rm so} & 0 & 0 & 0 & 0 & 0 & 0 \\ 0 & 0 & -\Delta'_{\rm so} & 0 & 0 & 0 & 0 & 0 \\ 0 & 0 & 0 & -\Delta'_{\rm so} & 0 & 0 & 0 & 0 \\ 0 & 0 & 0 & 0 & E_g & 0 & 0 & 0 \\ 0 & 0 & 0 & 0 & 0 & -\Delta'_{\rm so} & 0 & 0 \\ 0 & 0 & 0 & 0 & 0 & 0 & -\Delta'_{\rm so} & 0 \\ 0 & 0 & 0 & 0 & 0 & 0 & 0 & -\Delta'_{\rm so} \end{bmatrix}$$

E_g is the fundamental gap of the unstrained material, and $\Delta'_{\rm so} = \Delta_{\rm so}/3$, $\Delta_{\rm so}$ is the spin-orbit interaction energy.

$$H_{\rm so} = \Delta'_{\rm so} \begin{bmatrix} 0 & 0 & 0 & 0 & 0 & 0 & 0 & 0 \\ 0 & 0 & -i & 0 & 0 & 0 & 0 & 1 \\ 0 & i & 0 & 0 & 0 & 0 & 0 & -i \\ 0 & 0 & 0 & 0 & 0 & -1 & i & 0 \\ 0 & 0 & 0 & 0 & 0 & 0 & 0 & 0 \\ 0 & 0 & 0 & -1 & 0 & 0 & i & 0 \\ 0 & 0 & 0 & -i & 0 & -i & 0 & 0 \\ 0 & 1 & i & 0 & 0 & 0 & 0 & 0 \end{bmatrix}$$

The Hamilton matrix of the deformation potential interaction reads

$$D = \begin{bmatrix} a_c(\varepsilon_{xx} + \varepsilon_{yy} + \varepsilon_{zz}) & 0 & 0 & 0 \\ 0 & D_3 & 0 & 0 \\ 0 & 0 & a_c(\varepsilon_{xx} + \varepsilon_{yy} + \varepsilon_{zz}) & 0 \\ 0 & 0 & 0 & D_3 \end{bmatrix}$$

$$D_3 = \begin{bmatrix} l_\varepsilon \varepsilon_{xx} + m_\varepsilon(\varepsilon_{yy} + \varepsilon_{zz}) & n_\varepsilon \varepsilon_{xy} & n_\varepsilon \varepsilon_{xz} \\ n_\varepsilon \varepsilon_{yx} & l_\varepsilon \varepsilon_{yy} + m_\varepsilon(\varepsilon_{xx} + \varepsilon_{zz}) & n_\varepsilon \varepsilon_{yz} \\ n_\varepsilon \varepsilon_{zx} & n_\varepsilon \varepsilon_{zy} & l_\varepsilon \varepsilon_{zz} + m_\varepsilon(\varepsilon_{yy} + \varepsilon_{xx}) \end{bmatrix}$$

$$H_1(\boldsymbol{k}_\varepsilon) = \begin{bmatrix} H_4(\boldsymbol{k}_\varepsilon) & 0 \\ 0 & H_4(\boldsymbol{k}_\varepsilon) \end{bmatrix}$$

$$H_4(\boldsymbol{k}_\varepsilon) = \begin{bmatrix} 0 & ipk_{x\varepsilon} & ipk_{y\varepsilon} & ipk_{z\varepsilon} \\ -ipk_{x\varepsilon} & 0 & 0 & 0 \\ -ipk_{y\varepsilon} & 0 & 0 & 0 \\ -ipk_{z\varepsilon} & 0 & 0 & 0 \end{bmatrix}$$

where $\langle S|p_x|X\rangle = \langle S|p_y|Y\rangle = \langle S|p_z|Z\rangle = i(m_0/\hbar)p$, and p is real. This form was published by Enders *et al.*[32] The free-electron and second-order $\boldsymbol{k} \cdot \boldsymbol{p}$ interaction contributions are contained in the matrix

$$H_2 = \begin{bmatrix} Ak_\epsilon^2 & 0 & 0 & 0 \\ 0 & S(\boldsymbol{k}_\epsilon) & 0 & 0 \\ 0 & 0 & Ak_\epsilon^2 & 0 \\ 0 & 0 & 0 & S(\boldsymbol{k}_\epsilon) \end{bmatrix}$$

$k_\epsilon^2 = k_{x\epsilon}^2 + k_{y\epsilon}^2 + k_{z\epsilon}^2$. $S(\boldsymbol{k}_\epsilon)$ is the Shockley matrix

$$S(\boldsymbol{k}_\epsilon) = \begin{bmatrix} H & 0_{3\times3} \\ 0_{3\times3} & H \end{bmatrix}$$

$$H = \begin{bmatrix} Lk_x^2 + Mk_{yz}^2 & Nk_xk_y & Nk_xk_z \\ Nk_yk_x & Lk_y^2 + Mk_{zx}^2 & Nk_yk_z \\ Nk_zk_x & Nk_zk_y & Lk_z^2 + Mk_{xy}^2 \end{bmatrix}$$

where $k_{ij}^2 \equiv k_i^2 + k_j^2$, $0_{3\times3}$ is the 3×3 zero matrix, and L, M, and N are band parameters.

In the above equations, $A = \hbar^2/2m_e^*$, where m_e^* is normally referred to as the effective mass of electrons in the conduction band given by

$$\frac{m_0}{m_e^*} = 1 + 2F + \frac{E_p(E_g + 2\Delta_{\text{so}}/3)}{E_g(E_g + \Delta_{\text{so}})} \tag{1.32}$$

In the above equation, $E_p = p^2/2m_0$ is the energy parameter related to the momentum matrix element p, and F is the Kane parameter.

The confinement potential V_c for the electrons in a III–V semiconductor nanostructure is the conduction band offset between the barrier and well materials given by $0.65[E_g(\text{barrier}) - E_g(\text{well})]$, where 0.65 is the conduction band offset coefficient commonly used for III–V materials,[34] and $E_g(\text{barrier})$ and $E_g(\text{well})$ are the bandgap energies for the barrier and well materials, respectively. When calculating the bandgap energies of the constituent ternaries we account for the bandgap bowing effect. The bandgap energy of a relaxed $\text{In}_x\text{Ga}_{1-x}\text{As}$ alloy is given by:

$$E_g(\text{In}_x\text{Ga}_{1-x}\text{As}) = xE_g(\text{InAs}) + (1-x)E_g(\text{GaAs}) - x(1-x)C_{\text{InGaAs}} \tag{1.33}$$

where C is the bowing parameter. The recommended value for C_{InGaAs} is 0.477 eV.[34] Similar expressions are used for the other strain-relaxed ternary alloys, with $C_{\text{InGaN}} = 2.2\,\text{eV}$ for $\text{In}_x\text{Ga}_{1-x}\text{N}$ when $x = 0.36$,[40] $C_{\text{GaNAs}} = 20\,\text{eV}$ for $\text{GaN}_y\text{As}_{1-y}$ when $x = 0.01$ and 11 eV when $x = 0.03$,[34, 41] and the recommended bowing factor (C_{InNAs}) 4.22 eV for $\text{InN}_x\text{As}_{1-x}$.[34, 42] We also included the temperature dependence of the bandgap energies through the Varshni approximation:

$$E_g(T) = E_g(T=0) - \frac{\alpha T^2}{T + \beta} \tag{1.34}$$

where α and β are the empirical Varshni parameters and T is the absolute temperature.

To determine the confinement potential we have also applied the method where the top of the valence band of every binary III–V compound is measured with respect to a common reference energy level through a valence band offset (VBO) value.[43] The VBO values of the ternary alloys are obtained from the binaries using expressions similar to Eq. 1.33. Almost identical results are obtained using the two different band offset schemes, as can be expected for III–V materials.[34]

Other parameters in the model are

$$L = \frac{-\hbar^2}{m_0}(\gamma_1 + 4\gamma_2) \ , \ M = \frac{-\hbar^2}{m_0}(\gamma_1 - 2\gamma_2) \ , \ N = \frac{-\hbar^2}{m_0}(6\gamma_3) \tag{1.35}$$

where γ_1, γ_2, and γ_3 are Luttinger parameters. l, m, and n are defined by the deformation potential parameters a_v, b, and d as:

$$a_v = \frac{l + 2m}{3} \ , \ b = \frac{l - m}{3} \ , \ d = \frac{n}{\sqrt{3}} \tag{1.36}$$

For a layered structure such as a quantum well grown along the x direction, the epitaxial layer is biaxially strained in the yz-plane by an amount:

$$\epsilon_{yy} = \epsilon_{zz} = \frac{a_{yz}}{a_0} - 1 \tag{1.37}$$

where a_{yz} is the lattice constant of the relaxed epitaxial layer in the yz-plane and a_0 is the lattice constant of the substrate material. The epitaxial layer is also uniaxially strained in the x-direction by an amount:

$$\epsilon_{xx} = -\frac{2c_{12}}{c_{11}}\epsilon_{yy} \tag{1.38}$$

where c_{11} and c_{12} are the elastic constants. Other strain elements are zero, i.e., $\epsilon_{ij} = 0$ if $i \neq j$.

Table 1.3 Band structure parameters for zinc-blende GaAs, InP, InAs, GaN, AlN, and InN.

Parameters	GaAs	InP	InAs	GaN	AlN	InN
a [Å]	5.65325	5.8697	6.0583	4.50	4.38	4.98
E_g^Γ [eV]	1.519	1.4236	0.417	3.299	4.9	1.94
$\alpha(\Gamma)$ [meV/K]	0.5405	0.363	0.276	0.593	0.593	0.245
$\beta(\Gamma)$ [K]	204	162	93	600	600	624
Δ_{so} [eV]	0.341	0.108	0.39	0.017	0.019	0.006
m_e^*	0.067	0.0795	0.026	0.15	0.25	0.12
γ_1	6.98	5.08	20.0	2.67	1.92	3.72
γ_2	2.06	1.60	8.5	0.75	0.47	1.26
γ_3	2.93	2.10	9.2	1.10	0.85	1.63
F	−1.94	−1.31	−2.90	−0.92	0.76	−0.92
VBO [eV]	−0.80	−0.94	−0.59	−2.64	−3.44	−2.38
a_c [eV]	−7.17	−6.0	−5.08	−2.2	−6.0	−1.85
a_v [eV]	1.16	0.6	1.0	−5.2	−3.4	−1.5
b [eV]	−2.0	−2.0	−1.8	−2.2	−1.9	−1.2
d [eV]	−4.8	−5.0	−3.6	−3.4	−10	−9.3
c_{11} [GPa]	122.1	101.1	83.29	29.3	30.4	18.7
c_{12} [GPa]	56.6	56.1	45.26	15.9	16.0	12.5
E_p [eV]	28.8	20.7	21.5	25.0	27.1	25.0
ϵ[44]	12.40	12.5	15.15	10.4	9.14	6.55

Source: From Refs. [34,44].

Principal band structure parameters of the constituent binaries are listed in Table 1.3.[34] The permittivities ϵ are obtained from Ref.[44] Moreover, the permittivity of InN is derived from its refractive index of 2.56 listed in Ref.[16] Parameters for relaxed bulk ternary and quaternary alloys are obtained by interpolations similar to Eq. 1.33. Bowing effects are included whenever the values of the bowing parameters are available in Ref.[34] and references therein. Furthermore, note that in Ref.[34] the values of elastic constants c's are to be divided by a factor of 10.

1.5 HETEROSTRUCTURE MATERIALS

Among all possible combinations of semiconductors, the lattice-matched (or at least closely matched) combinations have attracted most interest. The reason is, of course, the possibility of ideal heterointerfaces if the lattice constants of the constituents are the same. It is interesting to note that perfect lattice match does not guarantee an deal interface, as can be seen from the results for the GaAs-Ge system.[45] In addition to lattice match condition, we also have to require that both systems be binary (or ternary, quaternary) in order to avoid, for instance, the site allocation problem.[46]

A picture of the effects of the heterojunctions on the band structure can be obtained by employing the effective-mass theorem. The simplest approach is to view the semiconductor as homogeneous with an additional potential superimposed on the crystal potential. The effective-mass theorem tells us that for electrons at band minima or maxima we can replace the Hamiltonian

$$\frac{-\hbar^2\nabla^2}{2m_0} + V_0(\boldsymbol{r}) + V(\boldsymbol{r}) \tag{1.39}$$

by

$$\frac{-\hbar^2\nabla^2}{2m^*} + V(\boldsymbol{r}) \tag{1.40}$$

and therefore transform away the periodic crystal potential $V_0(\boldsymbol{r})$ [provided that $V(\boldsymbol{r})$ and its Fourier components fulfil certain conditions]. In other words, we have reduced the heterojunction problem to a well-defined potential $V(\boldsymbol{r})$ problem for a quasi-free electron with effective mass m^*.

To complete the effective-mass theory we need to know the potential $V(\boldsymbol{r})$. Although it is difficult to determine $V(\boldsymbol{r})$ with respect to the vacuum level, a calculation of $V(\boldsymbol{r})$ relative to the bandedge of one component of the heterostructures seems to be easier. Harrison *et al.* proposed[47] that $V(\boldsymbol{r})$ can be determined by the linear combination of atomic orbitals method, which gives the valence bandedge E_v, as

$$E_v = \frac{E_a + E_c}{2} - \left[\left(\frac{E_a + E_c}{2}\right) + V_{xx}^2\right]^{1/2} \tag{1.41}$$

where E_a (E_c) are the p-orbital energies of the anion (cation) atoms, and V_{xx} is the interatomic matrix element of the crystal Hamiltonian between p-orbitals of adjacent atoms. The valence band offset ΔE_v for hole of, for example, an AlAs-GaAs heterojunction is therefore

$$\Delta E_v = E_v(\text{AlAs}) - E_v(\text{GaAs}) \tag{1.42}$$

and the conduction band discontinuity ΔE_c is

$$\Delta E_c = E_g(\text{AlAs}) - E_g(\text{GaAs}) - \Delta E_v \tag{1.43}$$

Here E_g is the energy bandgap of AlAs or GaAs. A comparison of these values with experimental results has been given by Margaritondo *et al.*,[48] who found an overall agreement between theory and experiments.

Note that it is equivalent to adopt the concept of VBO introduced in Section 1.4.

1.5.1 *Envelop Function*

We now seek to solve the wave equation

$$\Big[H_0 + V(\boldsymbol{r})\Big]\psi = E\psi \tag{1.44}$$

where

$$H_0 = \frac{p^2}{2m_0} + V_0(\boldsymbol{r})$$

is the Hamiltonian of electron in a perfect crystal, and $V(\boldsymbol{r})$ is a perturbing potential due to heterojunctions in the heterostructure system. m_0 is the free electron mass.

For the perfect crystal, the Bloch theorem says

$$H_0 u_{n\boldsymbol{k}} e^{i\boldsymbol{k}\cdot\boldsymbol{r}} = E_n(\boldsymbol{k}) u_{n\boldsymbol{k}} e^{i\boldsymbol{k}\cdot\boldsymbol{r}} \tag{1.45}$$

where $u_{n\boldsymbol{k}}(\boldsymbol{r}) = u_{n\boldsymbol{k}}(\boldsymbol{r} + \boldsymbol{R}_i)$, $\boldsymbol{R}_i$ is the position of cell i. $n = 1, 2, \ldots \mu$ is the band index.

For materials such as $Si_{1-x}Ge_x$, the bandgap is quite large so that the interaction between the valence band and the conduction band can be neglected and we only need to consider the mixings among the heavy hole, light hole, and spin-split-off band. By taking into account the spin-orbital interactions, the total number of bands involved, μ, is 6 for valence band holes. The subbands are no longer pure, neither heavy hole band, nor light hole band, nor spin-split-off band when $\boldsymbol{k}$ is moved away from the zone center because of the intermixings among them. For commonly used n-type materials, the conduction band is much simpler; it consists of either a single Γ-valley (GaAs), or six X-valleys (AlAs), or four L-valleys (AlGaSb). The X and L valleys are well separated in the $\boldsymbol{k}$-space so that interactions among them can be neglected. μ, in this case, is 2, including spin degeneracy. For III–V materials such as InAs the energy bandgap is small so that we have to include the interactions among the conduction-band electron and valence-band hole states for which $\mu = 8$, as discussed in the eight-band $\boldsymbol{k} \cdot \boldsymbol{p}$ model in Section 1.4.

When we investigate a crystal whose component composition is not uniform, it is more suitable to use the Wannier function

$$a_n(\boldsymbol{r}-\boldsymbol{R}_i)=\frac{1}{\sqrt{N}}\sum_{\boldsymbol{k}}u_{n\boldsymbol{k}}(\boldsymbol{r})e^{i\boldsymbol{k}\cdot(\boldsymbol{r}-\boldsymbol{R}_i)} \tag{1.46}$$

which is localized around the ith unit cell. Here we assume that the Bloch functions are normalized for a volume Ω containing N unit cells.

The perturbed wave function is expanded using Wannier functions in the form of

$$\psi=\sum_{i,n}\psi_n(\boldsymbol{R}_i)a_n(\boldsymbol{r}-\boldsymbol{R}_i) \tag{1.47}$$

In order to obtain the equations for $\psi_n(\boldsymbol{R}_i)$ we substitute the wave function given by Eq. 1.47 into Eq. 1.44, multiply by $a^*_{n'}(\boldsymbol{r}-\boldsymbol{R}_j)$, and integrate over the volume Ω. In this way we obtain the following equations

$$\sum_{i,n}\left[h_{n'}(\boldsymbol{R}_j-\boldsymbol{R}_i)\delta_{n'n}+V_{n'nji}\right]\psi_n(\boldsymbol{R}_i)=E\psi_{n'}(\boldsymbol{R}_j) \tag{1.48}$$

where

$$h_{n'}(\boldsymbol{R}_j-\boldsymbol{R}_i)\delta_{n'n}=\langle a_{n'}(\boldsymbol{r}-\boldsymbol{R}_j)|H_0|a_n(\boldsymbol{r}-\boldsymbol{R}_i)\rangle$$

because of the orthogonal property of wave functions from different bands and periodic property of the operator H_0,

$$V_{n'nji}=\langle a_{n'}(\boldsymbol{r}-\boldsymbol{R}_j)|V(\boldsymbol{r})|a_n(\boldsymbol{r}-\boldsymbol{R}_i)\rangle \tag{1.49}$$

$V_{n'nji}$ will be appreciable when there is considerable overlap between $a_{n'}(\boldsymbol{r}-\boldsymbol{R}_j)$ and $a_n(\boldsymbol{r}-\boldsymbol{R}_i)$. This will be so when $(\boldsymbol{R}_j-\boldsymbol{R}_i)$ is small. If $V(\boldsymbol{r})$ is a slowly varying function we may regard it as constant over the region where the product $\langle a_{n'}(\boldsymbol{r}-\boldsymbol{R}_j)|a_n(\boldsymbol{r}-\boldsymbol{R}_i)\rangle$ is significant and take it outside the integral for $V_{n'nji}$, which is equal to zero when $i\neq j$. Thus,

$$V_{n'nji}\approx V_{n'n}(\boldsymbol{R}_j)\delta_{ji} \tag{1.50}$$

$V(\boldsymbol{r})$ may not be a slowly varying function in situations such as interfaces between two different materials with respect to individual atomic potentials. However, as one consequence of surface kinetic processes, a certain degree of intermixing of the two materials is inevitable. Such intermixing is clearly indicated in the GaAs/AlGaAs superlattice by the fine structure emission lines in the photoluminescence spectra corresponding well to that calculated for one monolayer change in the well width. Therefore, it takes several atomic layers for the averaged atomic potential to change from one material over to the other. Since the Wannier function is quite localized, especially when the electronic states under investigation are close to the band edges, the approximation of $V_{n'nji}\approx V_{n'n}(\boldsymbol{R}_j)\delta_{ji}$ is reasonable.

Equation 1.48 is thus reduced to the form

$$\sum_i h_{n'}(\boldsymbol{R}_i)\psi_{n'}(\boldsymbol{R}_j-\boldsymbol{R}_i)+\sum_n V_{n'n}(\boldsymbol{R}_j)\psi_n(\boldsymbol{R}_j)=E\psi_{n'}(\boldsymbol{R}_j) \tag{1.51}$$

On the other hand, writing the energy $E_{n'}(\boldsymbol{k})$ as

$$E_{n'}(\boldsymbol{k}) = \langle u_{n'} \mathrm{e}^{i\boldsymbol{k}\cdot\boldsymbol{r}} | H_0 | u_{n'} \mathrm{e}^{i\boldsymbol{k}\cdot\boldsymbol{r}} \rangle \tag{1.52}$$

and substituting the Bloch functions in terms of Wannier functions, we obtain

$$E_{n'}(\boldsymbol{k}) = \frac{1}{N} \sum_{ij} \mathrm{e}^{-i\boldsymbol{k}\cdot(\boldsymbol{R}_j - \boldsymbol{R}_i)} h_{n'}(\boldsymbol{R}_j - \boldsymbol{R}_i) = \sum_j \mathrm{e}^{-i\boldsymbol{k}\cdot\boldsymbol{R}_j} h_{n'}(\boldsymbol{R}_j) \tag{1.53}$$

replacing $\boldsymbol{k}$ by $-i\nabla$ we see that

$$E_{n'}(-i\nabla)\psi_{n'}(\boldsymbol{r}) = \sum_j \mathrm{e}^{-\nabla\cdot\boldsymbol{R}_j} h_{n'}(\boldsymbol{R}_j)\psi_{n'}(\boldsymbol{r}) = \sum_j h_{n'}(\boldsymbol{R}_j)\psi_{n'}(\boldsymbol{r} - \boldsymbol{R}_j) \tag{1.54}$$

where we have used the relation

$$\mathrm{e}^{-\nabla\cdot\boldsymbol{R}_j}\psi_{n'}(\boldsymbol{r}) = \psi_{n'}(\boldsymbol{r} - \boldsymbol{R}_j) \tag{1.55}$$

The equation

$$\Big[E_{n'}(-i\nabla) - E\Big]\psi_{n'}(\boldsymbol{r}) + \sum_n V_{n'n}(\boldsymbol{r})\psi_n(\boldsymbol{r}) = 0 \tag{1.56}$$

is therefore seen to be equivalent to Eq. 1.51. Equation 1.56 is the differential equation for envelope function ψ_n.

To understand $\psi_n(\boldsymbol{r})$ more clearly, let us expand the wave function given by Eq. 1.47 in terms of Bloch functions:

$$\begin{aligned}\psi_n &= \frac{\sqrt{\Omega}}{N} \sum_i \int \psi_n(\boldsymbol{R}_i)\mathrm{e}^{-i\boldsymbol{k}\cdot\boldsymbol{R}_i} u_{n\boldsymbol{k}}(\boldsymbol{r})\mathrm{e}^{i\boldsymbol{k}\cdot\boldsymbol{r}}\mathrm{d}\boldsymbol{k} \\ &= \sqrt{\frac{\Omega}{N}} \sum_{\boldsymbol{k}} G_\nu(\boldsymbol{k})\mathrm{e}^{i\boldsymbol{k}\cdot\boldsymbol{r}} u_{n\boldsymbol{k}}(\boldsymbol{r})\end{aligned} \tag{1.57}$$

where

$$G_n(\boldsymbol{k}) = \frac{1}{\sqrt{N}} \sum_i \psi_n(\boldsymbol{R}_i)\mathrm{e}^{-i\boldsymbol{k}\cdot\boldsymbol{R}_i}$$

Replacing $\boldsymbol{R}_i$ by $\boldsymbol{r}$,

$$G_n(\boldsymbol{k}) = \sqrt{\frac{\Omega}{N}} \int \psi_n(\boldsymbol{r})\mathrm{e}^{-i\boldsymbol{k}\cdot\boldsymbol{r}}\mathrm{d}\boldsymbol{r}$$

we see that $G_n(\boldsymbol{k})$ is the Fourier transform of $\psi_n(\boldsymbol{r})$.

Assuming that $u_{n\boldsymbol{k}}(\boldsymbol{r})$ does not vary much with $\boldsymbol{k}$, Eq. 1.57 becomes

$$\psi_n \approx \sqrt{\frac{\Omega}{N}} u_n(\boldsymbol{r}) \int G_n(\boldsymbol{k})\mathrm{e}^{i\boldsymbol{k}\cdot\boldsymbol{r}}\mathrm{d}\boldsymbol{k} = u_n(\boldsymbol{r})\psi_n(\boldsymbol{r}) \tag{1.58}$$

where u_n is the Bloch function at the optimum point of interest. The function $\psi_n(\boldsymbol{r})$ therefore acts as a modulation function for the function u_n which may vary rapidly in a single cell, and is usually called the envelope function.

1.5.2 *Effective Mass Approximation*

We consider the conduction band of a semiconductor nanostructure composed of materials A and B that are described by energy dispersion relations of $E_n(\boldsymbol{k})$ in the form of

$$E_n(\boldsymbol{k}) = E_{n0} + \frac{\hbar^2 k_{xy}^2}{2m_{nt}^*} + \frac{\hbar^2 k_z^2}{2m_{n\ell}^*} \tag{1.59}$$

where m_{nt}^* and $m_{n\ell}^*$ are transverse and longitudinal effective mass along the z axis and in the xy plane, and E_{n0} is the conduction band edge of material n ($n = A$ and B). Replacing $\boldsymbol{k}$ by $-i\nabla$, we immediately come up to the well-known effective mass approximation from Eq. 1.56

$$\left[-\frac{\hbar^2}{2m_t^*(\boldsymbol{r})}\left(\frac{\partial^2}{\partial x^2} + \frac{\partial^2}{\partial y^2}\right) - \frac{\hbar^2}{2m_\ell^*(\boldsymbol{r})}\frac{\partial^2}{\partial z^2} + E_{n0}(\boldsymbol{r})\right]\psi(\boldsymbol{r}) = E\psi(\boldsymbol{r}) \tag{1.60}$$

where we have neglected the coupling of the conduction band with other bands (most importantly the valence band) in the above equation, which is valid for the most commonly used semiconductor materials.

More explicitly we concentrate on a one-dimensional heterostructure AB grown along the z direction with the heterointerface at $z = 0$. Assume $E_{A0} < E_{B0}$ (i.e., material A is the quantum well and B the barrier) and take E_{A0} as the energy reference. For energy $E \leq \Delta E_c = E_{B0} - E_{A0}$,

$$\left[-\frac{\hbar^2}{2m_{At}^*}\left(\frac{\partial^2}{\partial x^2} + \frac{\partial^2}{\partial y^2}\right) - \frac{\hbar^2}{2m_{A\ell}^*}\frac{\partial^2}{\partial z^2} + V(\boldsymbol{r})\right]\psi(\boldsymbol{r}) = E\psi(\boldsymbol{r}) \tag{1.61}$$

where $V(\boldsymbol{r}) = \Delta E_c$ for $z \geq 0$ and $V(\boldsymbol{r}) = 0$ otherwise. Since $V(\boldsymbol{r})$ is xy-independent,

$$\begin{aligned}
&\psi(\boldsymbol{r}) = \mathrm{e}^{i\boldsymbol{\rho}\cdot\boldsymbol{k}_{xy}}\psi(z) \\
&E = E_z + \frac{\hbar^2 k_{xy}^2}{2m_{At}^*} \\
&\left[-\frac{\hbar^2}{2m_{A\ell}^*}\frac{\partial^2}{\partial z^2} + V(z)\right]\psi(z) = E_z\psi(z)
\end{aligned} \tag{1.62}$$

where $\boldsymbol{\rho} = (x, y)$. It is noticed by the previous sub-section that the total wave function is $\psi(\boldsymbol{r})u(\boldsymbol{r})$, where $u(\boldsymbol{r})$ is the Bloch function.

In Eq. 1.61, for $E \leq \Delta E_c$ the effective mass of composite material A, $m_{A\ell}^*$, applies across the whole system, including the spatial region of material B, since the effective mass of material B, $m_{B\ell}^*$, is only defined for $E \geq \Delta E_c$. For $E \geq \Delta E_c$, the effective mass is spatial-position dependent, $m^*(z) = m_{B\ell}^*$ for $z \geq 0$ and $m^*(z) = m_{A\ell}^*$ otherwise. Extended discussions about the effective mass approximation were reported,[49–54] where the effective mass m^* is spatial-position dependent. It is generally accepted that

$$\left\{-\frac{\hbar^2}{2}\frac{\partial}{\partial z}\left[\frac{1}{m^*(z)}\frac{\partial}{\partial z}\right] + V(z)\right\}\psi(z) = E\psi(z) \tag{1.63}$$

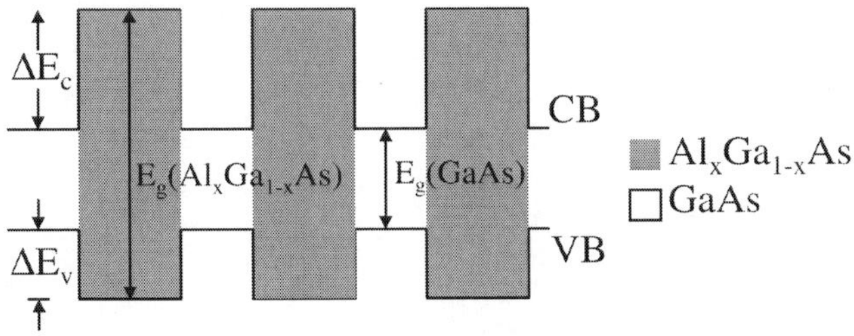

Fig. 1.8 A schematic illustration of one-dimensional square-well barrier model for conduction-band (CB) electrons and valence-band (VB) holes in GaAs/$Al_xGa_{1-x}As$ multiple quantum well system.

is the most reasonable choice, where

$$\psi(z) \text{ and } \frac{1}{m^*(z)}\frac{\partial\psi(z)}{\partial z} \tag{1.64}$$

are continuous across the system.

Practically Eq. 1.63 applies to both $E \leq \Delta E_c$ and $E \geq \Delta E_c$.

Conventionally the III–V heterostructure is referred to as type-I, as the quantum wells for both conduction-band electrons and valence-band holes locate at the same spatial regions, as schematically illustrated in Fig. 1.8. A band-offset coefficient is generally defined as

$$\frac{\Delta E_c}{\Delta E_c + \Delta E_v} = \frac{\Delta E_c}{\Delta E_g} \tag{1.65}$$

where $\Delta E_g = E_g(Al_xGa_{1-x}As) - E_g(GaAs)$ is the difference between energy band gaps of $Al_xGa_{1-x}As$ and GaAs. For III–V heterostructures, this coefficient is generally accepted as 0.65. The valence band offset between GaAs and $Al_xGa_{1-x}As$ is therefore $0.35\Delta E_g$. This expression is normally valid for direct band-gap materials. It becomes complicated when indirect-band-gap materials are involved.

For GaAs/$Al_xGa_{1-x}As$ heteromaterials, we have to consider the $\Gamma - X$-mixing effect when the Al mole fraction becomes larger than 0.35.[55] First of all, a valence band offset of $0.5x$ eV was obtained by the sp^3s^* tight-binding model, which fits very well with experiments.[56] The band offset between Γ bands of GaAs and $Al_xGa_{1-x}As$ is $1.247x + 1.147(x - 0.45)^2 - 0.5x$ eV when $x > 0.45$, as the Γ band gap of $Al_xGa_{1-x}As$ is $1.424 + 1.247x + 1.147(x - 0.45)^2$ eV.[57]

Group-IV hetero-materials, like Si/$Si_{1-x}Ge_x$, are much more complicated. While the conduction band consists of six ellipsoids described by longitudinal and transverse effective masses, the valence band must be described in the framework of the $\boldsymbol{k} \cdot \boldsymbol{p}$ theory due to the complications in the heavy-hole, light-hole, and spin-split-off bands, and their dependence on the strain due to lattice mismatch.

Lattice-mismatch-induced strain also occurs in the III–V system, an $In_xGa_{1-x}As$ layer grown on GaAs substrate is strained. The strain-induced conduction band

Table 1.4 Physical parameters for commonly used semiconductor materials in modern electronics and optoelectronics.

Parameter	Si	Ge	GaAs	InAs	AlAs
c_{11} [10^{11}dyn/cm^2]	16.56	12.853	11.88	8.33	12.02
c_{12} [10^{11}dyn/cm^2]	6.39	4.826	5.38	4.53	5.70
a_0 [Å]	5.4309	5.6561	5.6533	6.0536	5.660
a [eV]			-5.7	-6.0	
$E_{g,\mathrm{dir}}$ [eV] 300 K	4.135	0.805	1.424	0.354	3.03
$E_{g,\mathrm{ind}}$ [eV] 300 K	1.124	0.664			2.153
$E_{g,\mathrm{dir}}$ [eV] 4 K	4.185	0.898	1.519	0.418	3.13
$E_{g,\mathrm{ind}}$ [eV] 4 K	1.170	0.744			2.229
$\epsilon(0)/\epsilon_0$	11.9	16.0	12.4	15.15	10.06
m_ℓ^*/m_t^* [m_0]	0.19/0.98	1.64/0.082	0.067	0.0239	0.26
m_{lh}^*	0.16	0.044	0.082	0.026	0.153
m_{hh}^*	0.49	0.28	0.45	0.35	0.409

Source: From Refs. [57–62].

modification is given by[58, 59]

$$\delta E_c(x) = 2a(x)\left[1 - \frac{c_{12}(x)}{c_{11}(x)}\right]\left[\frac{a_0(\mathrm{GaAs})}{a_0(x)} - 1\right] \tag{1.66}$$

It is easy to obtain that $\delta E_c(0.15) = 64.86$ meV for $\mathrm{In_{0.15}Ga_{0.85}As}$ layer grown on a GaAs layer.

Table 1.4 lists important physical parameters of commonly used semiconductor materials in modern electronics and optoelectronics.

A good example of the one-dimensional heterostructure material is a GaAs/AlGaAs asymmetrically coupled quantum well system. Grown on the [001] oriented semi-insulated GaAs substrate, the sample consists of a 500 nm-GaAs buffer layer, a 50 nm-$\mathrm{Al_{0.53}Ga_{0.47}As}$ barrier, a 3 nm-GaAs well, a 2 nm-AlAs barrier, a 7 nm-GaAs well, a 50 nm-$\mathrm{Al_{0.53}Ga_{0.47}As}$ barrier, and finally a 20 nm-GaAs cap layer.

Ion implantations have been performed to enhance the interdiffusion of Al atoms and consequently modify the quantum potential energy for various electronics and optoelectronics applications. After the ion implantation, the quantum well is changed from a square well to an error-function-like well. Taking the center of the AlAs barrier (denoted as $\mathrm{Al}_{x_2}\mathrm{Ga}_{1-x_2}\mathrm{As}$, $x_2 = 1.0$) as $z = 0$, and the AlAs barrier width as $2h$, the two heterointerfaces between GaAs quantum wells and the AlAs barrier locate at $z = -h$ and $z = h$, $h = 10$ Å. The other two heterointerfaces between GaAs quantum wells and $\mathrm{Al}_{x_1}\mathrm{Ga}_{1-x_1}\mathrm{As}$ barriers ($x_1 = 0.53$) locate then at $z_1 = -40$ Å and $z_2 = 80$ Å. The degree of the Al diffusion across heterointerfaces is given by the diffusion equation of[63–66]

$$x(z) = \frac{x_1}{2}\left[2 - \mathrm{erf}\left(\frac{z - z_1}{2L}\right) + \mathrm{erf}\left(\frac{z - z_4}{2L}\right)\right] + \\ + \frac{x_2}{2}\left[\mathrm{erf}\left(\frac{z + h}{2L}\right) - \mathrm{erf}\left(\frac{z - h}{2L}\right)\right] \tag{1.67}$$

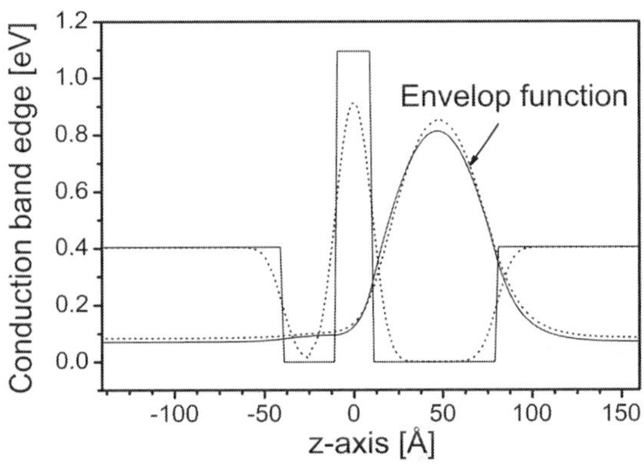

Fig. 1.9 Conduction band edge and envelop function (vertically shifted up by the energy position of the ground sublevel) of a GaAs/AlGaAs asymmetrically coupled quantum-well system. Solid lines: $L = 0.1$ Å; dotted lines: $L = 5.1$ Å.

where L is the diffusion length, and erf the error function. Here the same diffusion length is assumed at the GaAs/AlGaAs and GaAs/AlAs heterointerfaces.

$$\frac{1}{m^*(z)} = \frac{x(z)}{m^*_{\text{AlAs}}} + \frac{1 - x(z)}{m^*_{\text{GaAs}}} \tag{1.68}$$

in the virtual crystal approximation and

$$V(z) = 0.65 \times 1.247x(z) \text{ eV} \tag{1.69}$$

taking the conduction band edge of GaAs as zero reference energy. Similar equations can be written down for heavy holes. Figure. 1.9 shows the envelop function and energy position of the ground sublevel in the conduction band. By the position-dependent effective mass equation of Eq. 1.68, the ground sublevel in the conduction band is 59.1 meV when $L = 0.1$ Å; it becomes 75.2 meV when $L = 5.1$ Å. However, applying the effective mass of GaAs across the heterostructure, the ground sublevel becomes 70.0 and 83.5 meV, respectively, for $L = 0.1$ and 5.1 Å, whereas the envelop function is only slightly modified.

1.6 DIMENSIONALITY OF ENERGY DENSITY OF STATES

The density of energy states is the number of available electronic states per unit volume per unit energy. It is observed quantum mechanically that each electronic state can be occupied by two electrons (spin up and down). Here we exclude the magnetic field. In a three-dimensional bulk material, each electronic state is represented by its energy $E_{n\boldsymbol{k}}$ and wave vector $\boldsymbol{k}$ (Bloch theorem):

$$\psi^{3D}_{n\boldsymbol{k}}(\boldsymbol{r}) = u_{n\boldsymbol{k}}(\boldsymbol{r})e^{i\boldsymbol{k}\cdot\boldsymbol{r}} \tag{1.70}$$

where n is the band index and $u_{n\boldsymbol{k}}(\boldsymbol{r})$ is the Bloch function.

Assume periodic boundaries of the lattice with a period L_x in the x direction, L_y in the y direction, and L_z in the z direction,

$$\psi^{3\mathrm{D}}_{n\boldsymbol{k}}(x,y,z) = \psi^{3\mathrm{D}}_{n\boldsymbol{k}}(x+L_x,y,z) = \psi^{3\mathrm{D}}_{n\boldsymbol{k}}(x,y+L_y,z) = \psi^{3\mathrm{D}}_{n\boldsymbol{k}}(x,y,z+L_z)$$

we have the following expression for the wave vector k_i

$$k_i = \frac{2m_i\pi}{L_i} \tag{1.71}$$

where $i = x, y, z$ and m_i is an integer. When L_i is large, the spacing between the allowed $\boldsymbol{k}$ values is small so that we are able to discuss the volume in the $\boldsymbol{k}$-space that each electronic state occupies

$$\frac{(2\pi)^3}{L_xL_yL_z} = \frac{(2\pi)^3}{\Omega}$$

where $\Omega = L_xL_yL_z$ is the volume of the crystal. The total number of electronic states within $\mathrm{d}\boldsymbol{k}$ in the $\boldsymbol{k}$-space is

$$\frac{2\mathrm{d}\boldsymbol{k}}{\frac{(2\pi)^3}{\Omega}} \tag{1.72}$$

Here the factor of "2" in front of $\mathrm{d}\boldsymbol{k}$ denotes the spin degeneracy. We introduce the density of states per unit volume per energy interval

$$N_3(E)\mathrm{d}E = \int_{\boldsymbol{k}(E)}^{\boldsymbol{k}(E+\mathrm{d}E)} \frac{2\mathrm{d}\boldsymbol{k}}{(2\pi)^3} \tag{1.73}$$

by the energy dispersion relationship $E = E_n(\boldsymbol{k})$ for a three-dimensionally extended system.

For a spherical-parabolic conduction band described by

$$E = E_c + \frac{\hbar^2k^2}{2m^*} \tag{1.74}$$

where E_c is the conduction band edge and m^* is the effective mass of the conduction band electron (we concentrate on the conduction band, whereas the extension to the valence band is straightforward). We re-formulate Eq. 1.74

$$k^2 = \frac{2m^*(E-E_c)}{\hbar^2}$$

so that

$$\mathrm{d}\boldsymbol{k} = 4\pi^2k^2\mathrm{d}k = 2\pi^2k\mathrm{d}k^2 = 2\pi^2\left(\frac{2m^*}{\hbar^2}\right)^{3/2}\sqrt{E-E_c}\,\theta(E-E_c)\,\mathrm{d}E$$

$\theta(E)$ is the step function of E ($\theta = 1$ if $E \geq 0$; $\theta = 0$ when $E < 0$) which is introduced due to the requirement that $\boldsymbol{k}$ is real. The three-dimensional density of states $N_3(E)$ defined by Eq. 1.73 is

$$N_3(E) = \frac{1}{2\pi}\left(\frac{2m^*}{\hbar^2}\right)^{3/2}\sqrt{E-E_c}\,\theta(E-E_c) \tag{1.75}$$

When the system becomes confined (e.g., quantum well system) along, say, the z-axis, the translational symmetry in this direction is broken so that the corresponding wave vector k_z is replaced by the index i of discreet energy levels in this direction. The sublevel i is denoted as

$$\begin{aligned}\psi_{c\boldsymbol{k}i}^{2\mathrm{D}}(\boldsymbol{\rho}, z) &= \psi_i(z)u_c(\boldsymbol{r})\mathrm{e}^{i\boldsymbol{k}\cdot\boldsymbol{\rho}} \\ E_i(\boldsymbol{k}) &= E_i + E_c + \frac{\hbar^2 k^2}{2m^*}\end{aligned} \tag{1.76}$$

where $\boldsymbol{k}$ and $\boldsymbol{\rho}$ are wave vector and spatial coordinate in the xy plane. To make the equations simple, here we have approximated the energy band structure of the semiconductor as spherical and parabolic and the Bloch function $u_c(\boldsymbol{r})$ as $\boldsymbol{k}$-independent. The approach is valid when the energies of states under discussion are close to the band edges.

It is easy to obtain the two-dimensional density of states $N_2(E)$ as

$$N_2(E) = \frac{m^*}{\pi\hbar^2 L_z}\sum_i \theta(E - E_c - E_i) \tag{1.77}$$

where L_z is the effective width of the system in the quantum confinement direction (it is reminded that the density of states is defined as per unit volume).

When the dimensionality of the system is further decreased in the y-direction (so that the system is confined in the yz plane, for example, a quantum wire extended along the x direction), we come up with a one-dimensional system,

$$N_1(E) = \frac{1}{2\pi S_{yz}}\sqrt{\frac{2m^*}{\hbar^2}}\sum_i \frac{1}{\sqrt{E - E_c - E_i}}\theta(E - E_c - E_i) \tag{1.78}$$

where S_{yz} is the effective area of the system in the yz plane and E_i is the sublevel due to the confinement in the yz plane.

Finally we have the "zero" dimensional system, either an artificial quantum dot or a natural atom consisting of a set of discrete energy states E_i,

$$N_0(E) = \frac{1}{\Omega_{xyz}}\sum_i \delta(E - E_i) \tag{1.79}$$

where Ω_{xyz} is the effective volume of the quantum dot. Including the relaxation processes, the δ function is replaced by

$$\delta(E - E_i) \to \frac{\Gamma}{(E - E_i)^2 + \Gamma^2} \tag{1.80}$$

where Γ is the relaxation energy, which will be closely discussed in Chapter 2.

Figure. 1.10 shows the energy densities of conduction-band states of GaAs materials. We consider the conduction band structures of GaAs materials embedded in a barrier material (infinitely high barrier). 3D: bulk GaAs material. 2D: GaAs quantum well with a well width of 100 Å. 1D: GaAs quantum wire having a rectangular cross section of 100×75 Å^2. 0D: GaAs quantum dot of $100 \times 75 \times 50$ Å^3.$\Gamma = 1$ meV. We note that as the dimensionality of the system changes, the energy dependence

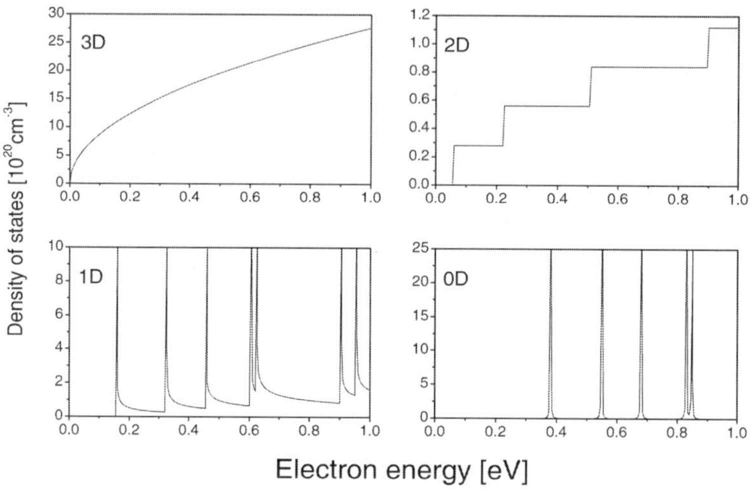

Fig. 1.10 Dimensionality of energy density of states.

of the density of states also changes. As shown in Fig. 1.10, for spherical-parabolic band, we have a $E^{1/2}$-dependence for a three-dimensionally extended system; for a two-dimensional system, the density of states is constant; and for a one-dimensional system, we have an $E^{-1/2}$-dependence. And finally we have the δ-function form for a zero-dimensional system.

References

[1] R. H. Dennard, F. H. Gaensslen, H. N. Yu, V. L. Rideout, E. Bassous, and A. R. LeBlanc, "Design of ion-implanted MOSFET's with very small physical (1974) dimensions. *IEEE J. Solid-State Circuits*, **sc9**, 256–268.

[2] G. A. Sai-Halasz, M. R. Wordeman, D. P. Kern, E. Ganin, S. Rishton, D. S. Zicherman, H. Schmidt, M. R. Polcari, H. Y. Ng, P. J. Restle, T. H. P. Chang, and R. H. Dennard. Design and experimental technology for 0.1-μm gate-length low-temperature operation FETs. IEEE Electron Device Lett, **EDL-8**, 463–466, 1987.

[3] T. Hashimoto, Y Sudoh, H. Kurino, A. Narai, S. Yokoyama, Y. Horiike, and M. Koyanagi. (1992) 3 V operation of 70 nm gate length MOSFET with new double punchthrough stopper structure," Ext. Abs. 1992 International Conferencion on Solid State Devices Materials, Tokyo, Japan, xxviii+772 pp. p.490–492, 1992.

[4] M. Ono, M. Saito, T. Yoshitomi, C. Fiegna, T. Ohguro, and H. Iwai. (1995) A 40 nm gate length n-MOSFET. *IEEE Trans. Electron Devices*, **42**, 1822–1830.

[5] H. Kawaura, T. Sakamoto, T. Baba, Y. Ochiai, J. Fujita, S. Matsui, and J. Sone. (1998) Transistor operation of 30 nm gate length EJ-MOSFETs. *IEEE Electron Device Lett*, **19**, 74–76.

[6] W. Langen, H. Hardtdegen, H. Luth, and P. Kordos, "Selectively grown vertical sub-100 nm dual-gate GaAs FETs," Compound Semiconductors 1994, Proceedings of 21st

International Symposium, IOP Publishing, Bristol, UK, 1995; xxvii+912 pp. p. 183–188.

[7] D. Behammer, L. Vescan, R. Loo, J. Moers, A. Mück, H. Lüth, and T. Grabolla. (1996) Selectively grown vertical Si-p MOS transistor with short channel lengths 130 nm Electron. Lett., **32**, 406–407,

[8] W. Langen, L. Vescan, R. Loo, H. Luth, and P. Kordos. (1996) Vertical 100 nm Si-p channel JFET grown by selective epitaxy. Appl. Surf. Sci., **102**, 252–254.

[9] D. Klaes, J. Moers, A. Tonnesmann, S. Wickenhauser, L. Vescan, M. Marso, T. Grabolla, M. Grimm, and H. Luth. (1998) Selectively grown vertical Si MOS transistor with reduced overlap capacitances. *Thin Solid Films*, **336**, 306–308.

[10] J. Moers, D. Klaes, A. Tonnesmann, L. Vescan, S. Wickenhauser, M. Marso, P. Kordos, and H. Luth. (1999) 19 GHz vertical Si p-channel MOSFET. Electron. Lett., **35**, 239–240.

[11] J. Moers, D. Klaes, A. Tonnesmann, V. Vescan, S. Wickenhauser, T. Grabolla, M. Marso, P. Kordos, and H. Luth. (1999) Vertical p-MOSFETs with gate oxide deposition before selective epitaxial growth. *Solid State Electron.*, **43**, 529–535.

[12] T. Murotani, I. Naritake, T. Matano, T. Ohtsuki, N. Kasai, H. Koga, K. Koyama, K. Nakajima, H. Yamaguchi, H. Watanabe, and T. Okuda, "A 4-level storage 4Gb DRAM," 1997 IEEE International Solid-State Circuits Conference, San Francisco, CA, Feb 1997, p. 74.

[13] H. Yasuda, "Multimedia impact on devices in the 21st century," 1997 IEEE International Solid-State Circuits Conference, San Francisco, CA, Feb 1997, p. 28.

[14] V. V. Mitin, Z. S. Gribnikov, and A. N. Korshak, "Quantum heterostructures for micro- and opto-electronics," Physics of Semiconductor Devices, Narosa Publishing House, Delhi, India, 1998; 2 vol. xxviii+1322 pp. p. 134–141 vol. 1.

[15] J. Appenzeller, C. H. Schroer, T. H. Schapers, A. van der Hart, A. Forster, B. Lengeler, and H. Luth, "Electron interference in a T-shaped quantum transistor based on Schottky-gate technology," Phys. Rev., **B53**, 9959–9963, 1996.

[16] C. G. Smith, M. Pepper, R. Newbury, H. Ahmed, D. G. Hasko, D. C. Peacock, J. E. P. Frost, D. A. Ritchie, G. A. C. Jones, and G. Hill, "One-dimensional quantised ballistics resistors in parallel configuration," J. Phys. Condens. Matter, **1**, 6763–6770, 1989.

[17] Y. Avishai, M. Kaveh, S. Shatz and Y. B. Band, "Ballistic electronic conductance of two parallel channels," J. Phys. Condens. Matter, **1**, 6907–6912, 1989.

[18] E. Castaño and G. Kirczenow, "Theory of the conductance of parallel ballistic constrictions," Phys. Rev., **B41**, 5055–5060, 1990.

[19] X. Baie, J. P. Colinge, V. Bayot, and E. Grivei, "Quantum-wire effects in thin and narrow SOI MOSFETs," 1995 IEEE International SOI Conference Proceedings (Cat.No.95CH35763), IEEE, New York, NY, USA, 1995; xiv+183 pp. p. 66–67; J. P. Colinge, X. Baie, V. Bayot, and E. Grivei, "A silicon-on-insulator quantum wire," Solid-State-Electron., **39**, 49–51, 1996.

[20] G. Kirczenow, A. S. Sachrajda, Y. Feng, R. P. Taylor, L. Henning, J. Wang, P. Zawadzki, and P. T. Coleridge, "Artificial impurities in quantum wires: from classical to quantum behaviour," Phys. Rev. Lett., **72**, 2069–2072, 1994.

[21] G. Kirczenow, "Scattering models of conduction around an antidot in a magnetic field," Phys. Rev., **B50**, 1649–1955, 1994; and references therein.

[22] J. P. Sun, G. I. Haddad, P. Mazumder, and J. N. Shulman, "Resonant tunneling diodes: Models and properties," Proc. IEEE, vol. 86, p. 641–661, 1998; P. Mazumder, S. Kulkarni, M. Bhattacharya, J. p. Sun, and G. I. Haddad, "Digital circuit applications of resonant tunneling diodes," Proc. IEEE, **86**, 664–686, 1998.

[23] J. Y. Andersson, J. Alverbro, J. Borglind, P. Helander, H. Martin, and M. Ostlund, "320*240 pixels quantum well infrared photodetector (QWIP) array for thermal imaging: fabrication and evaluation," Proc. SPIE, vol. 3061, p. 740–748, 1997; B. Brill and G. Sarusi, "QWIP research and development of 320*256 QWIP arrays in EL-OP," Proc. SPIE, **3061**, 781–788, 1997.

[24] D. A. Allee, A. N. Broers, and R. F. W. Pease, "Limits of nano-gate fabrication," Proc IEEE, **79**, 1093–1105, 1991.

[25] H. Ahmed, "Nanostructure fabrication," Proc. IEEE, **79**, 1140–1148, 1991.

[26] S. Brittain, K. Paul, X.-M. Zhao, and G. Whitesides, "Soft lithography and microfabrication," Phys. World, **11**, 31–36, 1998.

[27] H. Luth, "Tunneling in semiconductor nanostructures: physics and devices," Acta-Physica Polonica A, vol. 90, p. 667–679, 1996; "Semiconductor nanostructures for device applications," Physics of Semiconductor Devices, Narosa Publishing House, Delhi, India, 1998, 2 vol. xxviii+1322 pp. 18–25 vol. 1; H. Luth, "Semiconductor nanostructures: a new impact on electronics," Appl. Surf. Sci. **130–132**, 855–865, 1998.

[28] G. Dresselhaus, A. F. Kip, and C. Kittel, "Cyclotron resonance of electrons and holes in silicon and germanium crystals," Phys. Rev., **98**, 368–384, 1955.

[29] M. Tiersten, "Acoustic-mode scattering of holes," Ibbox J. Res. Dev., **5**, 122–131, 1961.

[30] G. L. Bir and G. E. Pikus, "Theory of the deformation potential for semiconductors with a complex band structure," Soviet Phys.-Solid State, **2**, 2039, 1960.

[31] E. O. Kane, Chapter 3 The $\boldsymbol{k} \cdot \boldsymbol{p}$ Method, in *Semiconductors and Semimetals*, ed. R. K. Willardson and A. C. Beer, vol. 1, Physics of III–V Compound, Academic Press, New York, 1966, 75–100.

[32] P. Enders, A. Bärwolff, M. Woerner, and D. Suisky, "$\boldsymbol{k} \cdot \boldsymbol{p}$ theory of energy bands, wave functions, and optical selection rules in strained tetrahedral semiconductors," Phys. Rev. B, **51**, 16695–16704, 1995.

[33] O. Stier, M. Grundmann, and D. Bimberg, "Electronic and optical properties of strained quantum dots modeled by 8-band $\boldsymbol{k} \cdot \boldsymbol{p}$ theory," Phys. Rev. B, **59**, 5688–5701, 1999.

[34] I. Vurgaftman, J. R. Meyer, and L. R. Ram-Mohan, "Band parameters for III-V compound semiconductors and their alloys," J. Appl. Phys., **89**, 5815–5875, 2001.

[35] K.-N. Tu, J. W. Mayer, and L. C. Feldman, *Electronic Thin Film Science: For Electrical Engineers and Materials Scientists*, New York, Macmillan, 1992.

[36] F. M. Ross, J. Tersoff, and R. M. Tromp, "Coarsening of self-assembled Ge quantum dots on Si(001)," Phys. Rev. Lett., **80**, 984–987, 1998.

[37] K. Brunner and G. Abstreiter, "Ordering and electronic properties of self-assembled Si/Ge quantum dots," Jpn. J. Appl. Phys., **40**, 1860–1865, 2001.

[38] J. Ibáñez, A. Patanè, M. Henini, L. Eaves, S. Hernández, R. Cuscó, L. Artús, Y. G. Musikhin, and P. N. Brounkov, Strain relaxation in stacked InAs/GaAs quantum dots studied by Raman scattering," Appl. Phys. Lett. **83**, 3069(3), 2003.

[39] K. Fu and Y. Fu, "Strain-induced Stranski-Krastanov three-dimensional growth mode of GaSb quantum dot on GaAs substrate," Appl. Phys. Lett., **94**, 181913(3), 2009.

[40] P. R. C. Kent, G. L. W. Hart, and A. Zunger, "Biaxial strain-modified valence and conduction band offsets of zincblende GaN, GaP, GaAs, InN, InP, and InAs, and optical bowing of strained epitaxial InGaN alloys," Appl. Phys. Lett., **81**, 4377–4379, 2002.

[41] A. Yu. Egorov, D. Bernklau, B. Borchert, S. Illek, D. Livshits, A. Rucki, M. Schuster, A. Kaschner, A. Hoffmann, Gh. Dumitras, M. C. Amann, and H. Riechert, "Growth

of high quality InGaAsN heterostructures and their laser application," Cryst. Growth, **227–228**, 545–552, 2001.

[42] T. Yang, S. Nakajima, and S. Sakai, "Tight-binding calculation of electronic structures of InNAs ordered alloys," Jpn. J. Appl. Phys., Part 2, **36**, 320–322, 1997.

[43] E. T. Yu, J. O. McCaldin, and T. C. McGill, "Band offsets in semiconductor heterojunctions," Solid State Phys., **46**, 1–146, 1992.

[44] *Semiconductors Group IV Elements and III–V Compounds*, ed. O. Madelung, Springer-Verlag, Berlin, 1991.

[45] H. Kroemer, K. J. Polasko, and S. C. Wight, "On the (110) orientation as the preferred orientation for the molecular beam epitaxial growth of GaAs or Ge, GaP on Si, and similar zincblende-on-diamond systems," Appl. Phys. Lett., **36**, 763–465, 1980.

[46] H. C. Casey, Jr and M. B. Panish, *Heterostructure Lasers, Part B: Materials and Operating Characteristics*, Academic, New York, 1978.

[47] W. A. Harrison, *Electronic Structure and the Properties of Solids*, Freeman, San Francisco, 1980.

[48] G. Margaritondo, A. D. Katnani, N. G. Stoffel, R. R. Daniels, and T.-X. Zhao, "Nature of the band discontinuities at semiconductor heterojunction interfaces," Solid State Commun., **43**, 163–166, 1982.

[49] S. R. White and J. L. Sham, "Electronic properties of flat-band semiconductor heterostructures," Phys. Rev. Lett., **47**, 879–882, 1981.

[50] T. Ando and S. Mori, "Effective-mass theory of semiconductor heterojunctions and superlattices," Surf. Sci., **113**, 124–130, 1982.

[51] Q. G. Zhu and H. Kroemer, "Interface connection rules for effective-mass wave functions at an abrupt heterojunction between two different semiconductors," Phys. Rev., **B27**, 3519–3527, 1983.

[52] K. B. Kahen and J. P. Leburton, "Optical constants of GaAs-$Al_xGa_{1-x}As$ superlattices and multiple quantum wells," Phys. Rev., **B33**, 5465–5472, 1986.

[53] R. A. Morrow, "Establishment of an effective-mass Hamiltonian for abrupt heterojunctions," Phys. Rev., **B35**, 8074–8079, 1987.

[54] Y. Fu and K. A. Chao, "Subband structures of GaAs/$Al_xGa_{1-x}As$ multiple quantum wells," Phys. Rev., vol. B40, p. 8349–8356, 1989; "Band offset in GaAs/$Al_xGa_{1-x}As$ multiple quantum wells calculated with the sp^3s^* tight-binding model," Phys. Rev., **B43**, 4119–4124, 1991.

[55] Y. Fu, M. Willander, E. L. Ivchenko, and A. A. Kiselev, "Valley mixing in GaAs/AlAs multilayer structures in the effective mass method," Phys. Rev., **B47**, 13498–13507, 1993.

[56] Y. Fu and K. A. Chao, "Band-offset in GaAs/AlGaAs multiple quantum wells calculated with the sp^3s^* tight-binding model," Phys. Rev., **B43**, 4119–4124, 1991.

[57] S. Adachi, "GaAs, AlAs, and $Al_xGa_{1-x}As$: Material parameters for use in research and device applications," J. Appl. Phys., **58**, R1–R29, 1985.

[58] T. G. Andersson, Z. G. Chen, V. D. Kulakovskii, A. Uddin, and J. T. Vallin, "Photoluminescence and photoconductivity measurements on band-edge offsets in strained molecular-beam-epitaxy-grown $In_xGa_{1-x}As$/GaAs quantum well," Phys. Rev., **B37**, 4032–4038, 1988.

[59] M. Beaudoin, A. Bensaada, R. Leonelli, P. Desjardins, R. A. Masut, L. Isnard, A. Chennouf, and G. L'Esperance, "Self-consistent determination of the band offsets in $InAs_xP_{1-x}$/InP strained-layer quantum wells and the bowing parameter of bulk $InAs_xP_{1-x}$," Phys. Rev., **B53**, 1990–1996, 1996.

[60] *Properties of Latticed-matched and Strained InGaAs*, ed. P. Bhattacharya, INSPEC, London, 1993.

[61] S. M. Sze, *Physics of Semiconductor Devices*, John-Wiley & Sons, New York, 1981.

[62] *Semiconductors, Group IV Elements and III-V Compounds*, ed. O. Madelung, Springer-Verlag, Berlin, 1991.

[63] J. Crank, *The Mathematics of Diffusion*, Clarendon, Oxford, 1956, pp. 9–14.

[64] G. F. Redinbo, H. G. Craighead, and J. M. Hong, "Proton implantation intermixing of GaAs/AlGaAs quantum well," J. Appl. Phys. **74**, 3099–3102, 1993.

[65] W. Feng, F. Chen, W. Q. Cheng, Q. Huang, and J. M. Zhou, "Influence of growth conditions on Al-Ga interdiffusion in low-temperature grown AlGaAs/GaAs quantum wells," Appl. Phys. Lett., **71**, 1676–1678, 1997.

[66] Na Li, Ning Li, W. Lu, X. Q. Liu, X. Z. Yuan, Z. F. Li, H. F. Dou, S. C. Shen, Y. Fu, M. Willander, L. Fu, H. H. Tan, C. Jagadish, M. B. Johnston, and M. Gal, "Proton implantation and rapid thermal annealing effects on GaAs/AlGaAs quantum well infrared photodetector," Superlattices and Microstructures, **26**, 317–324, 1999.

Chapter 2

Light-Matter Interactions

2.1 TIME-DEPENDENT PERTURBATION THEORY

We start the discussion by a formal theory of perturbation, which will be further elucidated by straightforward calculations.

2.1.1 *The Golden Rule*

Assume that H_0 is a time-independent Hamiltonian and all solutions of the eigenvalue problem $H_0|\boldsymbol{k}\rangle = E_{\boldsymbol{k}}|\boldsymbol{k}\rangle$ are known; it is straightforward to construct the solution of the equation of motion

$$i\hbar\frac{\partial|\psi(t)\rangle}{\partial t} = H_0|\psi(t)\rangle \tag{2.1}$$

which coincides at $t = 0$ with the given initial state $|\psi(0)\rangle$. Under these circumstances, the state vector

$$|\psi(t)\rangle = \mathrm{e}^{-iH_0t/\hbar}|\psi(0) \tag{2.2}$$

represents the general solution of the equation of motion, Eq. 2.1. Since

$$\sum_{\boldsymbol{k}} |\boldsymbol{k}\rangle\langle\boldsymbol{k}| = 1$$

we obtain

$$|\psi(t)\rangle = \sum_{\boldsymbol{k}} \mathrm{e}^{-iH_0t/\hbar}|\boldsymbol{k}\rangle\langle\boldsymbol{k}|\psi(0)\rangle = \sum_{\boldsymbol{k}} \mathrm{e}^{-iE_{\boldsymbol{k}}t/\hbar}|\boldsymbol{k}\rangle\langle\boldsymbol{k}|\psi(0)\rangle \tag{2.3}$$

We now calculate transition amplitudes between the relevant unperturbed eigenstates due to the appearance of an external perturbation V at $t = 0$. We transform the equation of motion in the Schrödinger picture

$$i\hbar\frac{\partial|\psi(t)\rangle}{\partial t} = (H_0 + V)|\psi(t)\rangle \tag{2.4}$$

to the interaction picture by a time-dependent unitary operator

$$|\hat{\psi}(t)\rangle = \exp{(iH_0t/\hbar)}|\psi(t)\rangle \tag{2.5}$$

so that in the interaction picture, the equation of motion for the state is

$$i\hbar\frac{\partial|\hat{\psi}(t)\rangle}{\partial t} = \hat{V}(t)|\hat{\psi}(t)\rangle \tag{2.6}$$

where the new interaction operator is given by

$$\hat{V}(t) = \exp(iH_0t/\hbar)V\exp(-iH_0t/\hbar) \tag{2.7}$$

Next we introduce a time development operator $\hat{T}(t)$ such that

$$|\hat{\psi}(t)\rangle = \hat{T}(t)|\hat{\psi}(0)\rangle \tag{2.8}$$

The time development operator satisfies the integral equation

$$\hat{T}(t) = 1 - \frac{i}{\hbar}\int_0^t \hat{V}(t)\hat{T}(t')\mathrm{d}t' \tag{2.9}$$

The relationship between the operator expressions in the two pictures is

$$\hat{T}(t) = \exp(iH_0t/\hbar)T(t)\exp(-iH_0t/\hbar) \tag{2.10}$$

The transition matrix elements of the time development operator between eigenstates of the unperturbed Hamiltonian are

$$\begin{aligned}\langle \boldsymbol{q}|\hat{T}(t)|\boldsymbol{k}\rangle &= \delta(\boldsymbol{q}-\boldsymbol{k}) - \frac{i}{\hbar}\int_0^t \langle \boldsymbol{q}|\hat{V}(t)\hat{T}(t')|\boldsymbol{k}\rangle\mathrm{d}t' \\ &= \delta(\boldsymbol{q}-\boldsymbol{k}) - \frac{i}{\hbar}\sum_{\boldsymbol{k}'}\int_0^t \langle \boldsymbol{q}|\hat{V}(t')|\boldsymbol{k}'\rangle\langle \boldsymbol{k}'|\hat{T}(t')|\boldsymbol{k}\rangle\mathrm{d}t'\end{aligned} \tag{2.11}$$

By Eq. 2.10, and note $H_0|\boldsymbol{k}\rangle = E_{\boldsymbol{k}}|\boldsymbol{k}\rangle$ and $H_0|\boldsymbol{q}\rangle = E_{\boldsymbol{q}}|\boldsymbol{q}\rangle$,

$$\langle \boldsymbol{q}|\hat{T}(t)|\boldsymbol{k}\rangle = \delta(\boldsymbol{q}-\boldsymbol{k}) - \frac{i}{\hbar}\sum_{\boldsymbol{k}'}\int_0^t \mathrm{e}^{i(E_{\boldsymbol{q}}-E_{\boldsymbol{k}'})t'/\hbar}\langle \boldsymbol{q}|V(t')|\boldsymbol{k}'\rangle\langle \boldsymbol{k}'|\hat{T}(t')|\boldsymbol{k}\rangle\mathrm{d}t' \tag{2.12}$$

The strategy of time-dependent perturbation theory is to assume that V is small and proceed with iterating Eq. 2.9 as a power series in terms of $\hat{V}$

$$\hat{T}(t) = 1 - \frac{i}{\hbar}\int_0^t \hat{V}'\mathrm{d}t' - \frac{1}{\hbar^2}\int_0^t \hat{V}'\mathrm{d}t'\int_0^{t'} \hat{V}''\mathrm{d}t'' + \cdots \tag{2.13}$$

and the transition amplitude from initial state $|\boldsymbol{k}\rangle$ to final state $|\boldsymbol{q}\rangle$ in the perturbation expansion becomes

$$\begin{aligned}\langle \boldsymbol{q}|\hat{T}(t)|\boldsymbol{k}\rangle = \delta(\boldsymbol{q}-\boldsymbol{k}) &- \frac{i}{\hbar}\int_0^t \mathrm{e}^{i(E_{\boldsymbol{q}}-E_{\boldsymbol{k}})t'/\hbar}\langle \boldsymbol{q}|V(t')|\boldsymbol{k}\rangle\mathrm{d}t' \\ &-\frac{1}{\hbar^2}\sum_{\boldsymbol{k}'\neq\boldsymbol{k}}\int_0^t \mathrm{e}^{i(E_{\boldsymbol{q}}-E_{\boldsymbol{k}'})t'/\hbar}\langle \boldsymbol{q}|V(t')|\boldsymbol{k}'\rangle \\ &\times\int_0^{t'} \mathrm{e}^{i(E_{\boldsymbol{k}'}-E_{\boldsymbol{k}})t''/\hbar}\langle \boldsymbol{k}'|V(t'')|\boldsymbol{k}\rangle\mathrm{d}t'\mathrm{d}t'' + \cdots\end{aligned} \tag{2.14}$$

It is equivalent to obtain the above equation by iteratively replacing the term $\langle \boldsymbol{k}'|\hat{T}(t')|\boldsymbol{k}\rangle$ on the right side of Eq. 2.12 by the whole right-side expression of

Eq. 2.12, i.e.,

$$\begin{aligned}
&\sum_{\boldsymbol{k}'}\int_0^{t'} \mathrm{e}^{i(E_{\boldsymbol{q}}-E_{\boldsymbol{k}'})t''/\hbar}\langle\boldsymbol{q}|V(t'')|\boldsymbol{k}'\rangle\langle\boldsymbol{k}'|\hat{T}(t'')|\boldsymbol{k}\rangle\mathrm{d}t'' \\
&= \int_0^{t'} \mathrm{e}^{i(E_{\boldsymbol{q}}-E_{\boldsymbol{k}})t''/\hbar}\langle\boldsymbol{q}|V(t'')|\boldsymbol{k}\rangle\langle\boldsymbol{k}|\hat{T}(t'')|\boldsymbol{k}\rangle\mathrm{d}t'' \\
&\quad + \sum_{\boldsymbol{k}'\neq\boldsymbol{k}}\int_0^{t'} \mathrm{e}^{i(E_{\boldsymbol{q}}-E_{\boldsymbol{k}'})t''/\hbar}\langle\boldsymbol{q}|V(t'')|\boldsymbol{k}'\rangle\langle\boldsymbol{k}'|\hat{T}(t'')|\boldsymbol{k}\rangle\mathrm{d}t''
\end{aligned} \tag{2.15}$$

By Eq. 2.12

$$\langle\boldsymbol{k}|\hat{T}(t')|\boldsymbol{k}\rangle = 1 - \frac{i}{\hbar}\sum_{\boldsymbol{k}''}\int_0^{t'} \mathrm{e}^{i(E_{\boldsymbol{k}}-E_{\boldsymbol{k}''})t''/\hbar}\langle\boldsymbol{k}|V(t'')|\boldsymbol{k}''\rangle\langle\boldsymbol{k}''|\hat{T}(t'')|\boldsymbol{k}\rangle\mathrm{d}t'' \tag{2.16}$$

The first term on the right side of the above equation results in the second term on the right side of Eq. 2.14. For $\boldsymbol{k}' \neq \boldsymbol{k}$ and by Eq. 2.12, the second term on the right side of Eq. 2.15 becomes

$$\langle\boldsymbol{k}'|\hat{T}(t')|\boldsymbol{k}\rangle = -\frac{i}{\hbar}\sum_{\boldsymbol{k}''}\int_0^{t'} \mathrm{e}^{i(E_{\boldsymbol{k}'}-E_{\boldsymbol{k}''})t''/\hbar}\langle\boldsymbol{k}'|V(t'')|\boldsymbol{k}''\rangle\langle\boldsymbol{k}''|\hat{T}(t'')|\boldsymbol{k}\rangle\mathrm{d}t'' \tag{2.17}$$

We use Eq. 2.15 to separate the term $\boldsymbol{k}'' = \boldsymbol{k}$ from $\boldsymbol{k}'' \neq \boldsymbol{k}$ in the above equation, which becomes the third term on the right side of Eq. 2.14.

Back to Eq. 2.14. If $\boldsymbol{k} \neq \boldsymbol{q}$, the transition amplitude from initial state $|\boldsymbol{k}\rangle$ to final state $|\boldsymbol{q}\rangle$ in the first-order perturbation theory is given by the simple and much-used formula

$$\langle\boldsymbol{q}|\hat{T}(t)|\boldsymbol{k}\rangle = -\frac{i}{\hbar}\int_0^{t} \mathrm{e}^{i(E_{\boldsymbol{q}}-E_{\boldsymbol{k}})t'/\hbar}\langle\boldsymbol{q}|V(t')|\boldsymbol{k}\rangle\mathrm{d}t' \tag{2.18}$$

If the perturbation is transient and sufficiently weak in magnitude, the transition amplitudes may remain small and the first-order approximation valid throughout all times. In this case, we may take the limit of $t \to \infty$ and calculate the transition amplitude from one unperturbed energy eigenstate to another by

$$\langle\boldsymbol{q}|\hat{T}(+\infty)|\boldsymbol{k}\rangle = -\frac{i}{\hbar}\int_0^{+\infty} \mathrm{e}^{i(E_{\boldsymbol{q}}-E_{\boldsymbol{k}})t/\hbar}\langle\boldsymbol{q}|V(t)|\boldsymbol{k}\rangle\mathrm{d}t \tag{2.19}$$

Note that $t = 0$ is a time preceding the onset of the perturbation.

If V does not depend on the time, or $V(t) = V_0^\dagger\mathrm{e}^{i\omega t} + V_0\mathrm{e}^{-i\omega t}$, where V_0 and ω are time-independent, the integration is trivial. We will discuss the second case in the next section.

Here we assume that V is time-independent,

$$\langle\boldsymbol{q}|\hat{T}(t)|\boldsymbol{k}\rangle = -\frac{i\langle\boldsymbol{q}|V|\boldsymbol{k}\rangle}{\hbar}\int_0^{t} \mathrm{e}^{i(E_{\boldsymbol{q}}-E_{\boldsymbol{k}})t'/\hbar}\mathrm{d}t' = \frac{\langle\boldsymbol{q}|V|\boldsymbol{k}\rangle}{\hbar}\,\frac{1-\mathrm{e}^{i(E_{\boldsymbol{q}}-E_{\boldsymbol{k}})t/\hbar}}{(E_{\boldsymbol{q}}-E_{\boldsymbol{k}})/\hbar} \tag{2.20}$$

and the probability for the system to be found in a different eigenstate of H_0 with energy $E_{\boldsymbol{q}}$ is

$$P_{\boldsymbol{q}\leftarrow\boldsymbol{k}}(t) = \left|\langle\boldsymbol{q}|\hat{T}(t)|\boldsymbol{k}\rangle\right|^2 = \frac{2}{\hbar^2}\left|\langle\boldsymbol{q}|V|\boldsymbol{k}\rangle\right|^2\frac{1-\cos\left[(E_{\boldsymbol{q}}-E_{\boldsymbol{k}})t/\hbar\right]}{(E_{\boldsymbol{q}}-E_{\boldsymbol{k}})^2/\hbar^2} \tag{2.21}$$

This is an oscillating function of t with a period $2\pi\hbar/|E_{\boldsymbol{q}} - E_{\boldsymbol{k}}|$, except at or near $E_{\boldsymbol{q}} - E_{\boldsymbol{k}} = 0$. Its derivation presupposes that the time development operator remains close to its initial value of unity in the time interval $(0, t)$. During such times, the probability $P_{\boldsymbol{k}\leftarrow\boldsymbol{k}}(t) \approx 1$, and the transition probabilities to states $\boldsymbol{q} \neq \boldsymbol{k}$ must stay small.

From Eq. 2.18 we see that as long as $|(E_{\boldsymbol{q}} - E_{\boldsymbol{k}})t|/\hbar \ll 1$, transitions to states with $E_{\boldsymbol{q}} \approx E_{\boldsymbol{k}}$ have amplitudes proportional to t, and Eq. 2.21 becomes approximately

$$P_{\boldsymbol{q}\leftarrow\boldsymbol{k}}(t) \approx \frac{1}{\hbar^2}\left|\langle\boldsymbol{q}|V|\boldsymbol{k}\rangle\right|^2 t^2 \tag{2.22}$$

We consider the total transition probability to all final states under consideration,

$$\sum_{\boldsymbol{q}} P_{\boldsymbol{q}\leftarrow\boldsymbol{k}}(t) = \frac{2}{\hbar^2}\sum_{\boldsymbol{q}}\left|\langle\boldsymbol{q}|V|\boldsymbol{k}\rangle\right|^2 \frac{1-\cos\left[(E_{\boldsymbol{q}} - E_{\boldsymbol{k}})t/\hbar\right]}{(E_{\boldsymbol{q}} - E_{\boldsymbol{k}})^2/\hbar^2} \tag{2.23}$$

In usual practice, $|\langle\boldsymbol{q}|V|\boldsymbol{k}\rangle|^2$ is sensibly constant over an energy range ΔE in the neighborhood of $E_{\boldsymbol{q}}$. By labeling collectively these states by f, we introduce the concept of density of states $N_{\mathrm{DOS}}(E_{\boldsymbol{q}})$ so that

$$\sum_{\boldsymbol{q}\in f} P_{\boldsymbol{q}\leftarrow\boldsymbol{k}}(t) = \frac{2}{\hbar^2}\int_{E_q-\Delta E/2}^{E_q+\Delta E/2}\left|\langle\boldsymbol{q}|V|\boldsymbol{k}\rangle\right|^2 \frac{1-\cos\left[(E_{\boldsymbol{q}} - E_{\boldsymbol{k}})t/\hbar\right]}{(E_{\boldsymbol{q}} - E_{\boldsymbol{k}})^2/\hbar^2} N_{\mathrm{DOS}}(E_{\boldsymbol{q}})\mathrm{d}E_{\boldsymbol{q}} \tag{2.24}$$

Again, in usual practice, $N_{\mathrm{DOS}}(E_{\boldsymbol{q}})$ is sensibly constant over an energy range ΔE in the neighborhood of $E_{\boldsymbol{q}}$. In this same interval, as a function of $E_{\boldsymbol{q}}$ the factor

$$\frac{1-\cos\left[(E_{\boldsymbol{q}} - E_{\boldsymbol{k}})t/\hbar\right]}{(E_{\boldsymbol{q}} - E_{\boldsymbol{k}})^2/\hbar^2} \tag{2.25}$$

varies rapidly for all t satisfying $t\Delta E \gg \hbar$ and has a pronounced peak at $E_{\boldsymbol{q}} - E_{\boldsymbol{k}} = 0$. The interval ΔE is generally comparable in magnitude to $E_{\boldsymbol{q}}$ itself. If this is so, $\hbar/\Delta E$ is a very short time, and there is a considerable range of t such that $t\Delta E \gg \hbar$ is fulfilled and yet the initial state $\boldsymbol{k}$ is not appreciably depleted. During such times, the perturbation method is expected to be valid, and Eq. 2.24 can be simplified to good approximation as

$$\sum_{\boldsymbol{q}\in f} P_{\boldsymbol{q}\leftarrow\boldsymbol{k}}(t) = \frac{2}{\hbar}\left|\langle\boldsymbol{q}|V|\boldsymbol{k}\rangle\right|^2 N_{\mathrm{DOS}}(E_{\boldsymbol{q}})\int_{-\infty}^{+\infty}\frac{1-\cos(\omega_{\boldsymbol{qk}}t)}{\omega_{\boldsymbol{qk}}^2}\,\mathrm{d}\omega_{\boldsymbol{qk}} \tag{2.26}$$

where $\hbar\omega_{\boldsymbol{qk}} = E_{\boldsymbol{q}} - E_{\boldsymbol{k}}$ and condition $t\Delta E \gg \hbar$ has been invoked in extending the integration limits. Under the condition stated, we can evaluate the integral for $t > 0$ as

$$\int_{-\infty}^{+\infty}\frac{1-\cos(\omega_{\boldsymbol{qk}}t)}{\omega_{\boldsymbol{qk}}^2}\,\mathrm{d}\omega_{\boldsymbol{qk}} = t\int_{-\infty}^{+\infty}\frac{\sin x}{x}\,\mathrm{d}x = \pi t \tag{2.27}$$

Hence, we can speak of a constant transition probability per unit time or transition rate $w_{\boldsymbol{q}\leftarrow\boldsymbol{k}}$

$$w_{\boldsymbol{q}\leftarrow\boldsymbol{k}} = \frac{\partial}{\partial t}\sum_{\boldsymbol{q}\in f} P_{\boldsymbol{q}\leftarrow\boldsymbol{k}}(t) = \frac{2\pi}{\hbar}\left|\langle\boldsymbol{q}|V|\boldsymbol{k}\rangle\right|^2 N_{\mathrm{DOS}}(E_{\boldsymbol{q}}) \tag{2.28}$$

which is normally addressed as the Golden rule.

2.1.2 *Exponential Decay*

Starting with an initial state $|\boldsymbol{k}\rangle$ at $t = 0$, we calculate the probabilities that after a time interval t the system still dwells in the initial state as well as the probability that the system has decayed into state $|\boldsymbol{q}\rangle$, due to a constant perturbation V switched on at $t = 0$.

Back to Eq. 2.12,

$$\langle \boldsymbol{q}|\hat{T}(t)|\boldsymbol{k}\rangle = -\frac{i}{\hbar}\langle \boldsymbol{q}|V|\boldsymbol{k}\rangle \int_0^t \mathrm{e}^{i(E_{\boldsymbol{q}}-E_{\boldsymbol{k}})t'/\hbar}\langle \boldsymbol{k}|\hat{T}(t')|\boldsymbol{k}\rangle \mathrm{d}t' \tag{2.29}$$

for $\boldsymbol{q} \neq \boldsymbol{k}$ and we keep the perturbation V time-independent. Here the transitions via intermediate states $\boldsymbol{k}' \neq \boldsymbol{k}$ are neglected, since under the first-order perturbation assumption, the possibility of a two-step process, from $\boldsymbol{k}$ to $\boldsymbol{k}'$ ($\boldsymbol{k}' \neq \boldsymbol{k}$), then from $\boldsymbol{k}'$ to $\boldsymbol{q}$ ($\boldsymbol{q} \neq \boldsymbol{k}$), is expected to be very small.

Furthermore, by Eq. 2.12, the equation of motion for $\langle \boldsymbol{k}|\hat{T}(t)|\boldsymbol{k}\rangle$ is (replacing $\boldsymbol{k}'$ by $\boldsymbol{q}$)

$$\langle \boldsymbol{k}|\hat{T}(t)|\boldsymbol{k}\rangle = 1 - \frac{i}{\hbar}\sum_{\boldsymbol{q}} \int_0^t \mathrm{e}^{i(E_{\boldsymbol{k}}-E_{\boldsymbol{q}})t'/\hbar}\langle \boldsymbol{k}|V(t')|\boldsymbol{q}\rangle\langle \boldsymbol{q}|\hat{T}(t')|\boldsymbol{k}\rangle \mathrm{d}t' \tag{2.30}$$

then, rigorously,

$$\frac{\partial}{\partial t}\langle \boldsymbol{k}|\hat{T}(t)|\boldsymbol{k}\rangle = -\frac{i}{\hbar}\sum_{\boldsymbol{q}} \mathrm{e}^{i(E_{\boldsymbol{k}}-E_{\boldsymbol{q}})t/\hbar}\langle \boldsymbol{k}|V|\boldsymbol{q}\rangle\langle \boldsymbol{q}|\hat{T}(t)|\boldsymbol{k}\rangle \tag{2.31}$$

Explicitly separate term $\boldsymbol{q} = \boldsymbol{k}$ from terms $\boldsymbol{q} \neq \boldsymbol{k}$,

$$\frac{\partial}{\partial t}\langle \boldsymbol{k}|\hat{T}(t)|\boldsymbol{k}\rangle = -\frac{i}{\hbar}\sum_{\boldsymbol{q}\neq\boldsymbol{k}} \mathrm{e}^{i(E_{\boldsymbol{k}}-E_{\boldsymbol{q}})t/\hbar}\langle \boldsymbol{k}|V|\boldsymbol{q}\rangle\langle \boldsymbol{q}|\hat{T}(t)|\boldsymbol{k}\rangle - \frac{i}{\hbar}\langle \boldsymbol{k}|V|\boldsymbol{k}\rangle\langle \boldsymbol{k}|\hat{T}(t)|\boldsymbol{k}\rangle \tag{2.32}$$

Substituting Eq. 2.29 into 2.32, we obtain the differential-integral equation for the probability amplitude that the system, at time $t \geq 0$, will still dwell in the initial state $|\boldsymbol{k}\rangle$

$$\begin{aligned}\frac{\partial}{\partial t}\langle \boldsymbol{k}|\hat{T}(t)|\boldsymbol{k}\rangle = &-\frac{1}{\hbar^2}\sum_{\boldsymbol{q}\neq\boldsymbol{k}} \left|\langle \boldsymbol{q}|V|\boldsymbol{k}\rangle\right|^2 \int_0^t \langle \boldsymbol{k}|\hat{T}(t')|\boldsymbol{k}\rangle e^{i(E_{\boldsymbol{q}}-E_{\boldsymbol{k}})(t'-t)/\hbar}\mathrm{d}t' \\ &-\frac{i}{\hbar}\langle \boldsymbol{k}|V|\boldsymbol{k}\rangle\langle \boldsymbol{k}|\hat{T}(t)|\boldsymbol{k}\rangle\end{aligned} \tag{2.33}$$

The solution of this equation demands care. We are interested in times t that imply rapid oscillations of the exponential factor in the integrand as a function of the final state energy $E_{\boldsymbol{q}}$. The slowly varying amplitude $\langle \boldsymbol{k}|\hat{T}(t')|\boldsymbol{k}\rangle$ can therefore be removed from the t' integrand, and the remaining integral can be evaluated by

$$\int_0^t \mathrm{e}^{-i(\omega-i\gamma)t'}\mathrm{d}t' = \frac{1}{i\omega+\gamma} = \pi\delta(\omega) - i\mathrm{P}\frac{1}{\omega} \tag{2.34}$$

for $\omega t \gg 1$. Here P denotes the Cauchy principal value. The resulting differential equation is solved for $t \geq 0$ by

$$\langle \boldsymbol{k}|\hat{T}(t)|\boldsymbol{k}\rangle = \exp\left[-\frac{\pi t}{\hbar}\sum_{\boldsymbol{q}\neq\boldsymbol{k}} \left|\langle \boldsymbol{q}|V|\boldsymbol{k}\rangle\right|^2\delta(E_{\boldsymbol{q}}-E_{\boldsymbol{k}}) - \frac{it}{\hbar}\Delta E_{\boldsymbol{k}}\right] \tag{2.35}$$

where we have denoted the perturbation energy shift of the level $|\boldsymbol{k}\rangle$, up to second order in V, by

$$\Delta E_{\boldsymbol{k}} = \langle \boldsymbol{k}|V|\boldsymbol{k}\rangle + \sum_{\boldsymbol{q}\neq\boldsymbol{k}} \frac{\left|\langle \boldsymbol{q}|V|\boldsymbol{k}\rangle\right|^2}{E_{\boldsymbol{q}} - E_{\boldsymbol{k}}} \tag{2.36}$$

Hence we see that

$$\langle \boldsymbol{k}|\hat{T}(t)|\boldsymbol{k}\rangle = \exp\left(-\frac{w_{\boldsymbol{k}}}{2}t - \frac{it}{\hbar}\Delta E_{\boldsymbol{k}}\right) \tag{2.37}$$

describing the exponential decay of the unstable state. Here

$$w_{\boldsymbol{k}} = \frac{2\pi}{\hbar}\sum_{\boldsymbol{q}\neq\boldsymbol{k}} \left|\langle \boldsymbol{q}|V|\boldsymbol{k}\rangle\right|^2 \delta(E_{\boldsymbol{q}} - E_{\boldsymbol{k}}) \tag{2.38}$$

describing the total transition rate from initial state $|\boldsymbol{k}\rangle$. We can see its relationship to the Golden rule of Eq. 2.28.

If Eq. 2.37 is substituted back into Eq. 2.29,

$$\langle \boldsymbol{q}|\hat{T}(t)|\boldsymbol{k}\rangle = -\frac{i}{\hbar}\langle \boldsymbol{q}|V|\boldsymbol{k}\rangle \int_0^t \mathrm{e}^{i(E_{\boldsymbol{q}}-E_{\boldsymbol{k}})t'/\hbar} \exp\left(-\frac{w_{\boldsymbol{k}}}{2}t - \frac{it}{\hbar}\Delta E_{\boldsymbol{k}}\right)\mathrm{d}t' \tag{2.39}$$

the integration can be carried out and we obtain for $t \geq 0$

$$\langle \boldsymbol{q}|\hat{T}(t)|\boldsymbol{k}\rangle = \langle \boldsymbol{q}|V|\boldsymbol{k}\rangle \frac{1 - \mathrm{e}^{-w_{\boldsymbol{k}}t/2}\mathrm{e}^{i(E_{\boldsymbol{q}}-E_{\boldsymbol{k}}-\Delta E_{\boldsymbol{q}})t/\hbar}}{E_{\boldsymbol{q}} - E_{\boldsymbol{k}} - \Delta E_{\boldsymbol{k}} + i\hbar w_{\boldsymbol{k}}/2} \tag{2.40}$$

Hence, the probability that the system has decayed into state $|\boldsymbol{q}\rangle$ is

$$P_{\boldsymbol{q}\leftarrow\boldsymbol{k}}(t) = \left|\langle \boldsymbol{q}|V|\boldsymbol{k}\rangle\right|^2 \frac{1 - 2\mathrm{e}^{-\Gamma_{\boldsymbol{k}}t/\hbar}\cos\left[(E_{\boldsymbol{q}} - E_{\boldsymbol{k}} - \Delta E_{\boldsymbol{k}})t/\hbar\right] + \mathrm{e}^{-2\Gamma_{\boldsymbol{k}}t/\hbar}}{(E_{\boldsymbol{q}} - E_{\boldsymbol{k}} - \Delta E_{\boldsymbol{k}})^2 + \Gamma_{\boldsymbol{k}}^2} \tag{2.41}$$

where $\Gamma_{\boldsymbol{k}} = \hbar w_{\boldsymbol{k}}/2$.

After a time that is very long compared with the lifetime $\hbar/\Gamma_{\boldsymbol{k}}$ of eigenstate $|\boldsymbol{k}\rangle$, we obtain the distribution

$$P_{\boldsymbol{q}\leftarrow\boldsymbol{k}}(+\infty) = \frac{\left|\langle \boldsymbol{q}|V|\boldsymbol{k}\rangle\right|^2}{(E_{\boldsymbol{q}} - E_{\boldsymbol{k}} - \Delta E_{\boldsymbol{k}})^2 + \Gamma_{\boldsymbol{k}}^2} \tag{2.42}$$

exhibiting the typical bell-shaped resonance behavior with a peak at $E_{\boldsymbol{q}} = E_{\boldsymbol{k}} + \Delta E_{\boldsymbol{k}}$ and a width $\Gamma_{\boldsymbol{k}}$. And the decay rate, i.e., the decay probability per unit time, is obtained by dividing $P_{\boldsymbol{q}\leftarrow\boldsymbol{k}}(+\infty)$ by the lifetime $\hbar/\Gamma_{\boldsymbol{k}}$

$$p_{\boldsymbol{q}\leftarrow\boldsymbol{k}} = \frac{\left|\langle \boldsymbol{q}|V|\boldsymbol{k}\rangle\right|^2}{\hbar} \frac{\Gamma_{\boldsymbol{k}}}{(E_{\boldsymbol{q}} - E_{\boldsymbol{k}} - \Delta E_{\boldsymbol{k}})^2 + \Gamma_{\boldsymbol{k}}^2} \tag{2.43}$$

Let us discuss $\Gamma_{\boldsymbol{k}}$. When $\langle \boldsymbol{q}|V|\boldsymbol{k}\rangle$ is constant in an energy range ΔE in the neighborhood of $E_{\boldsymbol{q}}$,

$$\begin{aligned}
\Gamma_{\boldsymbol{k}} = \frac{\hbar w_{\boldsymbol{k}}}{2} &= \pi \sum_{\boldsymbol{q}\neq\boldsymbol{k}} \left|\langle \boldsymbol{q}|V|\boldsymbol{k}\rangle\right|^2 \delta(E_{\boldsymbol{q}} - E_{\boldsymbol{k}}) \\
&\approx \pi \int_{\boldsymbol{q}\neq\boldsymbol{k}} \left|\langle \boldsymbol{q}|V|\boldsymbol{k}\rangle\right|^2 \delta(E_{\boldsymbol{q}} - E_{\boldsymbol{k}}) N_{\mathrm{DOS}}(E_{\boldsymbol{q}}) dE_{\boldsymbol{q}} \\
&= \pi \left|\langle \boldsymbol{q}_{\boldsymbol{k}}|V|\boldsymbol{k}\rangle\right|^2 N_{\mathrm{DOS}}(E_{\boldsymbol{q}_{\boldsymbol{k}}})
\end{aligned} \tag{2.44}$$

Here $|\boldsymbol{q}_{\boldsymbol{k}}\rangle$ is such a representative state that its eigenvalue equals $E_{\boldsymbol{k}}$, and $N_{\mathrm{DOS}}(E_{\boldsymbol{q}_{\boldsymbol{k}}})$ is the density of these states. Note that $|\boldsymbol{q}\rangle \neq |\boldsymbol{k}\rangle$!

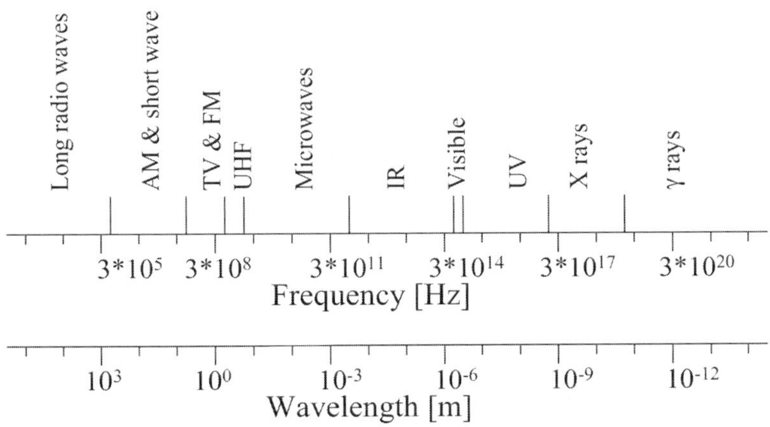

Fig. 2.1 Wavelength–frequency spectrum of electromagnetic wave (photon).

2.2 ELECTROMAGNETIC FIELD

Electromagnetic wave is associated with a combination of electric and magnetic fields that can be characterized by a wave function that varies as $e^{i(\boldsymbol{k}\cdot\boldsymbol{r}-\omega t)}$, where $\omega = 2\pi f$, f is the radiation frequency, $k = \omega/c$ is the wave vector, and c is the phase velocity, which in general differs from the group velocity $c_g = \mathrm{d}\omega/\mathrm{d}k$. The range of the electromagnetic wave frequency spans immensely, from very long-wavelength radio waves around 100 Hz to extremely high energy γ rays from space (as high as 10^{23} Hz). With the exception of the visible part of the electromagnetic spectrum shown in Fig. 2.1, the boundaries between the classifications are not sharp, which are based roughly on how the waves are generated and/or detected.

As we shall discuss the continuous interaction of electrons at energies E_1 and E_2 with photons of energy as $\hbar\omega = E_1 - E_2$, we first study the properties of photons, i.e., the electromagnetic wave, described in general by the Maxwell's equations in differential form

$$\begin{aligned}
&\nabla\times\boldsymbol{H} = \boldsymbol{J} + \frac{\partial \boldsymbol{D}}{\partial t}\\
&\nabla\times\boldsymbol{E} = -\frac{\partial \boldsymbol{B}}{\partial t}\\
&\nabla\cdot\boldsymbol{D} = \rho\\
&\nabla\cdot\boldsymbol{B} = 0
\end{aligned}\tag{2.45}$$

by MKS units (meter-kilogram-second), where $\boldsymbol{H}$ is the magnetic field intensity, $\boldsymbol{J}$ the conduction current density, $\boldsymbol{D}$ the electric flux density, $\boldsymbol{E}$ the electric field intensity vector, $\boldsymbol{B}$ the magnetic flux density, and ρ the volume charge density.

These are the basics of electromagnetic theory, which were formulated in the early 19th century and combined by Maxwell in 1864 into such a consistent set of equations. To the best of our knowledge, Maxwell's equations correctly describe the large-scale (macroscopic) electromagnetic phenomena that occur in nature. Ample experimental evidence is available to support this view, including the classical experiments of Coulomb, Ampére, and Faraday, which provided the original motivation for Maxwell to postulate the concept of the "electromagnetic theory."

In linear, isotropic, and non-dispersive materials (i.e., materials having field-independent, direction-independent, and frequency-independent electric and magnetic properties), we can relate $\boldsymbol{B}$ to $\boldsymbol{H}$ and $\boldsymbol{D}$ to $\boldsymbol{E}$ using simple proportions:

$$\boldsymbol{B} = \mu \boldsymbol{H} \quad , \quad \boldsymbol{D} = \epsilon \boldsymbol{E} \tag{2.46}$$

Here μ is the magnetic permeability and ϵ is the electric permittivity. In common molecular, bio, and semiconductor materials, we assume $\mu = \mu_0$, where μ_0 is the free-space permeability. In vacuum, $\epsilon = \epsilon_0$, which is the free-space permittivity. Now, permitting the possibility of electric losses that can dissipate electromagnetic fields in materials, we define an electric current to account for the electric loss mechanisms:

$$\boldsymbol{J} = \sigma \boldsymbol{E} \tag{2.47}$$

Here σ is the electric conductivity. Between $\boldsymbol{J}$ and the charge density ρ,

$$\nabla \cdot \boldsymbol{J} = -\frac{\partial \rho}{\partial t} \tag{2.48}$$

due to the charge conservation law. This relationship is also implicit in the Maxwell equations. The Maxwell equations reduce to

$$\nabla \times \boldsymbol{E} = -\mu \frac{\partial \boldsymbol{H}}{\partial t} \tag{2.49}$$

$$\nabla \times \boldsymbol{H} = \boldsymbol{J} + \epsilon \frac{\partial \boldsymbol{E}}{\partial t} \tag{2.50}$$

$$\nabla \cdot \boldsymbol{E} = \frac{\rho}{\epsilon} \tag{2.51}$$

$$\nabla \cdot \boldsymbol{H} = 0 \tag{2.52}$$

By introducing a vector potential, $\boldsymbol{A}$, and a scalar potential, ϕ, the electric and magnetic fields can be obtained from the relations of

$$\begin{aligned} \boldsymbol{E} &= -\nabla \phi - \frac{\partial \boldsymbol{A}}{\partial t} \\ \boldsymbol{B} &= \nabla \times \boldsymbol{A} \end{aligned} \tag{2.53}$$

The first and last Maxwell equations are automatically satisfied by the definitions. By the relation of

$$\nabla \times (\nabla \times \boldsymbol{A}) = \nabla(\nabla \cdot \boldsymbol{A}) - \nabla^2 \boldsymbol{A}$$

and in the Lorentz gauge of

$$\frac{1}{\mu} \nabla \cdot \boldsymbol{A} + \epsilon \frac{\partial \phi}{\partial t} = 0 \tag{2.54}$$

we have the following equations for the vector and scalar potentials

$$\nabla^2 \boldsymbol{A} - \epsilon\mu \frac{\partial^2 \boldsymbol{A}}{\partial t^2} = -\mu \boldsymbol{J} \tag{2.55}$$

$$\nabla^2 \phi - \epsilon\mu \frac{\partial^2 \phi}{\partial t^2} = -\frac{\rho}{\epsilon} \tag{2.56}$$

We consider the electric properties of common semiconductor materials compared with metals. Typically the free electron density of metallic elements is $10^{22}\,\text{cm}^{-3}$, whereas it is about $10^{19}\,\text{cm}^{-3}$ for heavily doped semiconductors. The resistivity of gold is $2.01 \times 10^{-8}\,\Omega\text{m}$ at room temperature. For silicon at 300 K, it is about $4 \times 10^{-5}\,\Omega\text{m}$ when the silicon is n-type doped at $10^{19}\,\text{cm}^{-3}$, which increases drastically to $4 \times 10^{-3}\,\Omega\text{m}$ when the doping level becomes $10^{16}\,\text{cm}^{-3}$.[1] For molecular and bio materials, the charge and current densities are even smaller. We can thus neglect the effects of the carrier concentration and current density on the electromagnetic field, i.e., $\boldsymbol{J} = 0$ and $\phi = 0$. In this case, the time-dependent solution for the vector potential becomes

$$\boldsymbol{A}(\boldsymbol{r}, t) = \boldsymbol{A}_0 \left[\mathrm{e}^{i(\boldsymbol{k}\cdot\boldsymbol{r} - \omega t)} + \text{c.c.} \right] \tag{2.57}$$

with $k^2 = \epsilon\mu\omega^2$. "c.c." in the above equation stands for complex conjugate. Note that in the MKS units, $\epsilon_0\mu_0 = 1/c_0^2$, where c_0 is the velocity of light in the vacuum. By the Lorentz gauge, it is observed that a constant scalar potential implies $\nabla \cdot \boldsymbol{A} = 0$, indicating $\boldsymbol{A} \cdot \boldsymbol{k} = 0$; i.e., the vector potential is perpendicular to the propagation direction of the electromagnetic wave.

It must be noticed that the scalar potential ϕ here is devoted solely for the electromagnetic field of the optical radiation. An equivalent scalar potential exists in the system due to the charge distribution and the ac and dc biases applied on the system. In almost all device applications, the frequencies of ac biases are very low compared with the frequency of the optical field, so that the vector field as well as the time dependence of the scalar field can be neglected, with which the second equation of Eq. 2.55 reduces to the standard steady-state Poisson equation.

The Poynting vector $\boldsymbol{S}$ representing the optical power of the electromagnetic field is

$$\boldsymbol{S} = \boldsymbol{E} \times \boldsymbol{H} = \frac{4ck^2 |\boldsymbol{A}_0|^2 \sin^2(\boldsymbol{k}\cdot\boldsymbol{r} - \omega t)}{\mu} \boldsymbol{k}_0 \tag{2.58}$$

where c is the velocity of light in the medium $c = 1/\sqrt{\epsilon\mu}$ and $\boldsymbol{k}_0$ is a unit vector in the direction of $\boldsymbol{k}$ ($\boldsymbol{k} = k\boldsymbol{k}_0$ and $k = |\boldsymbol{k}|$). In the linear, homogeneous, and isotropic medium, the phase velocity equals the group velocity, i.e., $c = c_g$.

The time-averaged value of the power is

$$\langle \boldsymbol{S} \rangle = 2c\epsilon\omega^2 |\boldsymbol{A}_0|^2 \boldsymbol{k}_0 \tag{2.59}$$

and the energy density is

$$\frac{|\langle \boldsymbol{S} \rangle|}{c} = 2\epsilon\omega^2 |\boldsymbol{A}_0|^2 \tag{2.60}$$

Here we have used the energy dispersion relation of the light:

$$k = \frac{\omega}{c} \tag{2.61}$$

On the other hand, taken the density of photons as n_{ph}, the photon energy density is

$$n_{\text{ph}}\hbar\omega$$

from which we obtain the amplitude of the vector potential

$$|\boldsymbol{A}_0|^2 = \frac{n_{\text{ph}}\hbar}{2\epsilon\omega} \tag{2.62}$$

The photon density n_{ph} is a physically measurable quantity that, according to this relation, tells us the strength of the vector potential $\boldsymbol{A}$.

Going back to the Maxwell equations and writing $\boldsymbol{J} = \sigma\boldsymbol{E}$ by Ohm's law, we obtain the wave equation for the electric field

$$\nabla^2\boldsymbol{E} = \epsilon\mu\frac{\partial^2\boldsymbol{E}}{\partial t^2} + \sigma\mu\frac{\partial\boldsymbol{E}}{\partial t} \tag{2.63}$$

This represents a wave propagating with dissipation. The general solution can be chosen as

$$\boldsymbol{E} = \boldsymbol{E}_0\mathrm{e}^{i(\boldsymbol{k}\cdot\boldsymbol{r}-\omega t)} \tag{2.64}$$

so that the amplitude of the wave vector, $k = |\boldsymbol{k}|$ is given by

$$k^2 = \epsilon\mu\omega^2 + i\sigma\mu\omega \tag{2.65}$$

compared with the energy dispersion, Eq. 2.61, in the non-conductive space ($\sigma = 0$).

A complex refractive index, n, is normally defined by dividing c_0 by the phase velocity in the medium, which can be written in terms of its real and imaginary parts so that

$$k = (n + i\beta)\frac{\omega}{c_0} \tag{2.66}$$

The solution of the electric field wave equation now becomes

$$\boldsymbol{E} = \boldsymbol{E}_0\mathrm{e}^{i\omega(nz/c_0-t)}\mathrm{e}^{-\omega\beta z/c_0} \tag{2.67}$$

The velocity of the wave is reduced from c_0 by n to c_0/n and its amplitude is damped exponentially by a fraction of

$$\exp\left(-2\beta/n\right)$$

per wavelength. The damping of the wave is associated with the absorption of the electromagnetic energy. The absorption coefficient α is defined by the energy intensity ($\propto |\boldsymbol{E}|^2$) decrease by a factor of 2.718282 so that

$$\alpha = \frac{2\beta\omega}{c} \tag{2.68}$$

2.3 GENERALIZED GOLDEN RULE

We now discuss the transition processes of electron states involving electromagnetic field. The processes include both the absorption and emission of electromagnetic energy by the electron to and from its surrounding environment.

We consider the perturbation of an electromagnetic field in the form of

$$V(t) = 2V \cos(\omega t) \tag{2.69}$$

where V and ω are time-independent. Rather importantly, V and ω are physical quantities so that they are real, and thus $V(t)$ is also real. Assuming that the electron occupies initially an eigenstate $|\boldsymbol{k}\rangle$ with energy $E_{\boldsymbol{k}}$ and starting from Eq. 2.18, the probability for the system to be found in a different eigenstate of $|\boldsymbol{q}\rangle$ with energy $E_{\boldsymbol{q}}$ is

$$\langle \boldsymbol{q}|\hat{T}(t,0)|\boldsymbol{k}\rangle = \langle \boldsymbol{q}|V|\boldsymbol{k}\rangle \left[\frac{\mathrm{e}^{i(E_{\boldsymbol{q}}-E_{\boldsymbol{k}}+\hbar\omega)t/\hbar} - 1}{E_{\boldsymbol{q}} - E_{\boldsymbol{k}} + \hbar\omega} + \frac{\mathrm{e}^{i(E_{\boldsymbol{q}}-E_{\boldsymbol{k}}-\hbar\omega)t/\hbar} - 1}{E_{\boldsymbol{q}} - E_{\boldsymbol{k}} - \hbar\omega} \right] \tag{2.70}$$

$$\begin{aligned} P_{\boldsymbol{q}\leftarrow\boldsymbol{k}}(t) = \frac{2}{\hbar^2}\big|\langle \boldsymbol{q}|V|\boldsymbol{k}\rangle\big|^2 \bigg\{ & \frac{1-\cos(\omega_{\boldsymbol{qk}-}t)}{\omega^2_{\boldsymbol{qk}-}} + \frac{1-\cos(\omega_{\boldsymbol{qk}+}t)}{\omega^2_{\boldsymbol{qk}+}} \\ & + \frac{2\cos(\omega t)\big[\cos(\omega t) - \cos(\omega_{\boldsymbol{qk}}t)\big]}{\omega^2_{\boldsymbol{qk}} - \omega^2} \bigg\} \end{aligned} \tag{2.71}$$

where $\hbar\omega_{\boldsymbol{qk}} = E_{\boldsymbol{q}} - E_{\boldsymbol{k}}$, $\omega_{\boldsymbol{qk}\pm} = \omega_{\boldsymbol{qk}} \pm \omega$.

If ω is finite and $E_{\boldsymbol{q}} > E_{\boldsymbol{k}}$,

$$P_{\boldsymbol{q}\leftarrow\boldsymbol{k},-}(t) = \frac{2}{\hbar^2}\big|\langle \boldsymbol{q}|V|\boldsymbol{k}\rangle\big|^2 \frac{1-\cos(\omega_{\boldsymbol{qk}-}t)}{\omega^2_{\boldsymbol{qk}-}} \tag{2.72}$$

by neglecting the rapidly oscillating term containing $\mathrm{e}^{i\omega_{\boldsymbol{qk}+}t}$ in Eq. 2.70. We then go through similar analysis from Eq. 2.23 to Eq. 2.28 to derive the generalized Golden rule:

$$w_{\boldsymbol{q}\leftarrow\boldsymbol{k},-} = \frac{2\pi}{\hbar}\big|\langle \boldsymbol{q}|V|\boldsymbol{k}\rangle\big|^2 N_{\mathrm{DOS}}(E_{\boldsymbol{k}} + \hbar\omega) \tag{2.73}$$

and the transition rate becomes

$$p_{\boldsymbol{q}\leftarrow\boldsymbol{k}} = \frac{\big|\langle \boldsymbol{q}|V|\boldsymbol{k}\rangle\big|^2}{\hbar} \frac{\Gamma_{\boldsymbol{q}\leftarrow\boldsymbol{k},-}}{(E_{\boldsymbol{q}} - E_{\boldsymbol{k}} - \hbar\omega - \Delta E_{\boldsymbol{k}})^2 + \Gamma^2_{\boldsymbol{q}\leftarrow\boldsymbol{k},-}} \tag{2.74}$$

where $\Delta E_{\boldsymbol{k}}$ is given by Eq. 2.36 and $\Gamma_{\boldsymbol{q}\leftarrow\boldsymbol{k},-} = \hbar w_{\boldsymbol{q}\leftarrow\boldsymbol{k},-}/2$.

It is easy to see that as long as ω is finite, we can write the total transition rate as

$$p_{\boldsymbol{q}\leftarrow\boldsymbol{k}} = \frac{\left|\langle\boldsymbol{q}|V|\boldsymbol{k}\rangle\right|^2}{\hbar}\left[\frac{\Gamma_{\boldsymbol{q}\leftarrow\boldsymbol{k},-}}{(E_{\boldsymbol{q}}-E_{\boldsymbol{k}}-\hbar\omega-\Delta E_{\boldsymbol{k}})^2+\Gamma^2_{\boldsymbol{q}\leftarrow\boldsymbol{k},-}} + \frac{\Gamma_{\boldsymbol{q}\leftarrow\boldsymbol{k},+}}{(E_{\boldsymbol{q}}-E_{\boldsymbol{k}}+\hbar\omega-\Delta E_{\boldsymbol{k}})^2+\Gamma^2_{\boldsymbol{q}\leftarrow\boldsymbol{k},+}}\right] \tag{2.75}$$

$$\Gamma_{\boldsymbol{q}\leftarrow\boldsymbol{k},\pm} = \pi\left|\langle\boldsymbol{q}|V|\boldsymbol{k}\rangle\right|^2 N_{\mathrm{DOS}}(E_{\boldsymbol{k}}\mp\hbar\omega) \tag{2.76}$$

The first term on the right side of Eq. 2.75 is normally referred to as the absorption of a quasi particle $\hbar\omega$ and the electron initially occupying a low-energy state $|\boldsymbol{k}\rangle$ transits to a high-energy state $|\boldsymbol{q}\rangle$, while the second one as the emission of $\hbar\omega$ and the electron jumps from a high-energy state $|\boldsymbol{q}\rangle$ to a low-energy state $|\boldsymbol{q}\rangle$.

We now discuss the transitions caused by

$$V(t) = 2V\cos(\omega t) = V\left(\mathrm{e}^{i\omega t}+\mathrm{e}^{-i\omega t}\right)$$

by following the steps of Section 2.1.2:

$$\langle\boldsymbol{q}|\hat{T}(t)|\boldsymbol{k}\rangle = -\frac{i}{\hbar}\langle\boldsymbol{q}|V|\boldsymbol{k}\rangle \int_0^t \left[\mathrm{e}^{i(E_{\boldsymbol{q}}-E_{\boldsymbol{k}}-\hbar\omega)t'/\hbar}+\mathrm{e}^{i(E_{\boldsymbol{q}}-E_{\boldsymbol{k}}+\hbar\omega)t'/\hbar}\right]\langle\boldsymbol{k}|\hat{T}(t')|\boldsymbol{k}\rangle\mathrm{d}t' \tag{2.77}$$

$$\frac{\partial}{\partial t}\langle\boldsymbol{k}|\hat{T}(t)|\boldsymbol{k}\rangle = -\frac{i}{\hbar}\sum_{\boldsymbol{q}\neq\boldsymbol{k}}\left[\mathrm{e}^{i(E_{\boldsymbol{k}}-E_{\boldsymbol{q}}-\hbar\omega)t/\hbar}+\mathrm{e}^{i(E_{\boldsymbol{k}}-E_{\boldsymbol{q}}+\hbar\omega)t/\hbar}\right]\langle\boldsymbol{k}|V|\boldsymbol{q}\rangle\langle\boldsymbol{q}|\hat{T}(t)|\boldsymbol{k}\rangle -\frac{2i}{\hbar}\cos(\omega t)\langle\boldsymbol{k}|V|\boldsymbol{k}\rangle\langle\boldsymbol{k}|\hat{T}(t)|\boldsymbol{k}\rangle \tag{2.78}$$

We expect to have three terms

$$\begin{aligned}&\int_0^t \mathrm{e}^{i(E_{\boldsymbol{k}}-E_{\boldsymbol{q}}\pm\hbar\omega)(t'-t)/\hbar}\mathrm{d}t' \to \delta(E_{\boldsymbol{k}}-E_{\boldsymbol{q}}\pm\hbar\omega)\\ &\int_0^t \mathrm{e}^{i(E_{\boldsymbol{k}}-E_{\boldsymbol{q}})(t'-t)/\hbar}\cos\left[\omega(t'+t)\right]\mathrm{d}t' \to 0\end{aligned} \tag{2.79}$$

so that

$$\frac{\partial}{\partial t}\langle\boldsymbol{k}|\hat{T}(t)|\boldsymbol{k}\rangle = -\langle\boldsymbol{k}|\hat{T}(t)|\boldsymbol{k}\rangle\left\{\sum_{\boldsymbol{q}\neq\boldsymbol{k}}\frac{\pi\left|\langle\boldsymbol{q}|V|\boldsymbol{k}\rangle\right|^2}{\hbar}\left[\delta(E_{\boldsymbol{k}}-E_{\boldsymbol{q}}-\hbar\omega) +\delta(E_{\boldsymbol{k}}-E_{\boldsymbol{q}}+\hbar\omega)\right]+\frac{2i}{\hbar}\cos(\omega t)\langle\boldsymbol{k}|V|\boldsymbol{k}\rangle\right\} \tag{2.80}$$

which results in a solution of

$$\langle\boldsymbol{k}|\hat{T}(t)|\boldsymbol{k}\rangle = \exp\left[-\frac{w_{\boldsymbol{k}}}{2}t-\frac{2i\sin(\omega t)}{\hbar\omega}\Delta E_{\boldsymbol{k}}\right] \tag{2.81}$$

where

$$w_{\boldsymbol{k}} = \frac{2\pi}{\hbar} \sum_{\boldsymbol{q} \neq \boldsymbol{k}} \left|\langle \boldsymbol{q}|V|\boldsymbol{k}\rangle\right|^2 \left[\delta(E_{\boldsymbol{q}} - E_{\boldsymbol{k}} - \hbar\omega) + \delta(E_{\boldsymbol{q}} - E_{\boldsymbol{k}} + \hbar\omega)\right] \tag{2.82}$$

$\Delta E_{\boldsymbol{k}}$ is given by its early expression of Eq. 2.36. It is easy to see that Eq. 2.37 is retrieved when $\omega \to 0$.

In practical systems and applications, the effective range of $\hbar\omega$ is at least in the order of meV so that $1/\omega$ is less than 10^{-12} s, while the time scale of most first-order perturbation processes is in the range of 10^{-15}. In this case we can neglect $\Delta E_{\boldsymbol{k}}$ and

$$\langle \boldsymbol{k}|\hat{T}(t)|\boldsymbol{k}\rangle \approx \mathrm{e}^{-w_{\boldsymbol{k}} t/2} \tag{2.83}$$

By Eq. 2.77,

$$\langle \boldsymbol{q}|\hat{T}(t)|\boldsymbol{k}\rangle = \langle \boldsymbol{q}|V|\boldsymbol{k}\rangle \frac{1 - \mathrm{e}^{i[(E_{\boldsymbol{q}} - E_{\boldsymbol{k}} - \hbar\omega)/\hbar + i w_{\boldsymbol{k}}/2]t}}{E_{\boldsymbol{q}} - E_{\boldsymbol{k}} - \hbar\omega + i\hbar w_{\boldsymbol{k}}/2} + \frac{1 - \mathrm{e}^{i[(E_{\boldsymbol{q}} - E_{\boldsymbol{k}} + \hbar\omega)/\hbar + i w_{\boldsymbol{k}}/2]t}}{E_{\boldsymbol{q}} - E_{\boldsymbol{k}} + \hbar\omega + i\hbar w_{\boldsymbol{k}}/2} \tag{2.84}$$

Note that $w_{\boldsymbol{k}} \neq 0$ and let $t \to \infty$

$$\langle \boldsymbol{q}|\hat{T}(+\infty)|\boldsymbol{k}\rangle = \frac{\langle \boldsymbol{q}|V|\boldsymbol{k}\rangle}{E_{\boldsymbol{q}} - E_{\boldsymbol{k}} - \hbar\omega + i\hbar w_{\boldsymbol{k}}/2} + \frac{\langle \boldsymbol{q}|V|\boldsymbol{k}\rangle}{E_{\boldsymbol{q}} - E_{\boldsymbol{k}} + \hbar\omega + i\hbar w_{\boldsymbol{k}}/2} \tag{2.85}$$

Finally, by defining $\Gamma_{\boldsymbol{k}} = \hbar w_{\boldsymbol{k}}/2$, we obtain

$$P_{\boldsymbol{q} \leftarrow \boldsymbol{k}} = \frac{|\langle \boldsymbol{q}|V|\boldsymbol{k}\rangle|^2 \left[(E_{\boldsymbol{q}} - E_{\boldsymbol{k}})^2 + \Gamma_{\boldsymbol{k}}^2\right]}{\left[(E_{\boldsymbol{q}} - E_{\boldsymbol{k}})^2 - \Gamma_{\boldsymbol{k}}^2 - \hbar^2\omega^2\right]^2 + 4(E_{\boldsymbol{q}} - E_{\boldsymbol{k}})^2\Gamma_{\boldsymbol{k}}^2} \tag{2.86}$$

It is observed here that we actually cannot distinguish the emission and absorption processes, as separately as in Eq. 2.75, which was obtained from Eq. 2.18, assuming a constant probability that the system stays at its initial state, i.e., $P_{\boldsymbol{k} \leftarrow \boldsymbol{k}}(t) = 1$. Only by setting $\Gamma_{\boldsymbol{k}} = 0$ in Eq. 2.86, can we observe immediately the clear separation of the emission and absorption processes.

2.4 LIGHT–MATTER INTERACTION

2.4.1 *Photons*

Photons are quantum mechanical particles of energy $\hbar\omega$ associated with an electromagnetic radiation. Unlike electrons, photons do not obey the Pauli exclusion principle: One can have as many photons in the same energy and wave vector as one can generate. At equilibrium, the probability of finding a photon in a state of energy $\hbar\omega$ is given by the Bose–Einstein distribution

$$b(\hbar\omega) = \frac{1}{\mathrm{e}^{\hbar\omega/k_B T} - 1} \tag{2.87}$$

Particles that obey this statistic distribution are bosons.

As discussed for electrons, it can be seen that within an enclosed box of volume Ω, a quantum state of photons corresponding to a wave function $\mathrm{e}^{i(\boldsymbol{k}\cdot\boldsymbol{r}-\omega t)}$ requires a volume $(2\pi)^3/\Omega$ in the $\boldsymbol{k}$ space in Section 1.6. Taking into account the fact that there are two independent polarizations for each $\boldsymbol{k}$ state, one can obtain the photon density (per unit volume) $n_{\mathrm{ph}}(\hbar\omega)\mathrm{d}\hbar\omega$ for photons between ω and $\omega + \mathrm{d}\omega$

$$n_{\mathrm{ph}}(\hbar\omega)\mathrm{d}(\hbar\omega) = 2b(\hbar\omega)\frac{4\pi k^2 \mathrm{d}k}{(2\pi)^3} = \frac{\omega^2}{\hbar\pi^2 c^2 c_g}\frac{\mathrm{d}(\hbar\omega)}{e^{\hbar\omega/k_B T} - 1} \tag{2.88}$$

which is Planck's law of radiation at equilibrium.

The photons are bosons with zero mass; in free space they carry linear momentum $\hbar\boldsymbol{s}$ (in the following we denote the photon momentum by $\hbar\boldsymbol{s}$ as to differ the momentum of electron, which is normally referred to as $\hbar\boldsymbol{k}$), and energy $\hbar\omega_{\boldsymbol{s}}$. Denoting the polarization vector $\boldsymbol{e}_{\boldsymbol{s}}$, the vector potential operator of the electromagnetic field can be expressed as

$$\boldsymbol{A}(\boldsymbol{r},t) = \sum_{\boldsymbol{s}}\sqrt{\frac{\hbar}{2\epsilon\omega_{\boldsymbol{s}}\Omega}}\left[b_{\boldsymbol{s}}\mathbf{e}_{\boldsymbol{s}}\mathrm{e}^{i(\boldsymbol{s}\cdot\boldsymbol{r}-\omega t)} + b_{\boldsymbol{s}}^{+}\boldsymbol{e}_{\boldsymbol{s}}\mathrm{e}^{-i(\boldsymbol{s}\cdot\boldsymbol{r}-\omega t)}\right] \tag{2.89}$$

where Ω is the spatial volume where the wave function of the photon in the coordinate representation is normalized as $\mathrm{e}^{i(\boldsymbol{s}\cdot\boldsymbol{r}-\omega t)}/\sqrt{\Omega}$.

The important physical operators for a system of free photons are the total energy, the total linear momentum, and the total number of photons

$$\sum_{\boldsymbol{s}}\hbar\omega_{\boldsymbol{s}}b_{\boldsymbol{s}}^{+}b_{\boldsymbol{s}} \quad , \quad \sum_{\boldsymbol{s}}\hbar\boldsymbol{s}b_{\boldsymbol{s}}^{+}b_{\boldsymbol{s}} \quad , \quad \sum_{\boldsymbol{s}}b_{\boldsymbol{s}}^{+}b_{\boldsymbol{s}} \tag{2.90}$$

In the momentum representation, the electromagnetic field is symbolizd by photon occupation numbers $|\ldots N_{\boldsymbol{s}}\ldots\rangle$

$$\begin{aligned} b_{\boldsymbol{s}}^{+}|\ldots N_{\boldsymbol{s}}\ldots\rangle &= \sqrt{N_{\boldsymbol{s}}+1}|\ldots N_{\boldsymbol{s}}+1\ldots\rangle \\ b_{\boldsymbol{s}}|\ldots N_{\boldsymbol{s}}\ldots\rangle &= \sqrt{N_{\boldsymbol{s}}}|\ldots N_{\boldsymbol{s}}-1\ldots\rangle \end{aligned} \tag{2.91}$$

2.4.2 *Electron in Electromagnetic Field*

The Hamiltonian H describing an electron with an electronic charge $-e$ in the electromagnetic field of $(\boldsymbol{A},\phi)$ [see Eqs. 2.55] is

$$H = \frac{1}{2m_0}(\boldsymbol{p} - e\boldsymbol{A})^2 - e\phi + V_0(\boldsymbol{r}) \tag{2.92}$$

where $V_0(\boldsymbol{r})$ is the potential energy of the system in the darkness, m_0 the free electron mass, and $\boldsymbol{p} = -i\hbar\nabla$ the electron momentum. By expanding the square term we have

$$H = \frac{p^2}{2m_0} + V_0(\boldsymbol{r}) - e\phi - \frac{e}{m_0}\boldsymbol{A}\cdot\boldsymbol{p} + \frac{ie\hbar}{2m_0}\nabla\cdot\boldsymbol{A} + \frac{e^2}{2m_0}\boldsymbol{A}^2 \tag{2.93}$$

In the above equation $H_0 = p^2/2m_0 + V_0(\boldsymbol{r})$ is the Hamiltonian of electron in the darkness. For electromagnetic field of optical radiation, $\phi = 0$ (Section 2.2), which results in $\nabla \cdot \boldsymbol{A} = 0$ in the Lorentz gauge; Four terms remain on the right side of Eq. 2.93. Consider the following ratios

$$\left|\frac{e\boldsymbol{A}\cdot\boldsymbol{p}}{m_0}\right| : \left|\frac{p^2}{2m_0}\right| = \frac{2e|\boldsymbol{A}|}{p}$$

$$\left|\frac{e^2\boldsymbol{A}^2}{2m_0}\right| : \left|\frac{e\boldsymbol{A}\cdot\boldsymbol{p}}{m_0}\right| = \frac{e|\boldsymbol{A}|}{2p}$$

Direct radiation from the Sun is about 120 watts/m^2 (W/m^2). The average optical power from today's light-emitting diode is in the order of mW. Consider an electromagnetic field of a radiation at 2.0 eV, i.e., the photon energy $\hbar\omega = 2.0\,\text{eV}$, corresponding to red light with wavelength 620 nm. For an optical power of $W = 1.0\,\text{W/cm}^2$ (solar power is about $0.12\,\text{W/cm}^2$),

$$\begin{aligned} W &= 1.0\,\text{W/cm}^2 = n_{\text{ph}}\hbar\omega c \\ &= n_{\text{ph}}\,[\text{cm}^{-3}] \times 2.0 \times 1.6 \times 10^{-19}[\text{J}] \times 3 \times 10^{10}\,[\text{cm/s}] \end{aligned}$$

for $c = 3 \times 10^{10}\,\text{cm/s}$, we obtain $n_{\text{ph}} = 1.04 \times 10^8\,\text{cm}^{-3}$ (in vacuum). By Eq. 2.62,

$$e|\boldsymbol{A}| = e\sqrt{\frac{n_{\text{ph}}\hbar}{2\epsilon\omega}} = 7.2 \times 10^{-32}\,\text{kg m/s} \tag{2.94}$$

when $\epsilon = \epsilon_0 = 8.85 \times 10^{-12}\,\text{F/m}$ in the free space. Using an electron speed of $10^4\,\text{m/s}$ (in common semiconductors) and $m_0 = 9.1 \times 10^{-31}\,\text{kg}$, one finds

$$\frac{e|\boldsymbol{A}|}{p} = \frac{7.2 \times 10^{-32}}{10^4 \times 9.1 \times 10^{-31}} = 0.8 \times 10^{-5} \tag{2.95}$$

We now increase the optical power W by a factor of 10^6, i.e., an optical beam carrying $10^6\,\text{W/cm}^2$. Since $|\boldsymbol{A}| \propto \sqrt{n_{\text{ph}}} \propto \sqrt{W}$, we will obtain a value of $e|\boldsymbol{A}|/p = 0.8 \times 10^{-2}$, which is still very small. It is thus concluded that the potential energy modification of the electron in the external electromagnetic field is normally rather small and only the term linear in $\boldsymbol{A}$ dominates in Eq. 2.93. The time-dependent Schödinger equation is now written as

$$\begin{aligned} i\hbar\frac{\partial\psi}{\partial t} &= (H_0 + H')\psi \\ H_0 &= -\frac{\hbar^2\nabla^2}{2m_0} + V_0(\boldsymbol{r}) \ , \ H' = \frac{ie\hbar}{m_0}\boldsymbol{A}\cdot\nabla \end{aligned} \tag{2.96}$$

where $\boldsymbol{A}(\boldsymbol{r},t)$ is given by Eq. 2.55.

In some cases, e.g., when the linear-$\boldsymbol{A}$ induced transitions are forbidden, the term in $\boldsymbol{A}^2$ has to be included.

2.4.3 *The Composite System*

We now discuss the energy dissipation to and from the electron by including the environment in the form of a composite system. As discussed, potential fields in the form of $V(\boldsymbol{r},t) = V(\boldsymbol{r})\cos(\omega t)$ can be quantized similarly as quasi boson particle systems. The quasi particle has an energy of $\hbar\omega$, and the density of the quasi particles is proportional to $\sqrt{V(\boldsymbol{r})}$. Typical examples of the field quantization are lattice vibration (phonon); and the electromagnetic field (photon); the latter was just discussed.

The electron is scattered from an initial state to a final one by gaining or losing energy; a boson is annihilated or created at the same time. Note, however, that the total energy of the composite system, including the electron and the boson particles, is conserved. We write the total Hamiltonian of the composite system generally as

$$H_0 + \sum_{\boldsymbol{s}} \left[V_{\boldsymbol{s}}(\boldsymbol{r}) b_{\boldsymbol{s}} + V_{\boldsymbol{s}}^{+}(\boldsymbol{r}) b_{\boldsymbol{s}}^{+} \right] + \sum_{\boldsymbol{s}} \hbar\omega_{\boldsymbol{s}} b_{\boldsymbol{s}}^{+} b_{\boldsymbol{s}} \tag{2.97}$$

where H_0 is the electron Hamiltonian, and b^+ and b are creation and annihilation operator of the boson field, respectively. We have neglected the zero-point energy $\sum_{\boldsymbol{s}} \hbar\omega_{\boldsymbol{s}}/2$ of the optical field.

For simplicity, we consider for the moment only one type of quasi particles so that

$$\begin{cases} H_0 & t \leq 0 \\ H_0 + \left[V(\boldsymbol{r}) b + V(\boldsymbol{r})^{+} b^{+} \right] + \hbar\omega b^{+} b & t > 0 \end{cases} \tag{2.98}$$

and the state of the composite system as $|\boldsymbol{k}N_{\boldsymbol{k}}\rangle$, where $|\boldsymbol{k}\rangle$ is the electron wave function while $N_{\boldsymbol{k}}$ denotes the number of the boson particles associated with $|\boldsymbol{k}\rangle$.

By defining $H_0 \equiv H_0 + \hbar\omega b^+ b$ and $V(\boldsymbol{r}, t) \equiv V(\boldsymbol{r}) b + V^{+}(\boldsymbol{r}) b^{+}$, and starting from Eq. 2.9, we obtain the transition matrix element between two eigenstates of H_0

$$\begin{aligned} &\langle \boldsymbol{q} N_{\boldsymbol{q}} | \hat{T}(t) | \boldsymbol{k} N_{\boldsymbol{k}} \rangle = \delta(\boldsymbol{q} - \boldsymbol{k}) \delta(N_{\boldsymbol{q}} - N_{\boldsymbol{k}}) \\ &- \frac{i}{\hbar} \sum_{\boldsymbol{k}'} \int_0^t \left[\langle \boldsymbol{q} | V | \boldsymbol{k}' \rangle e^{i(E_{\boldsymbol{q}} - E_{\boldsymbol{k}'} + \hbar\omega)t'/\hbar} \sqrt{N_{\boldsymbol{q}} + 1} \langle \boldsymbol{k}' N_{\boldsymbol{q}} + 1 | \hat{T}(t') | \boldsymbol{k} N_{\boldsymbol{k}} \rangle \right. \\ &\left. + \langle \boldsymbol{q} | V^{+} | \boldsymbol{k}' \rangle e^{i(E_{\boldsymbol{q}} - E_{\boldsymbol{k}'} - \hbar\omega)t'/\hbar} \sqrt{N_{\boldsymbol{q}}} \langle \boldsymbol{k}' N_{\boldsymbol{q}} - 1 | \hat{T}(t') | \boldsymbol{k} N_{\boldsymbol{k}} \rangle \right] \mathrm{d}t' \end{aligned} \tag{2.99}$$

And in its differential form

$$\begin{aligned} &\frac{\partial}{\partial t} \langle \boldsymbol{q} N_{\boldsymbol{q}} | \hat{T}(t) | \boldsymbol{k} N_{\boldsymbol{k}} \rangle \\ &= -\frac{i}{\hbar} \sum_{\boldsymbol{k}'} \left[\langle \boldsymbol{q} | V | \boldsymbol{k}' \rangle \mathrm{e}^{i(E_{\boldsymbol{q}} - E_{\boldsymbol{k}'} + \hbar\omega)t/\hbar} \sqrt{N_{\boldsymbol{q}} + 1} \langle \boldsymbol{k}' N_{\boldsymbol{q}} + 1 | \hat{T}(t) | \boldsymbol{k} N_{\boldsymbol{k}} \rangle \right. \\ &\left. + \langle \boldsymbol{q} | V^{+} | \boldsymbol{k}' \rangle \mathrm{e}^{i(E_{\boldsymbol{q}} - E_{\boldsymbol{k}'} - \hbar\omega)t/\hbar} \sqrt{N_{\boldsymbol{q}}} \langle \boldsymbol{k}' N_{\boldsymbol{q}} - 1 | \hat{T}(t) | \boldsymbol{k} N_{\boldsymbol{k}} \rangle \right] \end{aligned} \tag{2.100}$$

We can follow the following sequence of temporal development of Eq. 2.100 as in Table 2.1, where δt is the time interval of creating or annihilating a boson. Starting from an initial composite state $|\boldsymbol{k}_0 N_{\boldsymbol{k}_0}\rangle$, an electron state $|\boldsymbol{k}_1\rangle$ is generated after δt. At the same time, the boson field is modified by either creating or annihilating a boson. $2\delta t$ later, $|\boldsymbol{k}_0 N_{\boldsymbol{k}_0}\rangle \rightarrow |\boldsymbol{k}_1 N_{\boldsymbol{k}_0} \pm 1\rangle$, and $|\boldsymbol{k}_1 N_{\boldsymbol{k}_0} \pm 1\rangle \rightarrow |\boldsymbol{k}_2 N_{\boldsymbol{k}_0} + 2\rangle$, $|\boldsymbol{k}_2 N_{\boldsymbol{k}_0}\rangle$, and $|\boldsymbol{k}_2 N_{\boldsymbol{k}_0} - 2\rangle$. Naturally the time development is a multiple-channel

Table 2.1 Temporal development of Eq. 2.100.

	$t \le 0$	$0 < t \le \delta t$	$\delta t < t \le 2\delta$	$2\delta t < t \le 3\delta$
$N_{\boldsymbol{k}_0} - 3$				$\boldsymbol{k}_3$
$N_{\boldsymbol{k}_0} - 2$			$\boldsymbol{k}_2$	$\boldsymbol{k}_2$
$N_{\boldsymbol{k}_0} - 1$		$\boldsymbol{k}_1$	$\boldsymbol{k}_1$	$\boldsymbol{k}_1$, $\boldsymbol{k}_3$
$N_{\boldsymbol{k}_0}$	$\boldsymbol{k}_0$	$\boldsymbol{k}_0$	$\boldsymbol{k}_0$, $\boldsymbol{k}_2$	$\boldsymbol{k}_0$, $\boldsymbol{k}_2$
$N_{\boldsymbol{k}_0} + 1$		$\boldsymbol{k}_1$	$\boldsymbol{k}_1$	$\boldsymbol{k}_1$, $\boldsymbol{k}_3$
$N_{\boldsymbol{k}_0} + 2$			$\boldsymbol{k}_2$	$\boldsymbol{k}_2$
$N_{\boldsymbol{k}_0} + 3$				$\boldsymbol{k}_3$

process, as stated by $\sum_{\boldsymbol{k}'}$ in Eq. 2.100, whereas we have followed only one channel of $|\boldsymbol{k}_0\rangle \to |\boldsymbol{k}_1\rangle$, $|\boldsymbol{k}_1\rangle \to |\boldsymbol{k}_2\rangle$, and $|\boldsymbol{k}_2\rangle \to |\boldsymbol{k}_3\rangle$ in Table 2.1.

It is noticed that the transition from energy level $E_{\boldsymbol{k}}$ to $E_{\boldsymbol{q}}$ is determined by V. In addition, we do not observe the energy conservation requirement within the short time interval due to the uncertainty principle. However, the creation and annihilation of a boson particle are progressive.

The wave vector of a photon is negligibly small, so is the one of optical phonon, which results in $\langle \boldsymbol{q}|V|\boldsymbol{k}'\rangle = \langle \boldsymbol{q}|V^+|\boldsymbol{k}'\rangle$ by setting $\boldsymbol{s} = 0$. However, for acoustic phonon field, $\langle \boldsymbol{q}|V|\boldsymbol{k}'\rangle$ can be significantly different from $\langle \boldsymbol{q}|V^+|\boldsymbol{k}'\rangle$.

Thus far, we have not made any approximations. Equation 2.100 can be solved numerically by the standard iteration method. Here we follow our early approximate approach by assuming the dominance of one-step process so that for $|\boldsymbol{q}\rangle \ne |\boldsymbol{k}\rangle$,

$$\begin{aligned} &\langle \boldsymbol{q}N_{\boldsymbol{q}}|\hat{T}(t)|\boldsymbol{k}N_{\boldsymbol{k}}\rangle \\ &= -\frac{i}{\hbar}\int_0^t \Big[\langle \boldsymbol{q}|V|\boldsymbol{k}\rangle \mathrm{e}^{i(E_{\boldsymbol{q}}-E_{\boldsymbol{k}}+\hbar\omega)t'/\hbar}\sqrt{N_{\boldsymbol{q}}+1}\langle \boldsymbol{k}N_{\boldsymbol{q}}+1|\hat{T}(t')|\boldsymbol{k}N_{\boldsymbol{k}}\rangle \\ &\quad + \langle \boldsymbol{q}|V^+|\boldsymbol{k}\rangle \mathrm{e}^{i(E_{\boldsymbol{q}}-E_{\boldsymbol{k}}-\hbar\omega)t'/\hbar}\sqrt{N_{\boldsymbol{q}}}\langle \boldsymbol{k}N_{\boldsymbol{q}}-1|\hat{T}(t')|\boldsymbol{k}N_{\boldsymbol{k}}\rangle\Big]\mathrm{d}t' \end{aligned} \tag{2.101}$$

By inserting the above equation into Eq. 2.100, we obtain four terms:

$$\begin{aligned} &(1)\ \sqrt{(N_{\boldsymbol{k}}+1)(N_{\boldsymbol{k}}+2)}\langle \boldsymbol{k}|V|\boldsymbol{q}\rangle\langle \boldsymbol{q}|V|\boldsymbol{k}\rangle \\ &\qquad \times \mathrm{e}^{i(E_{\boldsymbol{q}}-E_{\boldsymbol{k}})(t'-t)/\hbar}\mathrm{e}^{i\omega(t'+t)}\langle \boldsymbol{k}N_{\boldsymbol{k}}+2|\hat{T}(t')|\boldsymbol{k}N_{\boldsymbol{k}}\rangle \\ &(2)\ (N_{\boldsymbol{k}}+1)|\langle \boldsymbol{k}|V|\boldsymbol{q}\rangle|^2\mathrm{e}^{i(E_{\boldsymbol{q}}-E_{\boldsymbol{k}}-\hbar\omega)(t'-t)/\hbar}\langle \boldsymbol{k}N_{\boldsymbol{k}}|\hat{T}(t')|\boldsymbol{k}N_{\boldsymbol{k}}\rangle \\ &(3)\ N_{\boldsymbol{k}}|\langle \boldsymbol{k}|V|\boldsymbol{q}\rangle|^2\mathrm{e}^{i(E_{\boldsymbol{q}}-E_{\boldsymbol{k}}+\hbar\omega)(t'-t)/\hbar}\langle \boldsymbol{k}N_{\boldsymbol{k}}|\hat{T}(t')|\boldsymbol{k}N_{\boldsymbol{k}}\rangle \\ &(4)\ \sqrt{N_{\boldsymbol{k}}(N_{\boldsymbol{k}}-1)}\langle \boldsymbol{k}|V^+|\boldsymbol{q}\rangle\langle \boldsymbol{q}|V^+|\boldsymbol{k}\rangle \\ &\qquad \times \mathrm{e}^{i(E_{\boldsymbol{q}}-E_{\boldsymbol{k}})(t'-t)/\hbar}\mathrm{e}^{-i\omega(t'+t)}\langle \boldsymbol{k}N_{\boldsymbol{k}}-2|\hat{T}(t')|\boldsymbol{k}N_{\boldsymbol{k}}\rangle \end{aligned}$$

We keep only the one-step process by neglecting $\langle \boldsymbol{k}N_{\boldsymbol{k}}+2|\hat{T}(t')|\boldsymbol{k}N_{\boldsymbol{k}}\rangle$ and $\langle \boldsymbol{k}N_{\boldsymbol{k}}-2|\hat{T}(t')|\boldsymbol{k}N_{\boldsymbol{k}}\rangle$

$$\begin{aligned} &\frac{\partial}{\partial t}\langle \boldsymbol{k}N_{\boldsymbol{k}}|\hat{T}(t)|\boldsymbol{k}N_{\boldsymbol{k}}\rangle = -\sum_{\boldsymbol{q}}\frac{|\langle \boldsymbol{q}|V|\boldsymbol{k}\rangle|^2}{\hbar^2}\int_0^t \mathrm{d}t'\langle \boldsymbol{k}N_{\boldsymbol{k}}|\hat{T}(t')|\boldsymbol{k}N_{\boldsymbol{k}}\rangle \\ &\quad \times\Big[(N_{\boldsymbol{k}}+1)\mathrm{e}^{i(E_{\boldsymbol{q}}-E_{\boldsymbol{k}}-\hbar\omega)(t'-t)/\hbar} + N_{\boldsymbol{k}}\mathrm{e}^{i(E_{\boldsymbol{q}}-E_{\boldsymbol{k}}+\hbar\omega)(t'-t)/\hbar}\Big]\mathrm{d}t \end{aligned} \tag{2.102}$$

Again, by assuming $\langle \boldsymbol{k}N_{\boldsymbol{k}}|\hat{T}(t)|\boldsymbol{k}N_{\boldsymbol{k}}\rangle$ is constant over the time scale t', we obtain

$$\frac{\partial}{\partial t}\langle \boldsymbol{k}N_{\boldsymbol{k}}|\hat{T}(t)|\boldsymbol{k}N_{\boldsymbol{k}}\rangle = -\langle \boldsymbol{k}N_{\boldsymbol{k}}|\hat{T}(t)|\boldsymbol{k}N_{\boldsymbol{k}}\rangle \times \sum_{\boldsymbol{q}} \frac{\pi|\langle \boldsymbol{q}|V|\boldsymbol{k}\rangle|^2}{\hbar}\Big[(N_{\boldsymbol{k}}+1)\delta(E_{\boldsymbol{q}}-E_{\boldsymbol{k}}-\hbar\omega)+N_{\boldsymbol{k}}\delta(E_{\boldsymbol{q}}-E_{\boldsymbol{k}}+\hbar\omega)\Big] \quad (2.103)$$

In the above equation, the sum over $|\boldsymbol{q}\rangle$ includes the term of $|\boldsymbol{q}\rangle = |\boldsymbol{k}\rangle$. However, the corresponding process is automatically switched off when ω is finite due to the δ functions, so that there is no energy level shift, i.e., $\Delta E_{\boldsymbol{k}}$ in Eq. 2.36. The first (second) term in the above equation indicates the transition from state $|\boldsymbol{q}\rangle$ to $|\boldsymbol{k}\rangle$ by emitting (absorbing) a boson.

By denoting

$$w_{\boldsymbol{k}N_{\boldsymbol{k}}} = \sum_{\boldsymbol{q}} \frac{2\pi|\langle \boldsymbol{q}|V|\boldsymbol{k}\rangle|^2}{\hbar}\Big[(N_{\boldsymbol{k}}+1)\delta(E_{\boldsymbol{q}}-E_{\boldsymbol{k}}-\hbar\omega)+N_{\boldsymbol{k}}\delta(E_{\boldsymbol{q}}-E_{\boldsymbol{k}}+\hbar\omega)\Big] \quad (2.104)$$

we obtain

$$\langle \boldsymbol{k}N_{\boldsymbol{k}}|\hat{T}(t)|\boldsymbol{k}N_{\boldsymbol{k}}\rangle = \mathrm{e}^{-w_{\boldsymbol{k}N_{\boldsymbol{k}}}t/2} \quad (2.105)$$

Then, $t \to \infty$,

$$\langle \boldsymbol{q}N_{\boldsymbol{q}}|\hat{T}(+\infty)|\boldsymbol{k}N_{\boldsymbol{k}}\rangle = \frac{\langle \boldsymbol{q}|V|\boldsymbol{k}\rangle\sqrt{N_{\boldsymbol{k}}}\delta_{N_{\boldsymbol{q}},N_{\boldsymbol{k}}-1}}{E_{\boldsymbol{q}}-E_{\boldsymbol{k}}+\hbar\omega+i\Gamma_{\boldsymbol{k}N_{\boldsymbol{k}}}} + \frac{\langle \boldsymbol{q}|V^{+}|\boldsymbol{k}\rangle\sqrt{N_{\boldsymbol{k}}+1}\delta_{N_{\boldsymbol{q}},N_{\boldsymbol{k}}+1}}{E_{\boldsymbol{q}}-E_{\boldsymbol{k}}-\hbar\omega+i\Gamma_{\boldsymbol{k}N_{\boldsymbol{k}}}} \quad (2.106)$$

where $\Gamma_{\boldsymbol{k}N_{\boldsymbol{k}}} = \hbar w_{\boldsymbol{k}N_{\boldsymbol{k}}}/2$.

For a finite ω, we can write explicitly that for $E_{\boldsymbol{q}} > E_{\boldsymbol{k}}$,

$$\langle \boldsymbol{q}N_{\boldsymbol{k}}+1|\hat{T}(+\infty)|\boldsymbol{k}N_{\boldsymbol{k}}\rangle_{E_{\boldsymbol{q}}>E_{\boldsymbol{k}}} = \frac{\langle \boldsymbol{q}|V^{+}|\boldsymbol{k}\rangle\sqrt{N_{\boldsymbol{k}}+1}}{E_{\boldsymbol{q}}-E_{\boldsymbol{k}}-\hbar\omega+i\Gamma_{\boldsymbol{k}N_{\boldsymbol{k}}}} \quad (2.107)$$

In other words, only the process involving the emission of a boson is significantly large as compared with the process of boson absorption. Similarly,

$$\langle \boldsymbol{q}N_{\boldsymbol{k}}-1|\hat{T}(+\infty)|\boldsymbol{k}N_{\boldsymbol{k}}\rangle_{E_{\boldsymbol{q}}<E_{\boldsymbol{k}}} = \frac{\langle \boldsymbol{q}|V|\boldsymbol{k}\rangle\sqrt{N_{\boldsymbol{k}}}}{E_{\boldsymbol{q}}-E_{\boldsymbol{k}}+\hbar\omega+i\Gamma_{\boldsymbol{k}N_{\boldsymbol{k}}}} \quad (2.108)$$

The relevant transition rates are

$$\begin{aligned} p_{\boldsymbol{q}N_{\boldsymbol{q}}\leftarrow\boldsymbol{k}N_{\boldsymbol{k}},E_{\boldsymbol{q}}>E_{\boldsymbol{k}}} &= \frac{|\langle \boldsymbol{q}|V^{+}|\boldsymbol{k}\rangle|^2}{\hbar}\frac{(N_{\boldsymbol{k}}+1)\Gamma_{\boldsymbol{k}N_{\boldsymbol{k}}}}{(E_{\boldsymbol{q}}-E_{\boldsymbol{k}}-\hbar\omega)^2+\Gamma^2_{\boldsymbol{k}N_{\boldsymbol{k}}}} \\ p_{\boldsymbol{q}N_{\boldsymbol{q}}\leftarrow\boldsymbol{k}N_{\boldsymbol{k}},E_{\boldsymbol{q}}<E_{\boldsymbol{k}}} &= \frac{|\langle \boldsymbol{q}|V|\boldsymbol{k}\rangle|^2}{\hbar}\frac{N_{\boldsymbol{k}}\Gamma_{\boldsymbol{k}N_{\boldsymbol{k}}}}{(E_{\boldsymbol{q}}-E_{\boldsymbol{k}}+\hbar\omega)^2+\Gamma^2_{\boldsymbol{k}N_{\boldsymbol{k}}}} \end{aligned} \quad (2.109)$$

These are identical to Eqs. 2.82, 2.85, and 2.86 in the last section, where we have assumed, implicitly, the boson field as a reservoir $N \gg 1$ (and N is included in V). In the last section, we have also neglected possible phase difference between V and V^{+}.

Eq. 2.109 describes the emission and absorption processes of one boson quasi particle during the transition from electron state $|\boldsymbol{k}\rangle$ to $|\boldsymbol{q}\rangle$. This is in accordance with what we have assumed the one-step process of $|\boldsymbol{k}\rangle \rightarrow |\boldsymbol{q}\rangle$ by neglecting $|\boldsymbol{k}\rangle \rightarrow |\boldsymbol{k}'\rangle \rightarrow |\boldsymbol{q}\rangle$, where $|\boldsymbol{k}'\rangle \neq |\boldsymbol{k}\rangle$. In order to study multiple-quasi-particle processes, we have to go beyond the one-step approximation by solving Eq. 2.100 directly in the numerical ways.

2.5 OPTICAL SPECTRUM

After the general discussions about the energy dissipation processes, we discuss in the following sections of two special optical processes, namely, optical absorption and emission. As before, we assume that the eigenvalues and eigenstates of H_0 have already been obtained. i.e., $H_0|\boldsymbol{k}\rangle = E_{\boldsymbol{k}}|\boldsymbol{k}\rangle$.

When the optical transition rates are relatively low so that the external optical field is more or less constant, the transition rate from an initial electron state $|\boldsymbol{k}\rangle$ to a final state $|\boldsymbol{q}\rangle$ induced by these photons is

$$p_{\boldsymbol{q}\leftarrow\boldsymbol{k}} = \frac{\mathrm{e}^2\hbar^2\Gamma_{\boldsymbol{k}}}{2m_0^2\epsilon\omega\Omega}\left|\frac{\sqrt{N_{\hbar\omega}+1}\,\langle\boldsymbol{q}|\boldsymbol{e_s}\cdot\nabla|\boldsymbol{k}\rangle}{E_{\boldsymbol{q}}-E_{\boldsymbol{k}}-\hbar\omega+i\Gamma_{\boldsymbol{k}}}+\frac{\sqrt{N_{\hbar\omega}}\,\langle\boldsymbol{q}|\boldsymbol{e_s}\cdot\nabla|\boldsymbol{k}\rangle}{E_{\boldsymbol{q}}-E_{\boldsymbol{k}}+\hbar\omega+i\Gamma_{\boldsymbol{k}}}\right|^2 \tag{2.110}$$

$$\Gamma_{\boldsymbol{k}} = \sum_{\boldsymbol{q}}\frac{\pi\mathrm{e}^2\hbar^3|\langle\boldsymbol{q}|\boldsymbol{e_s}\cdot\nabla|\boldsymbol{k}\rangle|^2}{2m_0^2\epsilon\omega\Omega}\Big[(N_{\hbar\omega}+1)\delta(E_{\boldsymbol{q}}-E_{\boldsymbol{k}}-\hbar\omega)+N_{\hbar\omega}\delta(E_{\boldsymbol{q}}-E_{\boldsymbol{k}}+\hbar\omega)\Big] \tag{2.111}$$

by Eqs. 2.104, 2.106, 2.107, 2.108. Here $N_{\hbar\omega}$ is the number of monochromatic photons having an energy of $\hbar\omega$ and, $\boldsymbol{e_s}$ is the polarization vector of the optical field. Note that $n_{\hbar\omega} = N_{\hbar\omega}/\Omega$ is the photon density and Ω is the normalization volume; see Eq. 2.89.

The first terms on the right sides of the above two equations represent the photon emission process, whereas the second ones correspond to the photon absorption when an electron transits from an initial state $|\boldsymbol{k}\rangle$ to the final state $|\boldsymbol{q}\rangle$. For finite $\hbar\omega$, the two processes are well distinguished by the relevant energy resonance requirements of $E_{\boldsymbol{q}} - E_{\boldsymbol{k}} \pm \hbar\omega$. Therefore, for $E_{\boldsymbol{q}} > E_{\boldsymbol{k}}$, only photon absorption is possible

$$\begin{aligned} p_{\boldsymbol{q}\leftarrow\boldsymbol{k}} &= \frac{\mathrm{e}^2\hbar^2 n_{\hbar\omega}|\langle\boldsymbol{q}|\boldsymbol{e_s}\cdot\nabla|\boldsymbol{k}\rangle|^2}{2m_0^2\epsilon\omega}\frac{\Gamma_{\boldsymbol{k}}}{\left(E_{\boldsymbol{q}}-E_{\boldsymbol{k}}-\hbar\omega\right)^2+\Gamma_{\boldsymbol{k}}^2} \\ \Gamma_{\boldsymbol{k}} &= \sum_{\boldsymbol{q}}\frac{\pi\mathrm{e}^2\hbar^3 n_{\hbar\omega}|\langle\boldsymbol{q}|\boldsymbol{e_s}\cdot\nabla|\boldsymbol{k}\rangle|^2}{2m_0^2\epsilon\omega}\,\delta(E_{\boldsymbol{q}}-E_{\boldsymbol{k}}-\hbar\omega) \end{aligned} \tag{2.112}$$

and the photon emission rate is

$$\begin{aligned} p_{\boldsymbol{k}\leftarrow\boldsymbol{q}} &= \frac{\mathrm{e}^2\hbar^2 n_{\hbar\omega}|\langle\boldsymbol{k}|\boldsymbol{e_s}\cdot\nabla|\boldsymbol{q}\rangle|^2}{2m_0^2\epsilon\omega}\frac{\Gamma_{\boldsymbol{q}}}{\left(E_{\boldsymbol{k}}-E_{\boldsymbol{q}}+\hbar\omega\right)^2+\Gamma_{\boldsymbol{q}}^2} \\ \Gamma_{\boldsymbol{q}} &= \sum_{\boldsymbol{k}}\frac{\pi\mathrm{e}^2\hbar^3 n_{\hbar\omega}|\langle\boldsymbol{k}|\boldsymbol{e_s}\cdot\nabla|\boldsymbol{q}\rangle|^2}{2m_0^2\epsilon\omega}\,\delta(E_{\boldsymbol{k}}-E_{\boldsymbol{q}}+\hbar\omega) \end{aligned} \tag{2.113}$$

Here since $N_{\hbar\omega} \gg 1$ due to the constant external optical field, $(N_{\hbar\omega}+1)/\Omega \approx N_{\hbar\omega}/\Omega = n_{\hbar\omega}$.

Eqs. 2.112 and 2.113 are very much alike by taking into account the facts that $|\langle \boldsymbol{q}|\boldsymbol{e_s}\cdot\nabla|\boldsymbol{k}\rangle| = |\langle \boldsymbol{k}|\boldsymbol{e_s}\cdot\nabla|\boldsymbol{q}\rangle|$ and $\delta(E_{\boldsymbol{q}} - E_{\boldsymbol{k}} - \hbar\omega) = \delta(E_{\boldsymbol{k}} - E_{\boldsymbol{q}} + \hbar\omega)$. However, $\Gamma_{\boldsymbol{q}} \neq \Gamma_{\boldsymbol{k}}$ because the densities of states at $E_{\boldsymbol{x}}$ and $E_{\boldsymbol{q}}$ can be very different. For example, in a quantum well infrared photodetector, the principal optical transition involves an extended state in the barrier conduction band and a localized quantum well sublevel.

We briefly introduce and relate the optical coefficient to the optical transition rate. If we consider a beam of photons travelling along the z-axis, we can write the continuity equation for the photon density

$$\frac{\mathrm{d}n_{\hbar\omega}(z)}{\mathrm{d}t} = \frac{\partial n_{\hbar\omega}(z)}{\partial t} - \frac{\partial\big[c\, n_{\hbar\omega}(z)\big]}{\partial z} \tag{2.114}$$

where the first term on the right side of the above equation represents the absorption rate of photons and the second term the photons leaving due to the photon current. Here c is the speed of light in the medium. At steady state we have, in general

$$n_{\hbar\omega}(z) = n_{\hbar\omega}(0)\mathrm{e}^{\alpha_{\hbar\omega} z} \tag{2.115}$$

which defines the optical coefficient $\alpha_{\hbar\omega}$.

We consider an optical transition of electron from state $\boldsymbol{k}$ to $\boldsymbol{q}$. We further assume $E_{\boldsymbol{q}} > E_{\boldsymbol{k}}$. More specifically, the electron absorbs a photon (so that the total number of photon decreases) and transits with a transition probability $p_{\boldsymbol{q}\leftarrow\boldsymbol{k}}$ from the initially occupied $\boldsymbol{k}$ with an occupation probability $f(E_{\boldsymbol{k}})$ to the final unoccupied $\boldsymbol{q}$ with an unoccupied probability $\big[1 - f(E_{\boldsymbol{q}})\big]$. The probability of the absorption process is:

$$p_{\boldsymbol{q}\leftarrow\boldsymbol{k}}\Big[1 - f(E_{\boldsymbol{q}})\Big]f(E_{\boldsymbol{k}})$$

Similarly we consider the photon emission case that the electron transits from $\boldsymbol{q}$ to $\boldsymbol{k}$ so that the number of photon increases. Thus,

$$\frac{\partial n_{\text{ph}}(\hbar\omega, z)}{\partial t} = \sum_{\boldsymbol{q},\boldsymbol{k}(E_{\boldsymbol{q}}>E_{\boldsymbol{k}})} \Big\{p_{\boldsymbol{k}\leftarrow\boldsymbol{q}}\Big[1 - f(E_{\boldsymbol{k}})\Big]f(E_{\boldsymbol{q}}) - p_{\boldsymbol{q}\leftarrow\boldsymbol{k}}\Big[1 - f(E_{\boldsymbol{q}})\Big]f(E_{\boldsymbol{k}})\Big\} \tag{2.116}$$

At the steady state, $\mathrm{d}n_{\hbar\omega}(z)/\mathrm{d}t = 0$,

$$\alpha_{\hbar\omega} = \sum_{\boldsymbol{q},\boldsymbol{k},(E_{\boldsymbol{q}}>E_{\boldsymbol{k}})} \frac{\mathrm{e}^2\hbar^2|\langle\boldsymbol{k}|\boldsymbol{e_s}\cdot\nabla|\boldsymbol{q}\rangle|^2}{2m_0^2 c\epsilon\omega} \times \left\{\frac{\Gamma_{\boldsymbol{q}}\Big[1 - f(E_{\boldsymbol{k}})\Big]f(E_{\boldsymbol{q}})}{\big(E_{\boldsymbol{q}} - E_{\boldsymbol{k}} - \hbar\omega\big)^2 + \Gamma_{\boldsymbol{q}}^2} - \frac{\Gamma_{\boldsymbol{k}}\Big[1 - f(E_{\boldsymbol{q}})\Big]f(E_{\boldsymbol{k}})}{\big(E_{\boldsymbol{q}} - E_{\boldsymbol{k}} - \hbar\omega\big)^2 + \Gamma_{\boldsymbol{k}}^2}\right\} \tag{2.117}$$

When $\alpha_{\hbar\omega} > 0$, the optical field enhances following its propagation through the medium. The situation is normally referred to as optical gain, for example, in light-emitting and laser devices. $\alpha_{\hbar\omega} < 0$ refers to the optical absorption, for example, in photodetectors.

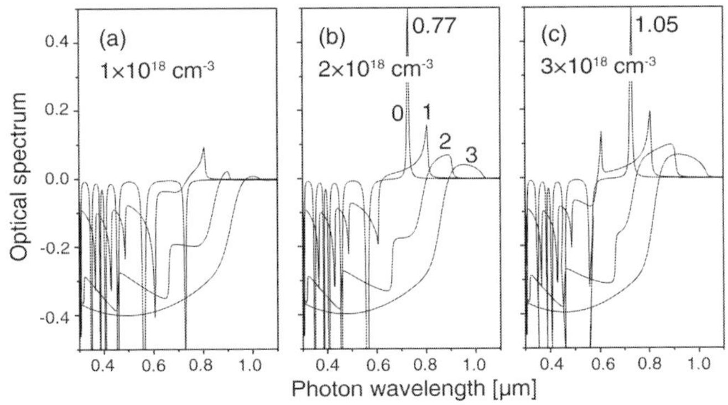

Fig. 2.2 Optical spectra of low-dimensional GaAs nanostructures. $L = 100$ Å, $T = 300$ K. Integer 0, 1, 2, and 3 indicate the dimensionality of the system.

Note that by Eqs. 2.112 and 2.113, $\Gamma_{\boldsymbol{k}} = \Gamma_{\boldsymbol{q}}$ when the two states are only coupled via optical transition and when their densities of states are similar (for example, when we consider optical transitions between two confined states in a quantum dot). In this case, Eq. 2.117 becomes

$$\alpha_{\hbar\omega} = \sum_{\boldsymbol{q},\boldsymbol{k},(E_{\boldsymbol{q}}>E_{\boldsymbol{k}})} \frac{\mathrm{e}^2\hbar^2|\langle\boldsymbol{k}|\boldsymbol{e_s}\cdot\nabla|\boldsymbol{q}\rangle|^2\Gamma_{\boldsymbol{kq}}}{2m_0^2c\epsilon\omega\big(E_{\boldsymbol{q}}-E_{\boldsymbol{k}}-\hbar\omega\big)^2+\Gamma_{\boldsymbol{kq}}^2}\Big[f(E_{\boldsymbol{q}})-f(E_{\boldsymbol{k}})\Big] \tag{2.118}$$

Here $\Gamma_{\boldsymbol{kq}} = \Gamma_{\boldsymbol{k}} = \Gamma_{\boldsymbol{q}}$. The above equation shows that the population inversion, $f(E_{\boldsymbol{q}}) > f(E_{\boldsymbol{k}})$ when $E_{\boldsymbol{q}} > E_{\boldsymbol{k}}$, is essential in the generation of laser radiation (i.e., $\alpha > 0$), where at equilibrium status, $f(E_{\boldsymbol{q}}) < f(E_{\boldsymbol{k}})$.

In Fig. 2.2 we show the optical spectra of various materials under various carrier injection conditions. Three-dimensionally extended bulk material (3D), 2D-extended quantum well, 1D quantum wire, and 0D quantum dot are considered. The extension of the quantum confinement is denoted as L (L is the quantum well width, $L \times L$ is then the cross-section of the quantum wire, and $L \times L \times L$ is the volume of the quantum dot). The carrier densities indicated in the figure are injected electron (hole) concentrations in the conduction (valence) bands. It is observed by Fig. 2.2 that the optical spectrum in general consists of two parts: the optical gain part in the long wavelength region, where the population inversion is achieved; and the optical absorption part in the short wavelength region, where high-energy electron states are largely unoccupied and high-energy hole states are occupied.

Reference

[1] S. M. Sze, *Physics of Semiconductor Devices*, 2nd edn, John Wiley & Sons, New York, 1981, p. 32.

Chapter 3

Exciton and Exciton Photogeneration

To precisely engineer light-matter interactions at the nanoscale with electron–hole pairs (excitons), and their coupling with photons (exciton-polaritons), is becoming increasingly important in contemporary research in nanoscience. Materials with nanoscale features have the potential to revolutionize optoelectronics, permitting new and interesting devices and system capabilities. One key design element is the geometry, i.e., size and shape, which defines the properties of the nanostructured material and which connects to the possibility to design new and significantly improved physical, electrical, and optical properties. Excitons represent one such important property that can be dictated by the size and the shape of a nanostructure in which they are confined, in addition to the material constitution of the nanostructure.

Excitons were intensively studied in the 1960s and 1970s, and their manifestations in various optical phenomena, including light reflection and transmission, photoluminescence (PL), and resonant light scattering, are by now well established and documented; see e.g. the dedicated volume[1] and references therein. Reference[2] gives a very intuitive description of excitons in nanoscale systems and a detailed review on excitons in polymers and nanotubes. Another key work is Ref.,[3] which reviews research and development of excitions and photons in confined structures, from three-dimensionally extended states (bulk) to atom-like quantum dots (QDs) and their applications in light emitting devices.

In the present chapter we outline a selected topic out of the broad set of world wide research activities in exciton science: We concentrate on low-dimensional semiconductor systems, where it has become possible during the last 20 years to handle and control single excitons. More specifically, we focus on excitons and exciton-polaritons in semiconductor QDs, which often are referred to as artificial atoms.

3.1 QUANTUM CONFINEMENT IN NANOPARTICLES

Optical spectroscopy deals with interactions of electromagnetic radiation with matter that occur at the ultraviolet (UV), visible (Vis), near-infrared (NIR), and

Table 3.1 Room-temperature energy band structure parameters of II–VI materials.[4–14]

Parameter	Wurtzite CdS	Hexagonal CdSe	Zincblende CdTe
E_g [eV]	2.485	1.751	1.49
Δ_{so} [eV]	0.062	0.416	0.80
m_c^*	0.18,[6] 0.2, 0.20[11]	0.12, 0.13[4, 11]	0.090, 0.096[11]
$m_{p,t}^*/m_{h,\ell}^*$	0.53,[6] 0.7/5, 0.8,[7] 1.02[8]	0.45/1, 0.41[4]	0.72
			0.12
$\epsilon(0)_t/\epsilon(0)_\ell$	8.45/9.12	9.29/10.16	10.2
$\epsilon(\infty)_t/\epsilon(\infty)_\ell$	5.32	6.3/6.2	7.1
E_p [eV]			20.7[11]
Δ_c [eV]	2.21,[9] 2.3–2.5[12]		
Δ_p [eV]	2.21[9]		
χ [eV]	4.79,[13] 4.22[14]	4.95[13]	4.28[13]
	Cubic ZnS	Wurtzite ZnS	Wurtzite ZnO
E_g [eV]	3.68	3.78	3.2
Δ_{so} [eV]		0.086	
m_c^*	0.184, 0.28,[11] 0.42[5]	0.28	0.24, 0.19[11]
$m_{h,t}^*/m_{h,\ell}^*$	1.76, 0.61[5]	0.49/1.4	0.59, 0.45[7]
	0.23		
$\epsilon(0)_t/\epsilon(0)_\ell$	8.0	9.6	7.8/8.75
$\epsilon(\infty)_t/\epsilon(\infty)_\ell$	5.1	5.13	3.70/3.75
E_p [eV]	20.4[11]		
χ [eV]	3.9[13]		4.57[13]

Source: Data are obtained from[10] if not otherwise stated.

infrared (IR) wavelengths. In the UV/Vis spectral regions (< 700 nm), light can penetrate only superficial tissue volumes (a few hundred microns to a millimeter in depth) due to the fact that biological medium is highly absorbing. However, in the NIR spectral region (700–900 nm) tissue is generally less absorbing and, furthermore, the number of elastic scattering events of light in tissues is approximately two orders of magnitude greater than the number of absorption events. This enables the light to propagate through tissue volumes that are up to several centimeters in depth. Furthermore, the success of bio application depends very much on the optical excitation and detection technologies, which are in the infrared regimes in most cases. In this chapter we concentrate mostly on II–VI QDs that fluorescence in the Vis–NIR region. Table 3.1 lists important physical parameters of common II–VI compounds, and physical parameters about exciton in relevant bulk semiconductors are listed in Table 3.2.

Electronic energy band structure in a nano-size particle is characterized by a series of discrete energy states, which has been focused on by many experimental and theoretical studies.[18] As one of the most important III–V semiconductors, GaAs has a large energy bandgap and a large exciton Bohr radius (11.6 nm, see Table 3.2) so that the quantum confinement effect in nano clusters is very pronounced, resulting in wide applications in photodetectors and long-wavelength lasers. GaAs nano

Table 3.2 Excitons in common semiconductor[10, 15–17] (low temperature for IV and III–V).

	Si	GaAs	AlAs	InAs	CdSe	ZnO
m_h (heavy hole)	0.537	0.51	0.409	0.35	0.45	0.59
m_e (electron DOS)	1.026	0.067	0.71	0.0239	0.13	0.28
E_g [eV]	1.170	1.519	2.229	0.418	1.842	3.435
a_{Br} [nm]	4.5	11.6	8.6	38.1	5.4	1.8
E_{10} [meV]	14.3	4.2	20.0	1.25	13.2	60.0

clusters have been prepared by different techniques including laser ablation, molecular beam epitaxy, radio frequency sputtering, and chemical method for various applications.[19, 20]

Quantitative understanding of structure–function relationships about quantum confinement and energy states is one of the key factors concerning the success of the nano system applications. However, semiconductor clusters in the nano-size are qualitatively different from bulk materials, suffering the breakdown of periodic boundary conditions, implying that the traditional solid physics approaches such as effective mass approximation and $\boldsymbol{k} \cdot \boldsymbol{p}$ theory are questionable in applying them to study nanostructures.[21] On the other hand, quantum chemistry theory is useful in describing the small-scale molecular systems and predicting spectra of many physical processes in them. Such technology has been applied to study structural and electronic properties of II–VI semiconductor clusters,[22] III–V compounds,[23, 24] as well as carbon nanoparticles.[25, 26] Figure 3.1 shows schematically a GaAs nano cluster with a diameter of 2.6 nm terminated with hydrogen atoms, whereas Fig. 3.2[27] shows that the electronic states of hydrogen terminated GaAs nano clusters up to 16.9 nm in diameter were studied using a large-scale quantum chemistry approach by which the quantum confinement effect is shown to quantitatively agree with experimental results.[28, 29]

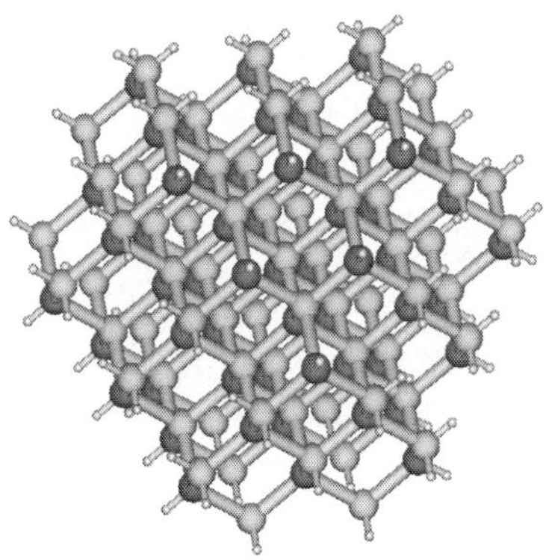

Fig. 3.1 Lattice structures of a GaAs nano cluster. See also Color Insert.

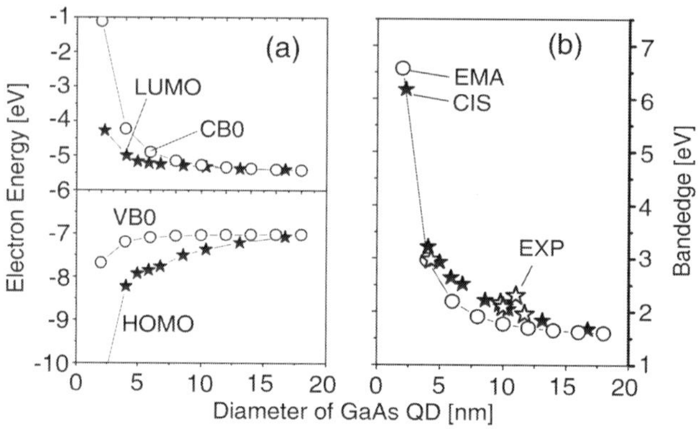

Fig. 3.2 Size-dependent (a) core-HOMO and core-LUMO, (b) band gap E_g between core-HOMO and core-LUMO of GaAs nano clusters. J. Jiang, B. Gao, T.-T. Han, and Y. Fu, Ab initio study of energy band structures of GaAs nanoclusters, Appl. Phys. Lett. vol. 94, p. 092110(3), 2009. Copyright (2009) by The American Institute of Physics.

As shown in Fig. 3.2, advanced theories such as advanced quantum chemistry,[27] effective bond-orbital model for valence-band holes,[30,31] sp^3s^* semi-empirical tight-binding theory,[4,5,32] and sp^3d^5 nearest-neighbor tight-binding model[6,33] improve very much the energy positions of peaks in the optical spectrum for small QDs. On the other hand, the size of QDs of practical interest is normally comparable with the exciton Bohr radius (see Table 3.2). We can well approximate the eigenvalue problem by the effective mass approximation of Ref.,[7–9,34–36] as shown in Fig. 3.2. Relevant energy band structure parameters are listed in Table 3.1. In this model, the total wave function of the conduction-band electron (valence-band hole) is written as $\psi_{ci}u_c$ ($\psi_{vj}u_v$), where ψ_{ci} (ψ_{vj}) is the envelop wave function and u_c (u_v) is the lattice-periodic Bloch function related to the conduction (valence) band.

We consider the one-electron Hamiltonian of a QD in the form of

$$H_c = -\frac{\hbar^2\nabla^2}{2m_c^*} + V_c(\boldsymbol{r}) \tag{3.1}$$

for conduction-band electrons; m_c^* is the effective mass of the electron,

$$V_c(\boldsymbol{r}) = \begin{cases} E_c & b \leq 1 \\ E_c + \Delta_c & \text{otherwise} \end{cases} \tag{3.2}$$

where

$$b = \frac{x^2}{a^2} + \frac{y^2}{a^2} + \frac{z^2}{a^2} \tag{3.3}$$

a is the QD radius. Subscript "c" stands for the conduction band, E_c is the conduction bandedge, and Δ_c is the band offset between the QD and the surrounding medium. When referring to vacuum, $\Delta_c = \chi$, which is the electron affinity of the material surface. Similar expressions can be written down for the valence-band holes.

For spherical QDs with $V_c(\boldsymbol{r}) = V_c(r)$, we pick eigenfunctions of the Schrödinger equation Eq. 3.1 in the form

$$\psi_{\ell m}(r,\theta,\phi) = R_\ell(r) Y_{\ell m}(\theta,\phi) \tag{3.4}$$

because of the rotational invariance, from which comes the radial Schrödinger equation

$$\frac{\mathrm{d}^2 R_\ell(r)}{\mathrm{d}r^2} + \frac{2}{r}\frac{\mathrm{d}R_\ell(r)}{\mathrm{d}r} - \frac{\ell(\ell+1)}{r^2} R_\ell(r) + \frac{2m_c^*\big[E - V_c(r)\big]}{\hbar^2} R_\ell(r) = 0 \tag{3.5}$$

where $Y_{\ell,m}(\theta,\phi)$ are the angular momentum eigenfunctions, which are also known as spherical harmonics. The first few normalized spherical harmonics $Y_{\ell m}(\theta,\phi)$ $(m = -\ell, -\ell+1, \ldots, \ell-1, \ell)$ are[37]

$$\begin{aligned}
Y_{00} &= \frac{1}{\sqrt{4\pi}} \\
Y_{11} &= -\sqrt{\frac{3}{8\pi}} \sin\theta \mathrm{e}^{i\phi} \quad Y_{10} = \sqrt{\frac{3}{4\pi}} \cos\theta \\
Y_{22} &= \sqrt{\frac{15}{32\pi}} \sin^2\theta \mathrm{e}^{i2\phi} \quad Y_{21} = -\sqrt{\frac{15}{8\pi}} \sin\theta\cos\theta \mathrm{e}^{i\phi} \\
Y_{20} &= \sqrt{\frac{5}{16\pi}} \left(3\cos^2\theta - 1\right) \\
Y_{33} &= -\sqrt{\frac{35}{64\pi}} \sin^3\theta \mathrm{e}^{3i\phi} \quad Y_{32} = \sqrt{\frac{105}{32\pi}} \sin^2\theta\cos\theta \mathrm{e}^{2i\phi} \\
Y_{31} &= -\sqrt{\frac{21}{64\pi}} \sin\theta \left(5\cos^2\theta - 1\right) \mathrm{e}^{i\phi} \\
Y_{30} &= \sqrt{\frac{7}{16\pi}} \left(5\cos^3\theta - 3\cos\theta\right)
\end{aligned} \tag{3.6}$$

It is convenient to introduce the function $u_\ell(r) = rR_\ell(r)$ so that

$$\frac{\mathrm{d}^2 u_\ell(r)}{\mathrm{d}r^2} + \frac{2m_c^*}{\hbar^2}\left[E - V_c(r) - \frac{\ell(\ell+1)\hbar^2}{2m_c^* r^2}\right] u_\ell(r) = 0 \tag{3.7}$$

This looks very much like a one-dimensional equation except that the definition of $u_\ell(r)$ and the finiteness of the wave function at the origin require that $u_\ell(0) = 0$. For a spherical square well of $V(r < a) = -V$ and $V(r \geq a) = 0$, the above equation also requires that

$$E \geq V_c(r) + \frac{\ell(\ell+1)\hbar^2}{2m_c^* r^2}$$

which also sets the maximal value of ℓ for E.

3.1.1 *Free Carrier and Infinite Potential Well*

For $V_c(r) = 0$ and $E > 0$, we denote $k^2 = 2m_c^* E/\hbar^2$; the regular solution of Eq. 3.5 is $j_\ell(\rho)$ known as spherical Bessel functions

$$j_\ell(\rho) = (-\rho)^\ell \left(\frac{1}{\rho}\frac{\mathrm{d}}{\mathrm{d}\rho}\right)^\ell \left(\frac{\sin\rho}{\rho}\right) \tag{3.8}$$

Table 3.3 Roots of ka for spherical infinite potentail well.

$\ell = 0$	1	2	3	4	5
3.1416 (π)	4.4934	5.7635	6.9879	8.1826	9.3558
6.2832 (2π)	7.7252	9.0950	10.4171	11.7049	12.9665
9.4248 (3π)	10.9042	12.3229	13.6980	15.0397	16.3547
12.5664 (4π)	14.0661	15.5146	16.9236	18.3013	19.6531
15.7080 (5π)	17.2208	18.6890	20.1218	21.5255	22.9046

where $\rho = kr$. And the irregular solution is called the spherical Neumann function $n_\ell(\rho)$

$$n_\ell(\rho) = -(-\rho)^\ell \left(\frac{1}{\rho}\frac{\mathrm{d}}{\mathrm{d}\rho}\right)^\ell \left(\frac{\cos\rho}{\rho}\right) \tag{3.9}$$

The first few values of these functions are

$$\begin{aligned}
j_0(\rho) &= \frac{\sin\rho}{\rho} \qquad j_1(\rho) = \frac{\sin\rho}{\rho^2} - \frac{\cos\rho}{\rho} \\
j_2(\rho) &= \left(\frac{3}{\rho^3} - \frac{1}{\rho}\right)\sin\rho - \frac{3}{\rho^2}\cos\rho \\
n_0(\rho) &= -\frac{\cos\rho}{\rho} \qquad n_1(\rho) = -\frac{\cos\rho}{\rho^2} - \frac{\sin\rho}{\rho} \\
n_2(\rho) &= -\left(\frac{3}{\rho^3} - \frac{1}{\rho}\right)\cos\rho - \frac{3}{\rho^2}\sin\rho
\end{aligned} \tag{3.10}$$

For infinite potential well of $V_c(r) = 0$ when $r < a$ and $V_c(r) = \infty$ when $r \geq a$, the solution that is regular at $r = 0$ is $R_\ell(r) = A_\ell j_\ell(kr)$ with the eigenvalues determined by the condition that the solution vanishes at $r = a$, i.e., $j_\ell(ka) = 0$. A_ℓ is the wave function normalization factor, and the roots of ka for a few values of ℓ are listed in Table 3.3.

It is easy to show by the comparison of the energies of the ground and first-excited electron levels obtained numerically and analytically that for the range of QD sizes of interest (diameter > 5 nm), the infinite-potential-well model describes very well the low-lying energy levels.

3.1.2 *Spherical Square Well*

Define the QD by a square potential well of $V_c(r) = -\Delta_c$ when $r < a$ and $V_c(r) = 0$ when $r \geq a$, where a is the radius of the QD. For continuum solutions $E > 0$, the solution for $r > a$ will be a combination of the regular and irregular solutions

$$R_\ell(r) = B_\ell\, j_\ell(kr) + C_\ell\, n_\ell(kr) \tag{3.11}$$

while the solution for $r \leq a$ must be the regular one of

$$R_\ell(r) = A_\ell\, j_\ell(qr) \tag{3.12}$$

where $q^2 = 2m_c^*(E + \Delta_c)/\hbar^2$, and $k^2 = 2m_c^* E/\hbar^2$ as before.

The two solutions must match at $r = a$ and so must the derivatives. This leads to

$$q\left[\frac{\mathrm{d}j_\ell(\rho)/\mathrm{d}\rho}{j_\ell(\rho)}\right]_{\rho=qa} = k\left[\frac{B_\ell \mathrm{d}j_\ell(\rho)/\mathrm{d}\rho + C_l \mathrm{d}n_\ell(\rho)/\mathrm{d}\rho}{B_\ell\, j_\ell(\rho) + C_\ell\, n_\ell(\rho)}\right]_{\rho=ka} \tag{3.13}$$

by which the ratio of B_ℓ/C_ℓ is determined. The ratio of A_ℓ/C_ℓ is

$$A_\ell/C_\ell = \frac{B_\ell/C_\ell\, j_\ell(ka) + n_\ell(ka)}{j_\ell(qa)} \tag{3.14}$$

and C_ℓ is obtained by the wave function normalization

$$|C_\ell|^2 \left\{|A_\ell/C_\ell|^2 \int_0^a j_\ell^2(kr) r^2 \mathrm{d}r + \int_a^\infty \Big[B_\ell/C_\ell\, j_\ell(kr) + n_\ell(kr)\Big]^2 r^2 \mathrm{d}r\right\} = 1 \tag{3.15}$$

For bound states $E \le 0$, we write $\alpha^2 = -2m_c^* E/\hbar^2$. The solution for $r \le a$ must be regular at the origin

$$R_\ell(r) = A_\ell\, j_\ell(qr) \tag{3.16}$$

and the solution for $r > a$ must vanish as $r \to \infty$,

$$R_\ell(r) = B_\ell\, h_\ell(i\alpha r) \tag{3.17}$$

where $h_\ell(\rho)$ is the spherical Hankel function

$$h_\ell(\rho) = j_\ell(\rho) + i n_\ell(\rho) \tag{3.18}$$

The first few spherical Hankel functions are

$$\begin{aligned} h_0(\rho) &= -\frac{i\mathrm{e}^{i\rho}}{\rho} \qquad h_1(\rho) = -\frac{\mathrm{e}^{i\rho}}{\rho}\left(1 + \frac{i}{\rho}\right) \\ h_2(\rho) &= \frac{i\mathrm{e}^{i\rho}}{\rho}\left(1 + \frac{3i}{\rho} - \frac{3}{\rho^2}\right) \end{aligned} \tag{3.19}$$

The solutions of bound states are given by the boundary conditions

$$q\left[\frac{\mathrm{d}j_\ell(\rho)/\mathrm{d}\rho}{j_\ell(\rho)}\right]_{\rho=qa} = i\alpha\left[\frac{\mathrm{d}h_\ell(\rho)/\mathrm{d}\rho}{h_\ell(\rho)}\right]_{\rho=i\alpha a} \tag{3.20}$$

A_ℓ/B_ℓ and B_ℓ are

$$\begin{aligned} A_\ell/B_\ell &= \frac{h_\ell(i\alpha a)}{j_\ell(qa)} \\ |B_\ell|^2 &\left[|A_\ell/B_\ell|^2 \int_0^a j_\ell^2(qr) r^2 \mathrm{d}r + \int_a^\infty |h_\ell(i\alpha r)|^2 r^2 \mathrm{d}r\right] = 1 \end{aligned} \tag{3.21}$$

The spherical Bessel functions have the following recurrence relation:

$$j_{\ell+1}(\rho) = \frac{\ell}{\rho}\, j_\ell(\rho) - \frac{\mathrm{d}j_\ell(\rho)}{\mathrm{d}\rho} = \frac{2\ell+1}{\rho}\, j_\ell(\rho) - j_{\ell-1}(\rho) \tag{3.22}$$

Alternatively, we can express the relationship as

$$\begin{bmatrix} j_{\ell+1}(\rho) \\ j'_{\ell+1}(\rho) \end{bmatrix} = \begin{bmatrix} \ell/\rho & -1 \\ 1 - \ell(\ell+2)/\rho^2 & (\ell+2)/\rho \end{bmatrix} \begin{bmatrix} j_\ell(\rho) \\ j'_\ell(\rho) \end{bmatrix} \tag{3.23}$$

The same recurrence relation holds true for $n_\ell(\rho)$ and $h_\ell(\rho)$. Recurrence relation in different forms:

$$\begin{aligned}
&j_{\ell-1}(\rho) + j_{\ell+1}(\rho) = \frac{2\ell+1}{\rho} j_\ell(\rho) \\
&\ell j_{\ell-1}(\rho) - (\ell+1) j_{\ell+1}(\rho) = (2\ell+1)\frac{\mathrm{d}j_\ell(\rho)}{\mathrm{d}\rho} \\
&\frac{\mathrm{d}}{\mathrm{d}\rho}\left[\rho^{\ell+1} j_\ell(\rho)\right] = x^{\ell+1} f_{\ell-1}(\rho) \\
&\frac{\mathrm{d}}{\mathrm{d}\rho}\left[\rho^{-\ell} j_\ell(\rho)\right] = x^{-\ell} f_{\ell+1}(\rho)
\end{aligned} \tag{3.24}$$

3.1.3 *Cylindrical Quantum Wire*

A cylindrical quantum wire positioned along the x axis is described by

$$\left[-\frac{\hbar^2}{2m_c^*}\left(\frac{\mathrm{d}^2}{\mathrm{d}y^2} + \frac{\mathrm{d}^2}{\mathrm{d}z^2}\right) + V_c(y,z)\right]\psi(y,z) = E\psi(y,z) \tag{3.25}$$

where

$$V_c(y,z) = \begin{cases} -\Delta_c & y^2+z^2 \le a^2 \\ 0 & y^2+z^2 > a^2 \end{cases} \tag{3.26}$$

where $\Delta_c \ge 0$. In the cylindrical coordinate of $(y,z) = (\rho,\phi)$ and by writing $\psi(y,z) = \mathrm{e}^{im\phi}R_m(\rho)$,

$$\frac{\mathrm{d}^2R(\rho)}{\mathrm{d}\rho^2} + \frac{1}{\rho}\frac{\mathrm{d}R(\rho)}{\mathrm{d}\rho} + \left\{\frac{2m_c^*[E - V_c(\rho)]}{\hbar^2} - \frac{m^2}{\rho^2}\right\}R(\rho) = 0 \tag{3.27}$$

where m is an integer (either negative or positive so the state of $|m|$ is two-fold degenerate).

For free electrons where $\Delta_c = 0$ and $E \ge 0$, Eq. 3.27 is the Bessel equation

$$\rho^2\frac{\mathrm{d}^2R(\rho)}{\mathrm{d}\rho^2} + \rho\frac{\mathrm{d}R(\rho)}{\mathrm{d}\rho} + \left(k^2\rho^2 - m^2\right)R(\rho) = 0 \tag{3.28}$$

where $k^2 = 2m_c^*E/\hbar^2$.

For finite Δ_c and for energy states confined in the quantum wire, i.e., $-\Delta_c \le E < 0$, the wave function outside of the quantum wire ($\rho > a$) is described by the Helmholtz equation

$$\rho^2\frac{\mathrm{d}^2R(\rho)}{\mathrm{d}\rho^2} + \rho\frac{\mathrm{d}R(\rho)}{\mathrm{d}\rho} - \left(\alpha^2\rho^2 + m^2\right)R(\rho) = 0 \tag{3.29}$$

where $\alpha^2 = -2m_c^*E/\hbar^2$. Inside the quantum wire ($\rho \le a$), we still have the Bessel equation

$$\rho^2\frac{\mathrm{d}^2R(\rho)}{\mathrm{d}\rho^2} + \rho\frac{\mathrm{d}R(\rho)}{\mathrm{d}\rho} + \left(q^2\rho^2 - m^2\right)R(\rho) = 0 \tag{3.30}$$

where $q^2 = 2m_c^*(E + \Delta_c)/\hbar^2$.

The Bessel function is

$$J_n(x) = \sum_{s=0}^{\infty} \frac{(-1)^s}{s!(n+s)!} \left(\frac{x}{2}\right)^{n+2s} = \frac{x^2}{2^n n!} - \frac{x^{n+2}}{2^{n+2}(n+1)!} + \cdots \tag{3.31}$$

which also holds for $n < 0$. And $J_{-n}(x) = (-1)^n J_n(x)$. Recurrence relations exist

$$\begin{aligned} J_{n-1}(x) + J_{n+1}(x) &= \frac{2n}{x} J_n(x) \\ J_{n-1}(x) - J_{n+1}(x) &= 2J_n'(x) \end{aligned} \tag{3.32}$$

Neumann functions, Bessel functions of the second kind

$$N_\nu(x) = \frac{\cos(\nu\pi) J_\nu(x) - J_{-\nu}(x)}{\sin(\nu\pi)} \tag{3.33}$$

is also a solution of Bessel's equation when $\nu = n$ is an integer. The most general solution for the Bessel's equation can therefore be written as

$$A_n J_n(x) + B_n N_n(x) \tag{3.34}$$

$J_n(x)$ and $N_n(x)$ are independent. $N_n(x)$ diverges at least logarithmically ($\ln x$) so that any boundary condition that requires the solution to be finite at the origin automatically excludes $N_n(x)$. Conversely, in the absence of such a requirement $N_n(x)$ must be considered. Neumann functions satisfy the recurrence relations

$$\begin{aligned} N_{n-1}(x) + N_{n+1}(x) &= \frac{2n}{x} N_n(x) \\ N_{n-1}(x) - N_{n+1}(x) &= 2N_n'(x) \end{aligned} \tag{3.35}$$

The solutions of the Helmholtz equation, Eq. 3.29, are Bessel functions of imaginary argument, often written as the modified Bessel functions

$$I_n(x) = i^{-n} J_n(ix) \tag{3.36}$$

In terms of infinite series,

$$I_\nu = \sum_{s=0}^{\infty} \frac{1}{s!(s+\nu)!} \left(\frac{x}{2}\right)^{2s+\nu} \tag{3.37}$$

$I_n(x) = I_{-n}(x)$. The recurrence relations for $I_\nu(x)$ are

$$\begin{aligned} I_{\nu-1}(x) - I_{\nu+1}(x) &= \frac{2\nu}{x} I_\nu(x) \\ I_{\nu-1}(x) + I_{\nu+1}(x) &= 2I_\nu^{'}(x) \end{aligned} \tag{3.38}$$

The second independent solution of Eq. 3.29 when ν is an integer is

$$K_\nu(x) = \frac{\pi}{2} \frac{I_{-\nu}(x) - I_\nu(x)}{\sin(\nu\pi)} \tag{3.39}$$

$K_\nu(x) = K_{-\nu}(x)$, which satisfies the recurrence relations

$$K_{\nu-1}(x) - K_{\nu+1}(x) = -\frac{2\nu}{x}K_\nu(x)$$
$$K_{\nu-1}(x) + K_{\nu+1}(x) = -2K_\nu^{'}(x) \tag{3.40}$$

Some values of the root of the equation $J_n(x) = 0$ are

	$n = 0$	$n = 1$	$n = 2$	$n = 3$	$n = 4$	$n = 5$
1	2.405	3.832	5.135	6.379	7.586	8.780
2	5.520	7.016	8.417	9.760	11.064	12.339
3	8.654	10.173	11.620	13.017	14.373	15.700
4	11.792	13.323	14.796	16.224	17.616	18.982
5	14.931	16.470	17.960	19.410	20.827	22.220

3.1.4 *Colloidal Quantum Dots*

Colloidal QDs (such as CdS, CdSe, ZnSe, and ZnO) can be prepared by sol-gel method at 250–300°C. Here, metallic oxide is selected as the source metal due to its low toxicity, whereas 1-Octadence (ODE) is selected as the only non-coordinating solvent because of its relatively low melting point (below 20°C), relatively high boiling point (about 320°C), low cost, low toxicity, low reactivity to precursors, and excellent solvation capability for many compounds at elevated temperatures. All these make ODE an ideal solvent for the growth of high-quality nanocrystals. This method is proposed by Peng *et al.* in 2002[38] and has been referred to as "organic green chemical method"[38–40] due to being environmentally friendly compared with other methods.

Figure 3.3 shows the schematic QD structure where E_{c0} and E_{v0} denote the ground-state energies measured from the conduction-band and valence-band edges,

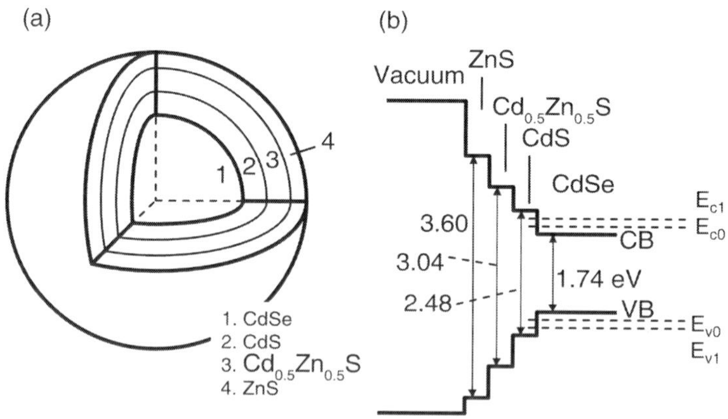

Fig. 3.3 Schematic (a) structure and (b) energy band of core-multishell CdSe QD.

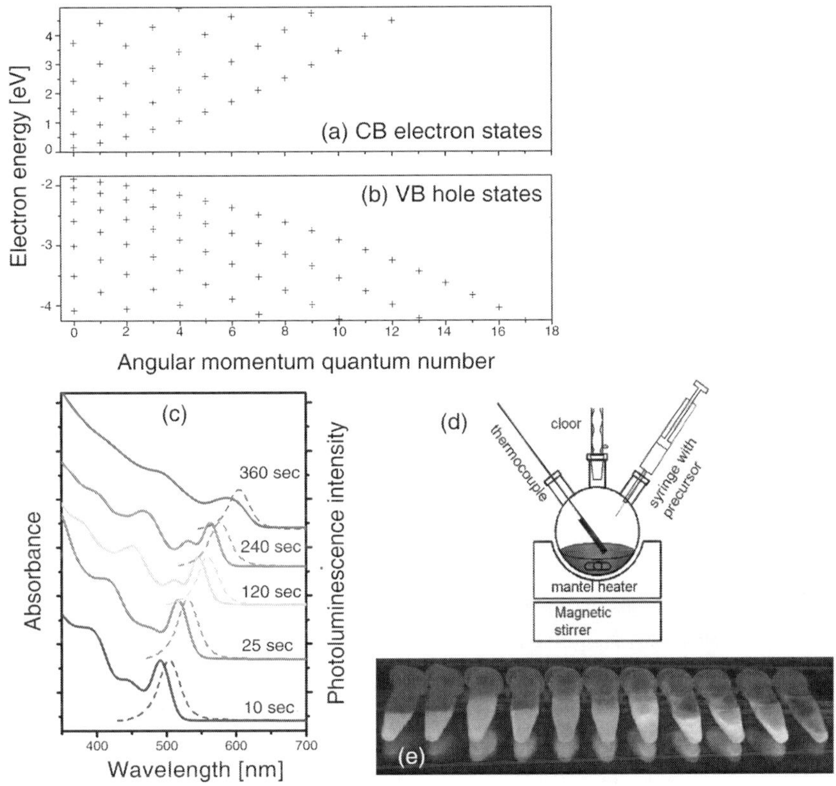

Fig. 3.4 (a) Conduction-band (CB) electron states and (b) valence-band (VB) hole states confined in a spherical CdSe QD with a radius of 4 nm as functions of angular momentum quantum number ℓ. (c) Absorbance (solid lines) and fluorescence (dashed lines), (d) basic chemical synthesis, and (e) fluorescence under UV radiation. See also Color Insert.

respectively, and $E = E_g + E_{c0} + E_{v0}$ is the ground-state transition energy, where E_g is the energy bandgap of the QD core material.

Figure 3.4a,b shows the conduction-band (CB) electron states and valence-band (VB) hole states confined in a spherical CdSe QD with a radius of 4 nm as functions of angular momentum quantum number ℓ. Note that states with $\ell \neq 0$ are $(2\ell + 1)$-fold degenerate. Absorption and fluorescence spectra of a series of CdSe QDs following the growth/reaction time of the chemical synthesis (schemcatically shown in Fig. 3.4d) are presented in Fig. 3.4c, while the fluorescence under UV radiation is shown in Fig. 3.4e.

As shown in Fig. 3.4a,b a QD has a discrete quantized energy spectrum due to the spatial confinements of the wave functions within the QD. In order to obtain a narrow bandwidth of the fluorescence from a QD assembly, the radii of the QDs must be controlled properly, in addition to other sources to the finite bandwidth such as the energy relaxation and carrier distribution in the energy sublevels of the QDs. For an infinite potential QD with a radius of r, $k^2 = 2m_c^* E_{c0}/\hbar^2$, $kr = \pi$ (see Table 3.3) so that

$$r^2 = \frac{\pi^2\hbar^2}{2m_c^* E_{c0}} \quad , \quad 2r\delta r = \frac{\pi^2\hbar^2}{2m_c^*}\frac{\delta E_{c0}}{E_{c0}^2} \tag{3.41}$$

where m_c^* is the electron effective mass, δE_{c0} is the variation in E_{c0} induced by δr. Similar equations can be written down for VB hole states E_{v0}. However, because of the large hole effective mass (about $0.5m_0$ for commonly used semiconductors, where m_0 is the electron rest mass), E_{v0} depends weakly on the QD radius r. Thus, $\delta E \approx \delta E_{c0}$ and

$$\frac{\delta r}{r} = \frac{\delta E_{c0}}{2E_{c0}} = \frac{\delta E_{c0}}{2(E - E_g - E_{v0})} \approx \frac{\delta E}{2(E - E_g)} \tag{3.42}$$

For large-size QDs where the quantum confinement effect is small, E is close to E_g. In this case, the tolerance in the radii of the QD assembly, i.e., $\delta r/r$, for the same emission bandwidth of $\delta E/E$, can be quite large. However, $\delta r/r$ must be small for small QDs. We also notice that E_{c0} is inversely proportional to the electron effective mass. When m_c^* is large, E_{c0} is small so that E is close to E_g, resulting in a much relaxed restriction for the QD size. The effective mass of electrons in commonly used III–V semiconductors is in the order of $0.05m_0$, while it is in the order of $0.5m_0$ in II–VI materials (Table 3.1). Here we understand immediately the advantages of using II–VI materials for fluorescence QDs from the fundamental physical point of view.

3.2 EFFECTIVE MASS THEORY OF EXCITON

In the absence of external fields, the wave function of an electron in a crystal is an eigenfunction of the Hamiltonian[41]

$$\sum_i H(\boldsymbol{r}_i) + \sum_{i<j} V(\boldsymbol{r}_i - \boldsymbol{r}_j) \tag{3.43}$$

where

$$H(\boldsymbol{r}_i) = -\frac{\hbar^2\nabla_i^2}{2m_0} + V_0(\boldsymbol{r}_i) \quad , \quad V(\boldsymbol{r}_i - \boldsymbol{r}_j) = \frac{e^2}{4\pi\epsilon|\boldsymbol{r}_i - \boldsymbol{r}_j|} \tag{3.44}$$

$\boldsymbol{r}_i$ is the spatial position of the ith electron and $V_0(\boldsymbol{r}_j)$ is the lattice potential. Here we have assumed that the nuclear and electronic motions can be separated.

In the Hartree–Fock approximation, the electronic wave function is written as a determinant single-electron function. The ground-state function can be written as

$$\psi_0 = \frac{1}{\sqrt{N!}} A\psi_{v_1\boldsymbol{k}_1}(\boldsymbol{r}_1)\cdots\psi_{v_1\boldsymbol{k}_n}(\boldsymbol{r}_n)\psi_{v_2\boldsymbol{k}_1}(\boldsymbol{r}_{n+1})\cdots\psi_{v_m\boldsymbol{k}_n}(\boldsymbol{r}_N) \tag{3.45}$$

for an N-electron system, where v_i runs over all occupied valence bands and $\boldsymbol{k}_i$ over the first Brillouin zone. The symbol A stands for the anti-symmetrizing operator. Consider an excited state of the system

$$\psi_{c\boldsymbol{k}_e,v\boldsymbol{k}_h} = \frac{1}{\sqrt{N!}} A\psi_{v_1\boldsymbol{k}_1}(\boldsymbol{r}_1)\cdots\psi_{c\boldsymbol{k}_e}(\boldsymbol{r}_j)\cdots\psi_{v_m\boldsymbol{k}_n}(\boldsymbol{r}_N) \tag{3.46}$$

in which a single electron has been transferred from the VB state $\psi_{v\boldsymbol{k}_h}$ to a CB state $\psi_{c\boldsymbol{k}_e}$. The excitation energy of this state above the ground state (E_0) is given by

$$E(c\boldsymbol{k}_e, v\boldsymbol{k}_h) - E_0 = E_c(\boldsymbol{k}_e) - E_v(\boldsymbol{k}_h) + V(c\boldsymbol{k}_e, v\boldsymbol{k}_h) \tag{3.47}$$

where $E_c(\boldsymbol{k}_e)$ and $E_v(\boldsymbol{k}_h)$ are the single-electron CB and VB energies given by

$$\begin{aligned}
E_c(\boldsymbol{k}_e) &= \langle c\boldsymbol{k}_e|H(\boldsymbol{r})|c\boldsymbol{k}_e\rangle + \sum_{v_i\boldsymbol{k}_i}\langle c\boldsymbol{k}_e, v_i\boldsymbol{k}_i|V(\boldsymbol{r}_1-\boldsymbol{r}_2)|c\boldsymbol{k}_e, v_i\boldsymbol{k}_i\rangle \\
&\quad - \sum_{v_i\boldsymbol{k}_i}\langle c\boldsymbol{k}_e, v_i\boldsymbol{k}_i|V(\boldsymbol{r}_1-\boldsymbol{r}_2)|v_i\boldsymbol{k}_i, c\boldsymbol{k}_e\rangle \\
E_v(\boldsymbol{k}_h) &= \langle v\boldsymbol{k}_h|H(\boldsymbol{r})|v\boldsymbol{k}_h\rangle + \sum_{v_i\boldsymbol{k}_i}\langle v\boldsymbol{k}_h, v_i\boldsymbol{k}_i|V(\boldsymbol{r}_1-\boldsymbol{r}_2)|v\boldsymbol{k}_h, v_i\boldsymbol{k}_i\rangle \\
&\quad - \sum_{v_i\boldsymbol{k}_i}\langle v\boldsymbol{k}_h, v_i\boldsymbol{k}_i|V(\boldsymbol{r}_1-\boldsymbol{r}_2)|v_i\boldsymbol{k}_i, v\boldsymbol{k}_h\rangle \\
V(c\boldsymbol{k}_e, v\boldsymbol{k}_h) &= -\langle c\boldsymbol{k}_e, v\boldsymbol{k}_h|V(\boldsymbol{r}_1-\boldsymbol{r}_2)|c\boldsymbol{k}_e, v\boldsymbol{k}_h\rangle \\
&\quad -\langle c\boldsymbol{k}_e, v\boldsymbol{k}_h|V(\boldsymbol{r}_1-\boldsymbol{r}_2)|v\boldsymbol{k}_h, c\boldsymbol{k}_e\rangle
\end{aligned} \tag{3.48}$$

where the sum goes over all VB states that are occupied in the ground state of the crystal. The last terms in Eq. 3.48 are of an exchange nature and are generally small and short-ranged. The correct value of the effective electron–hole interaction potential is rather complicated. When the electron and hole are widely separated and moving relatively slowly with respect to one another, the Coulomb interaction between them is screened by the static dielectric constant ϵ_b. The effective potential is given by

$$V(\boldsymbol{r}_e - \boldsymbol{r}_h) = -\frac{\mathrm{e}^2}{4\pi\epsilon_b|\boldsymbol{r}_e - \boldsymbol{r}_h|} \tag{3.49}$$

However as the electron and hole approach one another, their relative velocity increases until their motion is so rapid that the lattice polarization cannot follow it. This occurs when the frequency of motion is greater than the reststrahlen frequency of the crystal. In this region the interaction is given by Eq. 3.49 with ϵ_b replaced by the high-frequency dielectric constant ϵ_∞. Typical reststrahlen frequencies in III–V compound semiconductors are of order of $5 \times 10^{13}\,\mathrm{s}^{-1}$, yielding the result that Eq. 3.49 is invalid for $|\boldsymbol{r}|$ less than about 50 Å, in that region ϵ_b should be replaced by ϵ_∞.

The properties of a crystal with an electron missing from a VB state $\psi_{v\boldsymbol{k}_h}$ can be described by considering a single particle of positive charge in the state so that the exciton state of Eq. 3.46 is described by the two-particle wave function

$$\psi_{c\boldsymbol{k}_e}(\boldsymbol{r}_e)\psi_{v\boldsymbol{k}_h}(\boldsymbol{r}_h) \tag{3.50}$$

The total wave vector of the excited state is $\boldsymbol{K} = \boldsymbol{k}_e + \boldsymbol{k}_h$.

Because of the Coulomb interaction, Eq. 3.50 does not represent an eigenstate of the system. We approximate the eigenstate by a linear combination of electron-hole states

$$\Psi_{n\boldsymbol{K}}(\boldsymbol{r}_e, \boldsymbol{r}_h) = \sum_{c\boldsymbol{k}_e, v\boldsymbol{k}_h} A_{n\boldsymbol{K}}(c\boldsymbol{k}_e, v\boldsymbol{k}_h)\psi_{c\boldsymbol{k}_e}(\boldsymbol{r}_e)\psi_{v\boldsymbol{k}_h}(\boldsymbol{r}_h) \tag{3.51}$$

where n and $\boldsymbol{K}$ are indices of the exciton state. The energy difference E between this excited state and the ground state can be found as the solution of a set of simultaneous equations,

$$\begin{aligned}&\Big[E - E_c(\boldsymbol{k}_e) + E_v(\boldsymbol{k}_h)\Big] A_{n\boldsymbol{K}}(c\boldsymbol{k}_e, v\boldsymbol{k}_h)\\ &= \sum_{c'\boldsymbol{k}'_e, v'\boldsymbol{k}'_h} \langle c\boldsymbol{k}_e, v\boldsymbol{k}_h|V(\boldsymbol{r}_e - \boldsymbol{r}_h)|c'\boldsymbol{k}'_e, v'\boldsymbol{k}'_h\rangle A_{n\boldsymbol{K}}(c'\boldsymbol{k}'_e, v\boldsymbol{k}'_h)\end{aligned} \tag{3.52}$$

where $E_c(\boldsymbol{k}_e)$ and $E_v(\boldsymbol{k}_h)$ are given by Eq. 3.48. We have neglected the exchange term in Eq. 3.48.

The electron and hole states are Bloch functions and may be written in the form

$$\psi_{c\boldsymbol{k}}(\boldsymbol{r}) = \frac{1}{\sqrt{N}} \mathrm{e}^{i\boldsymbol{k}\cdot\boldsymbol{r}} u_{c\boldsymbol{k}}(\boldsymbol{r}) \tag{3.53}$$

where N is the number of unit cells and $u_{\boldsymbol{k}}(\boldsymbol{r})$ is the Bloch function with the period of the crystal lattice and is normalized in a unit cell of volume v_0. $\psi_{c\boldsymbol{k}}(\boldsymbol{r})$ is normalized in a large volume $\Omega = Nv_0$.

When $u_{\boldsymbol{k}}(\boldsymbol{r})$ varies slowly with $\boldsymbol{k}$ over the range of interest and this range is small compared to the dimensions of the Brillouin zone, we can write

$$u^*_{c\boldsymbol{k}_e}(\boldsymbol{r}_e)u_{c'\boldsymbol{k}'_e}(\boldsymbol{r}_e) = \frac{1}{v_0}\sum_{\nu} C_\nu(c\boldsymbol{k}_e, c'\boldsymbol{k}'_e)\mathrm{e}^{i\boldsymbol{K}_\nu\cdot\boldsymbol{r}_e} \tag{3.54}$$

and

$$u^*_{v\boldsymbol{k}_h}(\boldsymbol{r}_h)u_{v'\boldsymbol{k}'_h}(\boldsymbol{r}_h) = \frac{1}{v_0}\sum_{\mu} C_\mu(v\boldsymbol{k}_h, v'\boldsymbol{k}'_h)\mathrm{e}^{i\boldsymbol{K}_\mu\cdot\boldsymbol{r}_h} \tag{3.55}$$

where $\boldsymbol{K}_\nu$ and $\boldsymbol{K}_\mu$ are reciprocal lattice vectors. With these we obtain

$$\begin{aligned}&\langle c\boldsymbol{k}_e, v\boldsymbol{k}_h|V(\boldsymbol{r}_e - \boldsymbol{r}_h)|c'\boldsymbol{k}'_e, v'\boldsymbol{k}'_h\rangle\\ &= \frac{1}{\Omega}\sum_{\mu\nu} C_\nu(c\boldsymbol{k}_e, c'\boldsymbol{k}'_e)C_\mu(v\boldsymbol{k}_h, v'\boldsymbol{k}'_h)V(\boldsymbol{K}_\nu - \boldsymbol{q}_e)\delta_{\boldsymbol{q}_e+\boldsymbol{q}_h, \boldsymbol{K}_\nu+\boldsymbol{K}_\mu}\end{aligned} \tag{3.56}$$

where

$$V(\boldsymbol{q}) = \frac{1}{\Omega}\int \mathrm{e}^{i\boldsymbol{q}\cdot\boldsymbol{r}}V(\boldsymbol{r})d\boldsymbol{r} \tag{3.57}$$

is the Fourier transform of the electron–hole interaction potential. $\boldsymbol{q}_e = \boldsymbol{k}_e - \boldsymbol{k}'_e$ and $\boldsymbol{q}_h = \boldsymbol{k}_h - \boldsymbol{k}'_h$.

Since the total exciton wave vector $\boldsymbol{k}_e + \boldsymbol{k}_h$ can be restricted within the first Brillouin zone, the only possible solution of the δ-function is

$$\boldsymbol{K}_\nu + \boldsymbol{K}_\mu = 0 \tag{3.58}$$

which indicates the conservation of the wave vector of the exciton state

$$\boldsymbol{K} = \boldsymbol{k}_e + \boldsymbol{k}_h = \boldsymbol{k}'_e + \boldsymbol{k}'_h \tag{3.59}$$

Equation 3.56 can be rewritten as

$$\begin{aligned} &\langle c\boldsymbol{k}_e, v\boldsymbol{k}_h | V(\boldsymbol{r}_e - \boldsymbol{r}_h) | c'\boldsymbol{k}'_e, v'\boldsymbol{k}'_h \rangle \\ &= \frac{1}{\Omega} \sum_{\nu} C_\nu(c\boldsymbol{k}_e, c'\boldsymbol{k}'_e) C_{-\nu}(v\boldsymbol{k}_h, v'\boldsymbol{k}'_h) V(\boldsymbol{K}_\nu + \boldsymbol{q}) \end{aligned} \tag{3.60}$$

where $\boldsymbol{q} = \boldsymbol{q}_e = -\boldsymbol{q}_h$. If $V(\boldsymbol{r})$ is approximately Coulomb and $\boldsymbol{q}$ is small, the largest contribution is when $\boldsymbol{K}_\nu = 0$.

We now consider that the periodic functions $u_{\boldsymbol{k}}$ do not vary strongly with $\boldsymbol{k}$ so that we can write

$$u_{c'\boldsymbol{k}'_e}(\boldsymbol{r}_e) = u_{c'\boldsymbol{k}_e}(\boldsymbol{r}_e) + (\boldsymbol{k}'_e - \boldsymbol{k}_e)\nabla_{\boldsymbol{k}_e} u_{c'\boldsymbol{k}_e}(\boldsymbol{r}_e) + \cdots \tag{3.61}$$

Substituting this into Eq. 3.55 and integrating over one unit cell,

$$\begin{aligned} C_0(c\boldsymbol{k}_e, c'\boldsymbol{k}'_e) &= \delta_{cc'} - \boldsymbol{q} \cdot \boldsymbol{X}_{cc'}(\boldsymbol{k}_e) \\ C_0(v\boldsymbol{k}_h, v'\boldsymbol{k}'_h) &= \delta_{vv'} + \boldsymbol{q} \cdot \boldsymbol{X}_{vv'}(\boldsymbol{k}_h) \end{aligned} \tag{3.62}$$

where

$$\boldsymbol{X}_{nn'}(\boldsymbol{k}) = \langle u_{n\boldsymbol{k}} | \nabla_{\boldsymbol{k}} | u_{n'\boldsymbol{k}} \rangle_{v0} = \frac{m_0}{\hbar} \Big[E_{n'}(\boldsymbol{k}) - E_n(\boldsymbol{k}) \Big]^{-1} \langle \psi_{n\boldsymbol{k}} | \boldsymbol{p} | \psi_{n'\boldsymbol{k}} \rangle \tag{3.63}$$

Retaining only the $\boldsymbol{K}_\nu = 0$ term and with Eq. 3.62 we obtain

$$\begin{aligned} &\Big\langle c\boldsymbol{k}_e, v\boldsymbol{k}_h \Big| V(\boldsymbol{r}_e - \boldsymbol{r}_h) \Big| c'\boldsymbol{k}'_e, v'\boldsymbol{k}'_h \Big\rangle \\ &= \frac{V(\boldsymbol{q})}{\Omega} \Big[\delta_{cc'}\delta_{vv'} - \boldsymbol{q} \cdot (\delta_{vv'}\boldsymbol{X}_{cc'} - \delta_{cc'}\boldsymbol{X}_{vv'}) + \cdots \Big] \end{aligned} \tag{3.64}$$

to the first order in $\boldsymbol{q}$.

Retaining only the first term in the above equation, Eq. 3.52 becomes

$$\begin{aligned} &\Big[E - E_c(\boldsymbol{k}_e) + E_v(\boldsymbol{k}_h) \Big] A_{n\boldsymbol{K}}(c\boldsymbol{k}_e, v\boldsymbol{k}_h) \\ &= \frac{1}{\Omega} \sum_{\boldsymbol{q}} V(\boldsymbol{q}) A_{n\boldsymbol{K}}[c(\boldsymbol{k}_e - \boldsymbol{q}), v(\boldsymbol{k}_h + \boldsymbol{q})] \end{aligned} \tag{3.65}$$

The effective mass equation is obtained by introducing the Fourier transform of $A_{n\boldsymbol{K}}(c\boldsymbol{k}_e, v\boldsymbol{k}_h)$:

$$\psi_{n\boldsymbol{K}}(\boldsymbol{r}_e, \boldsymbol{r}_h) = \frac{1}{\Omega} \sum_{\boldsymbol{k}_e, \boldsymbol{k}_h} \mathrm{e}^{i\boldsymbol{k}_e \cdot \boldsymbol{r}_e} \mathrm{e}^{i\boldsymbol{k}_h \cdot \boldsymbol{r}_h} A_{n\boldsymbol{K}}(c\boldsymbol{k}_e, v\boldsymbol{k}_h) \tag{3.66}$$

Taking the Fourier transform of Eq. 3.65, it is easy to show that the functions $\psi_{n\boldsymbol{K}}(\boldsymbol{r}_e, \boldsymbol{r}_h)$ satisfy the differential equation

$$\Big[E_c(-i\nabla_e) - E_v(-i\nabla_h) + V(\boldsymbol{r}_e - \boldsymbol{r}_h) \Big] \psi_{n\boldsymbol{K}}(\boldsymbol{r}_e, \boldsymbol{r}_h) = E\psi_{n\boldsymbol{K}}(\boldsymbol{r}_e, \boldsymbol{r}_h) \tag{3.67}$$

where $E_c(-i\nabla_e)$ is the expression obtained by replacing $\boldsymbol{k}_e$ in the power-series expansion of $E_c(\boldsymbol{k}_e)$ by $-i\nabla_e$. The result is valid provided $E_c(\boldsymbol{k}_e)$ and $E_v(\boldsymbol{k}_h)$ are analytic and can be expanded in powers of $\boldsymbol{k}$.

Here it is noticed that if a function $f(\boldsymbol{k})$ is expandable in powers of $\boldsymbol{k}$,

$$\frac{1}{\Omega}\sum_{\boldsymbol{k}} \mathrm{e}^{i\boldsymbol{k}\cdot\boldsymbol{r}} f(\boldsymbol{k})G(\boldsymbol{k}) = f(-i\nabla)g(\boldsymbol{r})$$
$$g(\boldsymbol{r}) = \frac{1}{\Omega}\sum_{\boldsymbol{k}} \mathrm{e}^{i\boldsymbol{k}\cdot\boldsymbol{r}} G(\boldsymbol{k}) \tag{3.68}$$

Considering only an exciton formed from a hole in a given VB state and an electron in a given CB state, and $u_v(\boldsymbol{r}_h)$ and $u_c(\boldsymbol{r}_e)$ are independent of $\boldsymbol{k}_h$ and $\boldsymbol{k}_e$, the total wave function of such an exciton state is

$$\Psi_{n\boldsymbol{K}}(\boldsymbol{r}_e, \boldsymbol{r}_h) = v_0 \psi_{n\boldsymbol{K}}(\boldsymbol{r}_e, \boldsymbol{r}_h) u_c(\boldsymbol{r}_e) u_v(\boldsymbol{r}_h) \tag{3.69}$$

The exciton wave function is therefore made up of a product of the electron and hole band functions times a modulation function $\psi_{n\boldsymbol{K}}(\boldsymbol{r}_e, \boldsymbol{r}_h)$, which is in general referred to as the exciton envelop function.

3.3 EXCITON BINDING ENERGY

The fabrication of high-quality semiconductor heterostructures and multiple quantum wells (MQW) allows the study of the quantum size effect in the physical properties of these systems, which are quite different from those of the corresponding bulk materials. One of them is the exciton effect. The subband formation strongly influences the exciton effects, which dominate the optical absorption and the photoluminescence spectra of both single AlGaAs/GaAs QW and MQW. Even at room temperature the sharp exciton lines can be detected in the absorption spectra of AlGaAs/GaAs MQWs,[42] which can hardly be observed in bulk GaAs samples.

Variational method is commonly used to study the exciton effect in nanostructures. Here we will use a three-dimensional trial wave function to obtain the exciton binding energy E_b of an AlGaAs/GaAs MQW. Within the effective-mass approximation as discussed in the previous section, we need to solve the Schrödinger equation along the z-axis, i.e., the sample growth direction. The corresponding Hamiltonian of an exciton in a nanostructure has the form

$$H = -\frac{\hbar^2\nabla_e^2}{2m_e^*} - \frac{\hbar^2\nabla_h^2}{2m_h^*} - \frac{\mathrm{e}^2}{4\pi\epsilon|\boldsymbol{r}_e - \boldsymbol{r}_h|} + V_e(\boldsymbol{r}_e) + V_h(\boldsymbol{r}_h) \tag{3.70}$$

where $V_e(\boldsymbol{r}_e)$ and $V_h(\boldsymbol{r}_h)$ are confinement potentials for electron and hole, respectively, in the quantum well. ϵ is the dielectric constant of the crystal.

We build a three-dimensional trial wave function for the exciton ground state as

$$\psi(\boldsymbol{r}_e, \boldsymbol{r}_h) = \psi_e(z_e)\psi_h(z_h)\psi_{\mathrm{ex}}(\boldsymbol{r}_e, \boldsymbol{r}_h) \tag{3.71}$$

where $\boldsymbol{r}_e = (\boldsymbol{\rho}_e, z_e)$ and $\boldsymbol{r}_h = (\boldsymbol{\rho}_h, z_h)$ are spatial vectors of the electron and hole, respectively. $\psi_e(z_e)$ and $\psi_h(z_h)$ are eigenfunctions of the electron and hole in their corresponding confinement Hamiltonians

$$\begin{aligned} H_e &= -\frac{\hbar^2}{2m_e^*}\frac{d^2}{dz_e^2} + V_e(z_e) \\ H_h &= -\frac{\hbar^2}{2m_h^*}\frac{d^2}{dz_h^2} + V_h(z_h) \end{aligned} \tag{3.72}$$

The exciton envelope function $\psi_{\text{ex}}(\boldsymbol{r}_e, \boldsymbol{r}_h)$ is approximated by a three-dimensional hydrogenic-type wave function

$$\psi_{\text{ex}}(\boldsymbol{r}_e, \boldsymbol{r}_h) = C \exp\left[-\sqrt{\frac{|\boldsymbol{\rho}_e - \boldsymbol{\rho}_h|^2}{a_{xy}^2} + \frac{(z_e - z_h)^2}{a_z^2}}\right] \tag{3.73}$$

where a_{xy} and a_z are effective exciton Bohr radii to be determined by minimizing the total energy $\langle\psi|H|\psi\rangle$. C is the normalization factor. The binding energy of an exciton in its ground state is given by

$$\begin{aligned} E_b &= \langle\psi_e\psi_h|H_e + H_h|\psi_e\psi_h\rangle - \langle\psi|H|\psi\rangle \\ &= \int d\boldsymbol{r}_e d\boldsymbol{r}_h |\psi_e\psi_h|^2 \psi_{\text{ex}}^*(\boldsymbol{r}) \left(-\frac{\hbar^2\nabla_{\boldsymbol{r}}^2}{2m_r^*} + \frac{e^2}{4\pi\epsilon|\boldsymbol{r}|}\right) \psi_{\text{ex}}(\boldsymbol{r}) \end{aligned} \tag{3.74}$$

where $1/m_r^* = 1/m_e^* + 1/m_h^*$ is the reduced effective mass, $\boldsymbol{r} = \boldsymbol{r}_e - \boldsymbol{r}_h$.

It is easy to see that Eq. 3.73 reduces to the envelop function of an exciton in a bulk material, which is translationally symmetric in all the three dimensions ($a_{xy} = a_z$). In this case, the Bohr radius of the lowest exciton state is given by

$$a_3 = \frac{4\pi\epsilon\hbar^2}{m_r^* e^2} = \frac{\epsilon/\epsilon_0}{m_r^*/m_0}\, a_0 \tag{3.75}$$

where $a_0 = 4\pi\epsilon_0\hbar^2/m_0 = 0.529\,\text{Å}$ is the universal Bohr radius constant and the binding energy is

$$E_{3b} = \frac{m_r^*/m_0}{(\epsilon/\epsilon_0)^2}\, R_y \tag{3.76}$$

$R_y = \hbar^2/2m_0a_0^2 = 13.6058\,\text{eV}$. Note that subscript "3" in a_3 and E_{3b} stands for three-dimensions. Such hydrogenic spectra have been observed in the optical absorption that occurs below the interband threshold.

Furthermore, we can write down a general form of the envelope function in a nanostructure by introducing three different Bohr radii corresponding to the three spatial directions

$$\psi_{\text{ex}}(\boldsymbol{r}) = \exp\left(-\sqrt{\frac{x^2}{a_x^2} + \frac{y^2}{a_y^2} + \frac{z^2}{a_z^2}}\right) \tag{3.77}$$

with which it is easy to obtain

$$\nabla^2\psi_{\rm ex} = \frac{\psi_{\rm ex}(\boldsymbol{r})}{\left(\frac{x^2}{a_x^2}+\frac{y^2}{a_y^2}+\frac{z^2}{a_z^2}\right)^{3/2}}\left[\left(\frac{x^2}{a_x^4}+\frac{y^2}{a_y^4}+\frac{z^2}{a_z^4}\right)\sqrt{\frac{x^2}{a_x^2}+\frac{y^2}{a_y^2}+\frac{z^2}{a_z^2}} - \frac{1}{a_x^2}\left(\frac{y^2}{a_y^2}+\frac{z^2}{a_z^2}\right)-\frac{1}{a_y^2}\left(\frac{x^2}{a_x^2}+\frac{z^2}{a_z^2}\right)-\frac{1}{a_z^2}\left(\frac{x^2}{a_x^2}+\frac{y^2}{a_y^2}\right)\right] \tag{3.78}$$

which reduces to

$$\nabla^2\psi_{\rm ex} = \frac{{\rm e}^{-r/a}}{ar}\left(\frac{r}{a}-2\right)\psi_{\rm ex} \tag{3.79}$$

when $a_x = a_y = a_z = a$ for exciton in bulk material.

When the well width L_z is much larger than the exciton Bohr radius a_b, the exciton stays in a single well and its physical properties are the same as those of a free exciton in a bulk GaAs material. If L_z is comparable to or less than a_b, the exciton cannot be confined in a single well unless the barrier thickness d_b is so large that $L_z + d_b \gg a_b$. Furthermore, if $L_z + d_b$ is comparable with the Bohr radius, the exciton extends over more than a single well. Therefore, to calculate E_b from the above equation, the spatial extension in the numerical calculation must be sufficiently large in order to ensure a convergent result.

In the following we discuss E_b of an exciton consisting of a VB heavy hole and a CB electron. The binding energy of an exciton, which is a bound state of a VB light hole and a CB electron, can be similarly calculated. The resulted binding energy E_b of two sets of AlGaAs/GaAs samples is shown in Fig. 3.5.[43]

The samples have two values of the barrier thickness $d_b = 100$ and 300 Å, while the well width L_z varies from 10 to 400 Å. We see clearly the enhancement in the exciton binding energy when the well width is decreased. For a given barrier

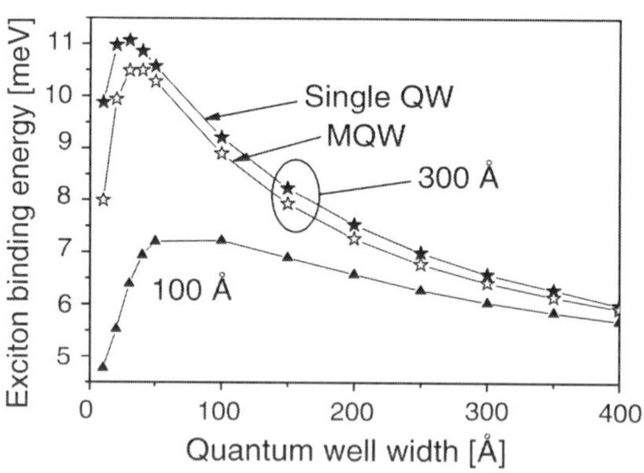

Fig. 3.5 Ground-state binding energy of a heavy-hole-electron exciton in AlGaAs/GaAs MQWs with barrier thickness of 100 and 300 Å, and in one single well sample.

thickness, E_b approaches the binding energy of an exciton in a bulk GaAs material as L_z approaches ∞; E_b approaches the exciton binding energy in a bulk AlGaAs material as L_z approaches 0.

We should mention that for most samples studied, the effective Bohr radius a_b in the envelope function has its optimal value around 100–180 Å. Hence, when the superlattice lattice constant $L_z + d_b$ is comparable with a_b, a substantial fraction of the exciton binding energy is due to the interaction between the electron in one well and the hole in another well (especially in an adjacent well). When $L_z + d_b \approx a_b$, a large contribution of about 1–4 meV is found from these inter-well interaction terms.

3.4 OPTICAL TRANSITION OF EXCITON

We now study the effect of exciton on the optical properties of nanostructure systems. Let $\boldsymbol{q}$ and $\boldsymbol{a}$ be, respectively, the wave vector and polarization vector of the photon. The probability that an electron in a nanostructure makes a transition from the ground state Ψ_0 to an excited exciton state $\Psi_{n\boldsymbol{K}}(\boldsymbol{r}_e, \boldsymbol{r}_h)$ is proportional to

$$|\langle \Psi_{n\boldsymbol{K}} | \mathrm{e}^{i\boldsymbol{q}\cdot\boldsymbol{r}} \boldsymbol{a} \cdot \boldsymbol{p} | \Psi_0 \rangle|^2 \tag{3.80}$$

where $\boldsymbol{p}$ is the momentum operator Eq. 2.96. By Eq. 3.51 we obtain

$$\langle \Psi_{n\boldsymbol{K}} | \mathrm{e}^{i\boldsymbol{q}\cdot\boldsymbol{r}} \boldsymbol{p} | \Psi_0 \rangle = \sum_{c\boldsymbol{k}_e, v\boldsymbol{k}_h} A_{n\boldsymbol{K}}(c\boldsymbol{k}_e, v\boldsymbol{k}_h) \langle \psi_{c\boldsymbol{k}_e} | \mathrm{e}^{i\boldsymbol{q}\cdot\boldsymbol{r}} \boldsymbol{p} | \theta \psi_{v\boldsymbol{k}_h} \rangle \tag{3.81}$$

where θ is the time-reversal operator and $\theta\psi_{v\boldsymbol{k}_h}$ is the unoccupied VB state out of which the electron was excited. Denote $\theta\psi_{v\boldsymbol{k}_h} = \psi_{v-\boldsymbol{k}_h}$ as the hole state. Furthermore, due to the momentum conservation,

$$\boldsymbol{K} = \boldsymbol{k}_e + \boldsymbol{k}_h = \boldsymbol{q}$$

while for optical radiation $\boldsymbol{q}$ is negligibly small so that $\boldsymbol{k}_e = -\boldsymbol{k}_h \equiv \boldsymbol{k}$,

$$\langle \Psi_{n0} | \boldsymbol{p} | \Psi_0 \rangle = \sum_{\boldsymbol{k}} A_{n0}(c\boldsymbol{k}, v - \boldsymbol{k}) \langle \psi_{c\boldsymbol{k}} | \boldsymbol{p} | \theta \psi_{v\boldsymbol{k}} \rangle \tag{3.82}$$

where we have neglected the contribution from all other bands except the VB (v) and the CB (c) in the vicinity of the Fermi level.

For commonly used semiconductor materials,

$$\langle \psi_{c\boldsymbol{k}} | \boldsymbol{p} | \psi_{v\boldsymbol{k}} \rangle \approx \boldsymbol{p}_{cv} \tag{3.83}$$

which depends very weakly on $\boldsymbol{k}$ over the range of $\boldsymbol{k}$ involved,

$$\langle \Psi_{n0} | \boldsymbol{p} | \Psi_0 \rangle = \boldsymbol{p}_{cv} \sum_{\boldsymbol{k}} A_{n0}(c\boldsymbol{k}, v - \boldsymbol{k}) \tag{3.84}$$

By Eq. 3.66 we obtain that the transition probability is therefore proportional to

$$|\langle \Psi_{n0} | \boldsymbol{p} | \Psi_0 \rangle|^2 \approx \boldsymbol{p}_{cv}^2 |\psi_{n0}(\boldsymbol{r}_e, \boldsymbol{r}_h)|^2_{\boldsymbol{r}_e = \boldsymbol{r}_h} \tag{3.85}$$

It is reminded that the quantum number $\boldsymbol{k}$ here denotes the wave vector for a three-dimensional Bloch state. For a heterostructure system grown along the z direction, translational symmetry is preserved in the xy plane, $\boldsymbol{k} = (\boldsymbol{k}_{xy}, i)$; here i is the index of subbands in the growth direction and $\boldsymbol{k}_{xy}$ is the wave vector in the xy plane. Similar discussions can be made for quantum wire and quantum dot cases.

To demonstrate briefly the exciton effect on the optical transition, we study a process in which an electric dipole transition creates an exciton from a filled VB in a bulk semiconductor material. For bulk material, we introduce the relative and center-of-mass coordinates

$$\boldsymbol{r} = \boldsymbol{r}_e - \boldsymbol{r}_h \ , \ \boldsymbol{R} = \frac{m_e^* \boldsymbol{r}_e + m_h^* \boldsymbol{r}_h}{m_e^* + m_h^*} \tag{3.86}$$

so that the Hamiltonian Eq. 3.70 describing the exciton becomes

$$H_{\text{ex}} = -\frac{\hbar^2 \nabla_{\boldsymbol{R}}^2}{2(m_e^* + m_h^*)} - \frac{\hbar^2 \nabla_{\boldsymbol{r}}^2}{2m_r^*} - \frac{\mathrm{e}^2}{4\pi\epsilon|\boldsymbol{r}|} \tag{3.87}$$

For bulk systems, the part of the wave function in $\boldsymbol{R}$ contains a factor of $\mathrm{e}^{i\boldsymbol{K}\cdot\boldsymbol{R}}$ and the part in the relative coordinates contains a hydrogenic $\psi_{n\boldsymbol{K}}(\boldsymbol{r}_e, \boldsymbol{r}_h) = \psi_{n\boldsymbol{K}}(\boldsymbol{r}_e - \boldsymbol{r}_h)$, so that the exciton wave function and its energy are

$$\begin{aligned} &\mathrm{e}^{i\boldsymbol{K}\cdot\boldsymbol{R}} \psi_{n\boldsymbol{K}}(\boldsymbol{r}_e, \boldsymbol{r}_h) \psi_e(\boldsymbol{r}_e) \psi_h(\boldsymbol{r}_h) \\ E_{n\boldsymbol{K}} &= \frac{\hbar^2 K^2}{2(m_e^* + m_h^*)} - \frac{m_r^* \mathrm{e}^4}{2\epsilon^2 n^2} \end{aligned} \tag{3.88}$$

and the wave function of the ground exciton state corresponding to the last two terms of H_{ex} in Eq. 3.87 is

$$\psi_{n0}(\boldsymbol{r}_e, \boldsymbol{r}_h) = \frac{1}{\sqrt{\pi a_{\text{Br}}^3}} \mathrm{e}^{-|\boldsymbol{r}|/a_{\text{Br}}} \tag{3.89}$$

where $a_{\text{Br}} = 4\pi\epsilon_\infty \hbar^2 / m_r^* \mathrm{e}^2$ is the Bohr radius of the exciton in the semiconductor bulk materials (Table 3.2). Note that the energy is often referred to as the exciton binding energy, E_{10}, or more commonly, E_b.

By Eqs. 3.10 and 3.85 we can observe that for spherical bands, $|\psi_{n0}(\boldsymbol{r}_e, \boldsymbol{r}_h)|^2_{\boldsymbol{r}_e = \boldsymbol{r}_h}$ is nonzero only for s states ($\ell = 0$) which are

$$|\psi_{n0}(\boldsymbol{r}_e, \boldsymbol{r}_h)|^2_{\boldsymbol{r}_e = \boldsymbol{r}_h} \propto \frac{1}{n^3} \tag{3.90}$$

where n is the principal quantum number in Eq. 3.88. The recombination process of the electron–hole pair is discussed in a similar way by considering the electric dipole transition from an occupied CB state to an empty VB state. It is thus concluded that only low-energy exciton states (small n) can be probed optically for bulk materials.

The exciton binding energy for the ground exciton state ($n = 1$) in commonly used semiconductors is less than 10 meV, so the room temperature luminescence is very weak (exciton thermally dissolved so that n is large). The thermal energy at room temperature is $k_B T = 25\,\text{meV}$ when $T = 300\,\text{K}$. Moreover, the exciton Bohr radii, i.e., the effective spatial extensions of the excitons, in bulk materials are also quite extended (Table 3.2).

In high-quality semiconductor nanostructures, quantum size effects prevail and subband formation strongly influences the exciton as the electron and hole are forced to stay in a very confined spatial area. The Coulomb interaction between the electron and hole, i.e., the exciton binding energy, then becomes much increased. Even at room temperature the sharp exciton lines can be detected in the absorption spectra of nanostructures such as AlGaAs/GaAs MQW.[42] Interestingly, one of the reasons for the fast development of ZnO and its related materials is their large exciton binding energy of about 60 meV, which binds the electron and hole strongly at room temperature.[44] A photonic switch operating by controlling exciton excitations in ZnO quantum wells (QWs) via an optical near field was reported, where ZnO QWs were in the form of disks with a disk diameter of 80 nm and the heights of 2 nm (in a structure composed of one single-QW and three double-QWs) and 3.5 nm in the one double-QW case.[45, 46] Another development based on the large exciton binding energy of ZnO is the room-temperature polariton laser,[47] which features very low threshold energies. The laser operates by Bose–Einstein condensation of exciton–polaritons in one or multiple QWs inside a microcavity at room temperatures and does not need population inversion.

Excitons are normally categorized into two types. The Frenkel exciton is an excited electron state where the electron and the hole are situated in the same molecule or atom. Because of their small radius, Frenkel excitons usually are considered to be local. The large-radius Wannier-Mott exciton in bulk semiconductors is weakly bound with large exciton Bohr radii. A hybrid state of Wannier-Mott and Frenkel excitons in different heterostructure configurations involving QDs was found to exist at the interfaces of QDs and the surrounding organic medium with complimentary properties of both kinds of excitons.[48]

In heterostructures both electrons in the CB and holes in the VB will experience extra potential energies such as band offsets. For low-energy Frenkel excitons in a QD, the electrons and holes are confined within the same QD volume (type-I exciton) so that we will have a common potential energy $V(|\boldsymbol{r}-\boldsymbol{a}|)$ that confines the exciton, where $\boldsymbol{a}$ denotes the center of the QD. For common semiconductor QDs, the radius of the QD, R, is in the order of the exciton Bohr radius, a_{Br}, so that one may neglect the internal motion of the electron–hole pair in the QD, i.e., $\boldsymbol{K} = 0$ in Eq. 3.88. For the ground state of the exciton we can approximate $V(|\boldsymbol{r}-\boldsymbol{a}|) = 0$ when $|\boldsymbol{r}-\boldsymbol{a}| < r$ and $V(|\boldsymbol{r}-\boldsymbol{a}|) = \infty$ elsewhere so that the wave function corresponding to the motion of the center-of-mass becomes

$$\frac{1}{\sqrt{2\pi r}\,|\boldsymbol{r}-\boldsymbol{a}|}\sin\left(\frac{\pi|\boldsymbol{r}-\boldsymbol{a}|}{R}\right) \tag{3.91}$$

for $|\boldsymbol{r}-\boldsymbol{a}| \leq R$, i.e., the first function in Eq. 3.10. The wave function is zero elsewhere. We thus obtain the total wave function of the exciton ground state inside the QD as

$$\frac{1}{\sqrt{2\pi r}\,|\boldsymbol{r}-\boldsymbol{a}|}\sin\left(\frac{\pi|\boldsymbol{r}-\boldsymbol{a}|}{R}\right)\frac{1}{\sqrt{\pi a_{\mathrm{Br}}^3}}\,\mathrm{e}^{-|\boldsymbol{r}_e-\boldsymbol{r}_h|/a_{\mathrm{Br}}}u_c(\boldsymbol{r}_e)u_v(\boldsymbol{r}_h) \tag{3.92}$$

Here the Bloch functions of the conduction- and valence-band edge states are included.

3.5 RADIATIVE AND NONRADIATIVE RECOMBINATIONS

In practical devices, efficient creation of excitons and long lifetimes are desirable to take advantage of the exciton effects. A good understanding of the exciton dynamics is thus vital and much work has been carried out in this area. Comparable recombination rates of approximately 300 ps for excitons and biexcitons in one single CdSe/ZnSe QD embedded in ZnSe were obtained and related to the spatial wave function distribution and the spin structure of the particles and their coupling to the photon field.[49] Decay time of the PL for cubic CuCl QD pairs was in the order of 5 ns, and the signal rise time, which corresponds to the energy transfer time, was from 25 to 180 ps.[50] For colloidal QDs, the radiative lifetime can be of the order of nanoseconds at room temperature.[51] The decay times of III–V QDs in solid state matrices are normally found to be much shorter. It was observed to be only 9.5 ps for a single InAs QD in Ref.[52] Similar results have been obtained for CdSe/ZnSe QDs in GaAs[53] and CdTe QDs in ZnTe.[54]

Dekel *et al.* quantitatively explained the long radiative time in terms of the reduced spatial coherence between the confined exciton dipole moment and the radiating electromagnetic field.[55] Cooney *et al.* reported via femtosecond experiments that the exciton relaxation processes were in the form of state-to-state transitions.[56] It is, however, notable that the theoretical analysis of Cooney *et al.* was based on the Fermi Golden rule and rate equations that were derived for steady state situations. For the femtosecond region with an energy scale of 1 eV under investigation, $\mathrm{d}t\,\mathrm{d}E$ is in the magnitude of $\hbar$ whereas the uncertainty principle says $\mathrm{d}t\mathrm{d}E \geq \hbar$. A nonperturbative theoretical analysis should thus be more appropriate.

High-order nonlinear transient spectroscopies show the nature of multi-particle dynamics on ultra-short time scales in colloidal CdSe QDs.[57] Multi-exciton dynamics has also been identified by photon correlation spectroscopy.[58] In InAs QDs in InGaAs QW, carrier capture and relaxation were found to be dominated by Auger carrier–carrier scattering at low temperatures, with thermal emission playing an increasing role with temperature.[59] Electron–phonon scattering in $\mathrm{In_{0.7}Ga_{0.3}As}$ QDs in GaAs was responsible for the fast dephasing at high temperature (above 150 K); only at low temperatures ($<$ 30 K) does population relaxation of multi-excitons determine the dephasing.[60] Electron–phonon interaction was attributed to be the principal energy dissipation channel of high-energy carriers in the colloidal

II–VI QDs via dynamic quantum Monte Carlo simulations.[61] Phonon scattering processes are thus concluded to be predominant in room-temperature situations.

3.5.1 *Spontaneous Emission*

Denote the eigensolution of Eq. 3.70 as exciton state j

$$H_0(\boldsymbol{r}_e, \boldsymbol{r}_h)\psi_j(\boldsymbol{r}_e, \boldsymbol{r}_h) = E_j\psi_j(\boldsymbol{r}_e, \boldsymbol{r}_h) \tag{3.93}$$

$\boldsymbol{A}$ as the vector potential of the electromagnetic field; the Hamiltonian describing the light–matter interaction is $H' = e\boldsymbol{A}\cdot\boldsymbol{p}/m_0$ (Eq. 2.96). Let $\boldsymbol{r} = (\boldsymbol{r}_e, \boldsymbol{r}_h)$; the wave function of the total system composed of electrons and photons is

$$\Psi(\boldsymbol{r}, t) = \sum_j C_j(t)\psi_j(\boldsymbol{r})\exp\left(-iE_jt/\hbar\right)|N_{\hbar\omega}\rangle \tag{3.94}$$

where $|N_{\hbar\omega}\rangle$ describes the photon field with $N_{\hbar\omega}$ as the number of photons at energy $\hbar\omega$. The following equation is obtained for the wave function coefficient

$$\begin{aligned} i\hbar\frac{\mathrm{d}C_j(t)}{\mathrm{d}t} = \frac{e}{m_0}\sqrt{\frac{\hbar}{2\omega\epsilon\Omega}}&\sum_i\langle\psi_j(\boldsymbol{r})|\boldsymbol{a}\cdot\boldsymbol{p}|\psi_i(\boldsymbol{r})\rangle C_i(t) \\ &\times\left\{\sqrt{N_{\hbar\omega}}\exp\left[\frac{i(E_j - E_i - \hbar\omega)t}{\hbar}\right]\right. \\ &\left.+\sqrt{N_{\hbar\omega}+1}\exp\left[\frac{i(E_j - E_i + \hbar\omega)t}{\hbar}\right]\right\} \end{aligned} \tag{3.95}$$

by inserting Eq. 3.94 into the time-dependent Schrödinger equation. The term headed by $\sqrt{N_{\hbar\omega}}$ corresponds to the photon absorption, while the one with $\sqrt{N_{\hbar\omega}+1}$ is the photon emission. $N_{\hbar\omega}/\Omega$ is the photon density.

Let ψ_0 be the vacuum state, i.e., the VB is completely filled while the CB is totally empty. When the spatial variation of the excitation field inside the QD is small, the matrix element of radiative recombination between ψ_j and ψ_0 becomes[15]

$$\langle\psi_j(\boldsymbol{r}_e, \boldsymbol{r}_h)|\boldsymbol{a}\cdot\boldsymbol{p}|\psi_0(\boldsymbol{r}_e, \boldsymbol{r}_h)\rangle = \boldsymbol{a}\cdot\boldsymbol{p}_{cv}\int\psi_j^*(\boldsymbol{r}, \boldsymbol{r})\,\mathrm{d}\boldsymbol{r} \tag{3.96}$$

when $\boldsymbol{p}_{cv} = \langle u_c(\boldsymbol{r})|\boldsymbol{p}|u_v(\boldsymbol{r})\rangle$, referred to as the dipole momentum between CB and VB is approximated as constant over the range of $\boldsymbol{k}$ involved. $u_c(\boldsymbol{r}_e)$ and $u_v(\boldsymbol{r}_h)$ are Bloch functions of the CB and VB edge states normalized per lattice cell. We will derive this relationship closely in Chapter 4.

By the generalized Golden rule (Section 2.1.2) and denoting $\langle\psi_j|\hat{T}(t)|\psi_j\rangle \approx \mathrm{e}^{-w_jt/2}$ as the temporal development of state j, it is easy to obtain

$$w_j = \frac{2\pi}{\hbar}\sum_{i\neq j}\left|\langle\psi_i|V|\psi_j\rangle\right|^2\left[\delta(E_i - E_j - \hbar\omega) + \delta(E_i - E_j + \hbar\omega)\right] \tag{3.97}$$

$\Gamma_j = \hbar w_j/2$ is the energy relaxation and $1/w_j$ the decaying time. In numerical calculations the δ functions in the above equation are replaced by $\Gamma_j/[\Gamma_j^2 + (E_i - E_j \pm \hbar\omega)^2]$. A simple estimation of the interaction energy

$$\frac{ep_{cv}}{m_0}\sqrt{\frac{\hbar}{2\omega\epsilon\Omega}}\int\psi_j^*(\boldsymbol{r}, \boldsymbol{r})\,\mathrm{d}\boldsymbol{r} = 6.5\times10^{-2}\,\mathrm{meV} \tag{3.98}$$

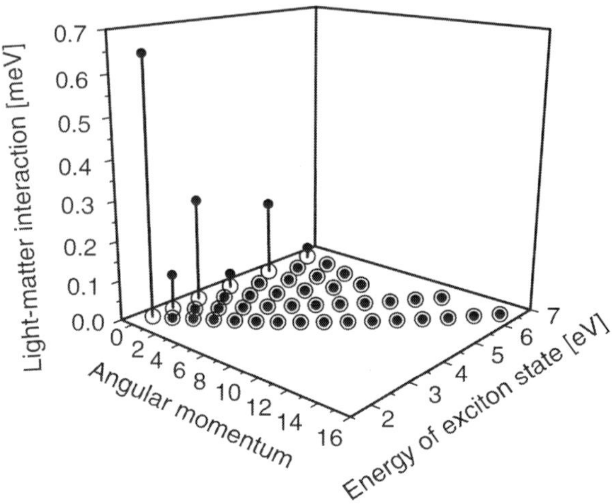

Fig. 3.6 Exciton states (open circles) confined in the QD and their corresponding light-matter interactions (solid circles). Y. Fu, H. Ågren, J. M. Kowalewski, H. Brismar, J. Wu, Y. Yue, N. Dai, and L. Thylén, Radiative and nonradiative recombination of photoexcited excitons in multi-shell-coated CdSe/CdS/ZnS quantum dots, EuroPhysics Lett. vol. 86, p. 37003(6), 2009. Copyright (2009) by the European Physical Society.

when $a = 10\,\text{nm}$, $E_p = p_{cv}^2/2m_0 = 20\,\text{eV}$, $\epsilon = 12$, and $\hbar\omega = 1.0\,\text{eV}$ for the ground exciton state. Here we take the space occupied by emitted photon Ω to be the size of the QD. It is thus obtained that the relaxation energy and the decaying time of a ground exciton state are about $6.5 \times 10^{-2}\,\text{meV}$ and $10\,\text{ps}$, respectively.

We note that the QDs are spherical so that the exciton wave function can be expressed by a radial function and spherical harmonics. The degeneracy of each exciton state is $2(2\ell+1)$, where ℓ is the quantum number of the angular momentum. Figure 3.6 shows the exciton states (open circles) confined in the QD as well as their corresponding light–matter interactions (solid circles). For spherical QDs, only exciton states having zero angular momenta ($\ell = 0$) have nonzero relaxation energies. Moreover, only confined exciton states (i.e., both the electron and hole states that compose the exciton state under discussion) have significant light–matter interactions. The optical coupling between confined state and extended state is weak because of the small wave function overlapping. Optical coupling between two extended states is also small; see Eq. 3.90.

3.5.2 *Electron–Phonon Interaction*

Nonradiative phonon–electron scattering processes for electrons in high-energy states are believed to be a major source of optical dephasing and damping in semiconductor QDs.[62] It was shown that the phonon spectra of the CdSe QD are modified by the boundary conditions. The modifications are, however, very much confined at the surface of the QD so that we expect a bulk-like phonon system to interact with the photoexcited excitons confined inside the QD.[63] The electron–phonon interaction normally does not allow one to incorporate both incoherent relaxation phenomena and dephasing due to fluctuations of the environment. Here we express the scattering rate for an exciton to relax from state i to state j by emitting a phonon as[61, 64, 65]

$$P_{ij}(t) = \frac{\pi(D_t K)^2}{\rho \Omega \omega_p} (N_p + 1)\, \delta[E_i - E_j + \hbar\omega_p] \tag{3.99}$$

where $D_t K$ is the optical deformation potential and $\rho = 5.8\,\mathrm{g/cm^3}$ is the mass density of the CdSe material. $\hbar\omega_p = 37\,\mathrm{meV}$ is the optical phonon energy of CdS at which the phonon density of states is maximal.[66, 67] At room temperature, all phonons are basically thermally excited so that $N_p \gg 1$. In the following numerical calculations, the above expression is re-formatted by replacing the δ function with a Lorentzian function of $\Gamma/[\Gamma^2 + (E_i - E_j + \hbar\omega_p)^2]$; Γ, the relaxation energy, is obtained similarly by Eq. 3.97, which is about 2.04 meV by adopting $D_t K = 10^8\,\mathrm{eV/cm}$ for most commonly used semiconductors,[65] which gives a scattering rate of $3.09 \times 10^{12}\,\mathrm{s^{-1}}$ at phonon emission resonance by including the phonon density of states. Such a rate is typical in common semiconductor bulk materials[68] as well as in QD-based light-emitting devices.[69, 70] Note that the relaxation energy in three-dimensionally extended bulk materials is typically about 50 meV due to the large density of electronic states.[64]

3.5.3 *Radiative Lifetime*

We now study the room-temperature measurements of radiative lifetime of II–VI CdSe core ZnS/CdS multi-shell QDs by using a Sarastro 2000 confocal microscope. Here the excitation was done using a 405 nm diode laser with 20 MHz pulses and emission from QDs was collected using a PMH-100-0 photomultiplier and SPC-140 single photon counting software (Becker & Hickl). The temporal profile of the laser pulse was collected during 30 s using reflected light from a mirror. Lifetime measurements were made using dry and pure QDs fixated and mounted on a glass slide. Ten repetitive measurements were done at the same point in the sample. Each fluorescence decay measurement was performed with a time duration of 10 s. The laser pulse profile was de-convoluted from the fluorescence decay of each measurement, and a triple exponential was fitted to the data using SPC image (Becker & Hickl). The ten repetitive measurements confirmed the stability of the curve fitting. Figure 3.7a shows the temporal development of the fluorescence of the QDs following one laser pulse excitation. The vertical axis shows photon counts per measurement. An average lifetime of 3.6 ns with a standard deviation of 6.2% was obtained. Note that a slight amount of ringing was present in the laser pulse and resulted in a small bump at about 12 ns in the fluorescence decay.

The numerical simulation starts with the photoexcitation of the QDs by the laser pulse where optical absorption, emission, and phonon scattering are included in a kinetic Monte Carlo scheme[61] by which we obtain the distribution of excited

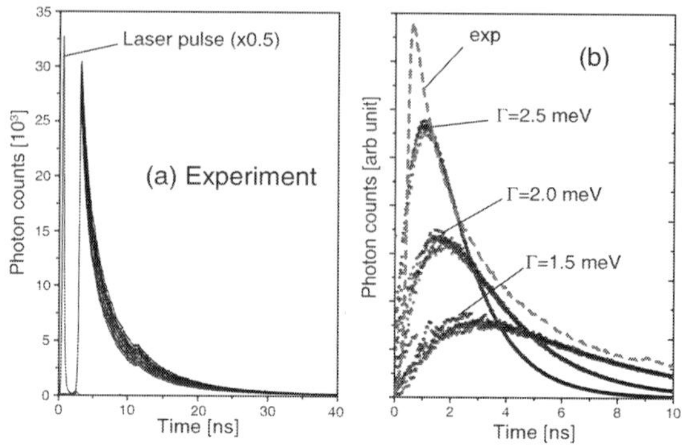

Fig. 3.7 (a) Ten repetitive measurement results of the temporal development of luminescence from II-VI CdSe core ZnS/CdS multi-shell QDs. The small bump at about 12 ns in the fluorescence decay was caused by a ringing in the laser pulse. (c) Monte Carlo simulations of spontaneous luminescence of an assembly of 5 million QDs for three different electron-phonon interactions Γ. Y. Fu, H. Ågren, J. M. Kowalewski, H. Brismar, J. Wu, Y. Yue, N. Dai, and L. Thylén, Radiative and nonradiative recombination of photoexcited excitons in multi-shell-coated CdSe/CdS/ZnS quantum dots, EuroPhysics Lett. vol. 86, p. 37003(6), 2009. Copyright (2009) by the European Physical Society. See also Color Insert.

excitons in the assembly of the QDs. In the Monte Carlo scheme the time interval is chosen in such a way that maximally only one event can occur. After calculating the probability for every possible process within the time interval by Eqs. 3.97 and 3.99, a random number is generated to determine the occurrence of events. The results of Monte Carlo simulations for the spontaneous luminescence from a QD assembly containing 5 million QDs are shown in Fig. 3.7b. Note that the general features of the theoretical and experimental spectra agree quite well, while detailed peak shapes differ quite significantly. One major factor is the value of D_tK used in calculating the phonon relaxation energy. In Fig. 3.7, the value of bulk material is adopted while modifications are expected in the nanoscale QD case, since the number of atoms at the surface of the QD is comparable with the total number of atoms in the QD. Furthermore, there are additional nonradiative relaxation processes such as surface states and defects, which are expected to have noticeable impact on the high-energy exciton states in the multi-shell coated QDs. Theoretical luminescence spectrum as a function of Γ (in the vicinity of pure-bulk material value of 2.04 meV) is shown in Fig. 3.7d to demonstrate the effect of the phone relaxation energy. By comparing with the experimental spectrum we can conclude that there are multiple phonon relaxation channels in the QDs. This is also reflected in the experimental data, which can be very well fitted by a triple exponential using SPC image software from Becker & Hickl.

In spite of the complications due to fluctuations in the physical parameters about the QDs, it can be firmly concluded that the general physical processes about the dynamic exciton excitation and recombination are the following: exciton states of $\ell = 0$ (see Fig. 3.4a) are initially photogenerated by the excitation laser pulse.

Basically all the exciton states confined in the QD are excited via the combination of the photoexcitation by the excitation laser (in Fig. 3.7 the wavelength of the laser is 405 nm, which can excite all the $\ell = 0$ exciton states[71]) and electron–phonon interactions. They decay to other exciton states of $\ell \neq 0$ largely via electron–phonon interactions (a few radiative relaxations have been observed in the simulation but the total counts are very small within the time duration of the laser pulse). After switching off the excitation laser pulse, exciton states of $\ell \neq 0$ need to relax to $\ell = 0$ states via electron–phonon interactions before the electron–hole pair recombines in the form of spontaneous radiative emission.

For $\Gamma = 2.0\,\text{meV}$, the principal peak at about 2 ns in Fig. 3.7 is due to the following physical processes. As stated before, the wavelength of the excitation laser is 405 nm, by which the exciton states at about 3.06 eV are excited resonantly. They need to go through about 30 electron–phonon interaction processes (phone emissions) before they can reach the ground exciton state. Since the energy separations between adjacent exciton states are not exactly the same as the phonon energy, the electron–phonon interactions are not resonant so that the average electron–phonon scattering rate is significantly lower than the resonant rate. This conforms with the observed decrease of the nonradiative electron–phonon scattering rate in QD-based laser systems, i.e., the so-called phonon bottleneck; see Refs.[69, 70] In the QDs under discussion this rate is more than two orders of magnitude smaller than its resonant value (the energy difference between the lowest exciton state of $\ell = 1$ and ground exciton state of $\ell = 0$ is 204 meV, while $\hbar\omega_p = 37\,\text{meV}$ by which the Lorentzian function of

$$\frac{\Gamma}{\Gamma^2 + (E_i - E_j + \hbar\omega_p)^2} \approx \frac{\Gamma}{(E_i - E_j + \hbar\omega_p)^2} \approx \frac{2}{150^2}\ \text{meV}^{-1} \tag{3.100}$$

while at resonance it is $1/\Gamma = 1/2\,\text{meV}^{-1}$, which finally results in the fluorescence peak at about 2 ns in Fig. 3.7.

It is thus concluded that the large number of exciton states confined in the colloidal CdSe QDs is responsible for the long luminescence lifetime of the order of nanoseconds. Many solid-state-substrate based QDs contain a limited number of confined exciton states (see e.g., Ref.[72]) so that the luminescence lifetime is largely determined by the lifetime of the exciton ground state in the QD, which is of the order of picoseconds. There are hundreds of exciton states confined in the core-shell colloidal CdSe QDs. The radiative lifetimes of these states are normally long as they gradually decay into the ground state via electron–phonon scattering. It is these long nonradiative decay processes that determine the long luminescence lifetime shown in Fig. 3.7b.

Further simulations on CdSe/ZnSe QD embedded in a ZnSe substrate predict a fluorescence peak at about 200 ps, quite close to the experimental data of Ref.[49] The difference in the fluorescence decay spectra of these QDs and the current QDs is again the number of exciton states confined in the QD. For CdSe/ZnSe QDs embedded in ZnSe, the potential barrier for the exciton states (i.e., the difference between the electron affinities of CdSe/ZnSe and ZnSe) is significantly lower compared with their electron affinities (the energy separation between the CB edge and the vacuum state). In self-assembled InAs QDs in GaAs matrices for laser device applications, the number of confined exciton states is even lower so that the decay time is determined by the radiative light–matter interaction of the ground excitonic state. Furthermore, phonon bottleneck is expected due to the limited number of confined exciton states as well as large energy separations among the exciton states.[69, 70]

More delicate nonlinear effects in QDs have also been investigated. For more or less the same volumes, CdSe nanorods clearly show fast recombinations of high-energy exciton states as compared with CdSe nanodots as a result of the decrease in symmetry.[73]

3.6 MULTIPHOTON PROCESSES

Another important aspect concerning the nonlinear optical properties of QDs is the multiphoton process, which using molecular chromophores has been widely applied in recent years in biological and medical imaging after the pioneer work of Goeppert-Mayer.[74–77] The multiphoton process has largely been treated theoretically by steady-state perturbation approaches, e.g., the scaling rules of multiphoton absorption by Wherrett[78] and the analysis of two-photon excitation spectroscopy of CdSe QDs by Schmidt *et al.*[79] The nonperturbative time-dependent Schrödinger equation was solved to analyze the ultrafast (as fast as fs) and ultraintense (in many experiments the optical power of the laser pulse peak can reach GW/cm^2) dynamics of multiphoton processes in QDs. With respect to the commonly used multiphoton molecular chromophores, QDs have the advantage of being highly fluorescent with excellent photochemical stability, extreme brightness, and a broad excitation spectral range.[71, 80]

We now discuss details about multiphoton optical transitions in a spherical QD. As indicated by Eq. 3.85, the matrix element of optical transition between a VB state and a CB one, denoted as interband transition, is

$$\begin{aligned}\langle\psi_j u_c|\boldsymbol{a}\cdot\boldsymbol{p}|\psi_i u_v\rangle &= \boldsymbol{a}\cdot\langle\psi_j u_c|\boldsymbol{p}|\psi_i u_v\rangle\\ &= \boldsymbol{a}\cdot\Big(\langle\psi_j|\psi_i\rangle\langle u_c|\boldsymbol{p}|u_v\rangle + \langle\psi_j|\boldsymbol{p}|\psi_i\rangle\langle u_c|u_v\rangle\Big)\\ &= \boldsymbol{a}\cdot\langle\psi_j|\psi_i\rangle\langle u_c|\boldsymbol{p}|u_v\rangle = \boldsymbol{a}\cdot\boldsymbol{p}_{cv}\langle\psi_j|\psi_i\rangle \end{aligned} \tag{3.101}$$

where we have used the wave function orthogonality properties $\langle u_c|u_v\rangle = 0$, and $\boldsymbol{p}_{cv} = \langle u_c|\boldsymbol{p}|u_v\rangle$. Thus,

$$\begin{aligned}\langle\psi_j u_c|\boldsymbol{a}\cdot\boldsymbol{p}|\psi_i u_v\rangle &= \boldsymbol{a}\cdot\boldsymbol{p}_{cv}\langle\psi_j|\psi_i\rangle\\ &= \boldsymbol{a}\cdot\boldsymbol{p}_{cv}\int_0^\infty r^2\mathrm{d}r\int_0^\pi \sin\theta\mathrm{d}\theta\\ &\quad\times\int_0^{2\pi}\mathrm{d}\phi\, R^*_{\ell_j}(r)Y^*_{\ell_j m_j}(\theta,\phi)R_{\ell_i}(r)Y_{\ell_i m_i}(\theta,\phi)\\ &= \boldsymbol{a}\cdot\boldsymbol{p}_{cv}\delta_{\ell_j\ell_i}\delta_{m_j m_i}\int_0^\infty r^2\mathrm{d}r\, R^*_{\ell_j}(r)R_{\ell_i}(r)\end{aligned} \tag{3.102}$$

Between two states within the same band, either the CB or VB, referred as intraband transition, the optical matrix element is

$$\langle\psi_j u_c|\boldsymbol{a}\cdot\boldsymbol{p}|\psi_i u_c\rangle = \boldsymbol{a}\cdot\Big(\langle\psi_j|\psi_i\rangle\langle u_c|\boldsymbol{p}|u_c\rangle + \langle\psi_j|\boldsymbol{p}|\psi_i\rangle\langle u_c|u_c\rangle\Big) \tag{3.103}$$

Since $\langle u_c|u_c\rangle = 1$ and $\langle u_c|\boldsymbol{p}|u_c\rangle = 0$, and $\boldsymbol{p} = m_0 \mathrm{d}\boldsymbol{r}/\mathrm{d}t$,

$$\begin{aligned}\langle\psi_j u_c|\boldsymbol{a}\cdot\boldsymbol{p}|\psi_i u_c\rangle &= m_0\boldsymbol{a}\cdot\left\langle\psi_j\left|\frac{\mathrm{d}\boldsymbol{r}}{\mathrm{d}t}\right|\psi_i\right\rangle\\ &= \frac{im_0}{\hbar}\,\boldsymbol{a}\cdot\langle\psi_j|[H_0,\boldsymbol{r}]|\psi_i\rangle\\ &= \frac{im_0(E_j-E_i)}{\hbar}\,\langle\psi_j|\boldsymbol{a}\cdot\boldsymbol{r}|\psi_i\rangle\end{aligned} \tag{3.104}$$

$$\begin{aligned}\langle\psi_j|\boldsymbol{a}\cdot\boldsymbol{r}|\psi_i\rangle \;&= \int_0^\infty r^2\mathrm{d}r\int_0^\pi \sin\theta\mathrm{d}\theta\\ &\times\int_0^{2\pi}\mathrm{d}\phi\, R^*_{\ell_j}(r)Y^*_{\ell_j m_j}(\theta,\phi)\boldsymbol{a}\cdot\boldsymbol{r}R_{\ell_i}(r)Y_{\ell_i m_i}(\theta,\phi)\end{aligned} \tag{3.105}$$

Writing $\boldsymbol{r} = r\boldsymbol{r}_0$, where $\boldsymbol{r}_0$ is the unit vector, $\boldsymbol{a} = a_x\boldsymbol{x}_0 + a_y\boldsymbol{y}_0 + a_z\boldsymbol{z}_0$, and $\boldsymbol{x}_0$, $\boldsymbol{y}_0$ and $\boldsymbol{z}_0$ are also unit vectors along x-, y-, and z-directions, respectively,

$$\begin{aligned}\boldsymbol{a}\cdot\boldsymbol{r}_0 &= a_x\sin\theta\cos\phi + a_y\sin\theta\sin\phi + a_z\cos\phi\\ &= \sqrt{\frac{4\pi}{3}}\left(\frac{-a_x+ia_y}{\sqrt{2}}Y_{11} + a_zY_{10} + \frac{a_x+ia_y}{\sqrt{2}}Y_{1\bar{1}}\right)\end{aligned} \tag{3.106}$$

so that the angular integral in Eq. 3.105 involves

$$\int_0^\pi \sin\theta\mathrm{d}\theta\int_0^{2\pi}\mathrm{d}\phi\, Y^*_{\ell_j m_j}(\theta,\phi)Y_{1,m}(\theta,\phi)Y_{\ell_i m_i}(\theta,\phi) \tag{3.107}$$

The azimuthal integration yields

$$\int_0^{2\pi}\mathrm{e}^{-im_j\phi}\mathrm{e}^{im\phi}\mathrm{e}^{im_i\phi}\,\mathrm{d}\phi = 2\pi\delta_{m,m_j-m_i} \tag{3.108}$$

If we define the z-axis to lie along the photon momentum direction $\boldsymbol{s}$, i.e., $a_z = 0$, the above equation indicates $m_j - m_i = \pm 1$.

By using the addition theorem for spherical harmonics

$$\begin{aligned}Y_{\ell_1 m_1}Y_{\ell_2 m_2} &= \sum_{\ell=|\ell_1-\ell_2|}^{\ell_1+\ell_2}\sqrt{\frac{(2\ell_1+1)(2\ell_2+1)}{4\pi(2\ell+1)}}\\ &C(\ell 0|\ell_1 0;\ell_2 0)C(\ell m_1+m_2|\ell_1 m_1;\ell_2 m_2)Y_{\ell m_1+m_2}\end{aligned} \tag{3.109}$$

so that the angular integration in the optical matrix element vanishes unless $\ell_j = \ell_i + 1, \ell_i, |\ell_i - 1|$, which is the general form of the electric dipole radiation selection rule $\Delta\ell = 1, 0, -1$. In the above equation, $C(\ell m|\ell_1 m_1;\ell_2 m_2)$ are the Clebsch–Gordan (CG) coefficients, which are nonzero only when $m = m_1 + m_2$. The expression below for the CG-coefficient is the most symmetric one of the

Table 3.4 CG-coefficient $C(\ell m|\ell_1 m_1; 1 m_2)$ $(m = m_1 + m_2)$ and $C(\ell m + m_2|1m; \ell_2 m_2)$

ℓ	$m_2 = 1$	$m_2 = 0$	$m_2 = -1$
ℓ_1+1	$\sqrt{\frac{(\ell_1+m)(\ell_1+m+1)}{(2\ell_1+1)(2\ell_1+2)}}$	$\sqrt{\frac{(\ell_1-m+1)(\ell_1+m+1)}{(2\ell_1+1)(\ell_1+1)}}$	$\sqrt{\frac{(\ell_1-m)(\ell_1-m+1)}{(2\ell_1+1)(2\ell_1+2)}}$
ℓ_1	$-\sqrt{\frac{(\ell_1+m)(\ell_1-m+1)}{2\ell_1(\ell_1+1)}}$	$\sqrt{\frac{m^2}{\ell_1(\ell_1+1)}}$	$\sqrt{\frac{(\ell_1-m)(\ell_1+m+1)}{2\ell_1(\ell_1+1)}}$
ℓ_1-1	$\sqrt{\frac{(\ell_1-m)(\ell_1-m+1)}{2\ell_1(2\ell_1+1)}}$	$-\sqrt{\frac{(\ell_1-m)(\ell_1+m)}{\ell_1(2\ell_1+1)}}$	$\sqrt{\frac{(\ell_1+m)(\ell_1+m+1)}{2\ell_1(2\ell_1+1)}}$
ℓ_1+1	$\sqrt{\frac{(\ell_1+m_1+1)(\ell_1+m_1+2)}{(2\ell_1+1)(2\ell_1+2)}}$	$\sqrt{\frac{(\ell_1-m_1+1)(\ell_1+m_1+1)}{(2\ell_1+1)(\ell_1+1)}}$	$\sqrt{\frac{(\ell_1-m_1+1)(\ell_1-m_1+2)}{(2\ell_1+1)(2\ell_1+2)}}$
ℓ_1	$-\sqrt{\frac{(\ell_1+m_1+1)(\ell_1-m_1)}{2\ell_1(\ell_1+1)}}$	$\sqrt{\frac{m_1^2}{\ell_1(\ell_1+1)}}$	$\sqrt{\frac{(\ell_1-m_1+1)(\ell_1+m_1)}{2\ell_1(\ell_1+1)}}$
ℓ_1-1	$\sqrt{\frac{(\ell_1-m_1-1)(\ell_1-m_1)}{2\ell_1(2\ell_1+1)}}$	$-\sqrt{\frac{(\ell_1-m_1)(\ell_1+m_1)}{\ell_1(2\ell_1+1)}}$	$\sqrt{\frac{(\ell_1+m_1-1)(\ell_1+m_1)}{2\ell_1(2\ell_1+1)}}$
	$m = 1$	$m = 0$	$m = -1$
ℓ_2+1	$\sqrt{\frac{(\ell_2+m_2+1)(\ell_2+m_2+2)}{(2\ell_2+1)(2\ell_2+2)}}$	$\sqrt{\frac{(\ell_2-m_2+1)(\ell_2+m_2+1)}{(2\ell_2+1)(\ell_2+1)}}$	$\sqrt{\frac{(\ell_2-m_2+1)(\ell_2-m_2+2)}{(2\ell_2+1)(2\ell_2+2)}}$
ℓ_2	$\sqrt{\frac{(\ell_2+m_2+1)(\ell_2-m_2)}{2\ell_2(\ell_2+1)}}$	$-\frac{m_2}{\sqrt{\ell_2(\ell_2+1)}}$	$-\sqrt{\frac{(\ell_2-m_2+1)(\ell_2+m_2)}{2\ell_2(\ell_2+1)}}$
ℓ_2-1	$\sqrt{\frac{(\ell_2-m_2-1)(\ell_2-m_2)}{2\ell_2(2\ell_2+1)}}$	$-\sqrt{\frac{(\ell_2-m_2)(\ell_2+m_2)}{\ell_2(2\ell_2+1)}}$	$\sqrt{\frac{(\ell_2+m_2-1)(\ell_2+m_2)}{2\ell_2(2\ell_2+1)}}$

various existing forms[81, 82]

$$\begin{aligned} &C(\ell m|\ell_1 m_1; \ell_2 m_2) \\ &= \delta_{m,m_1+m_2}\sqrt{\frac{(\ell_1+\ell_2-\ell)!(\ell_1-\ell_2+\ell)!(-\ell_1+\ell_2+\ell)!}{(\ell_1+\ell_2+\ell+1)!}} \\ &\times \sum_k \frac{(-1)^k\sqrt{(2\ell+1)(\ell_1+m_1)!(\ell_1-m_1)!(\ell_2+m_2)!}}{k!(\ell_1+\ell_2-\ell-k)!(\ell_1-m_1-k)!(\ell_2+m_2-k)!} \\ &\times \frac{\sqrt{(\ell_2-m_2)!(\ell+m)!(\ell-m)!}}{(\ell-\ell_2+m_1+k)!(\ell-\ell_1-m_2+k)!} \end{aligned} \tag{3.110}$$

and the sum runs over all values of integer k, which do not lead to negative factorials. Table 3.4 lists $C(\ell m_1+m_2|\ell_1 m_1; 1m_2)$ and $C(\ell m+m_2|1m; \ell_2 m_2)$.

By Section 2.4, the total Hamiltonian of the electron in the electromagnetic field is

$$H(\boldsymbol{r},t) = H_0(\boldsymbol{r}) + \frac{e}{m_0}\boldsymbol{A}\cdot\boldsymbol{p} + \sum_{\boldsymbol{s}} \hbar\omega_{\boldsymbol{s}}\left(b_{\boldsymbol{s}}^+ b_{\boldsymbol{s}} + \frac{1}{2}\right) \tag{3.111}$$

where H_0 is the Hamiltonian of the electron in the unperturbed QD, and Ω is the spatial volume of the QD. $b_{\boldsymbol{s}}$ and $b_{\boldsymbol{s}}^+$ are annihilation and creation operators, respectively of the photon of state $\boldsymbol{s}$ that carries a momentum $\hbar\boldsymbol{s}$ and an energy $\hbar\omega_{\boldsymbol{s}}$. Denote

$$H_0(\boldsymbol{r})\psi_i(\boldsymbol{r}) = e_i\psi_i(\boldsymbol{r}) \tag{3.112}$$

to characterize the CB electrons and VB holes, where ψ and e_i are eigenfunction and eigenvalue of single-electron state i, respectively. Consider that the QD contains n electrons distributed in a number of single-electron states. The electron system at time t occupies one of the following system states

$$\Psi_{\boldsymbol{k}} = \sum_i C_{\boldsymbol{k}i}|\psi_i\rangle \tag{3.113}$$

$|C_{\boldsymbol{k}i}| = 1$ (0) if the single-electron state ψ_i is occupied (empty), and $\sum_i |C_{\boldsymbol{k}i}|^2 = n$. The energy of system state $\boldsymbol{k}$ is $E_{\boldsymbol{k}} = \sum_i |C_{\boldsymbol{k}i}|^2 e_i$, i.e., the sum of energies of the occupied single-electron states.

After the switch on of the electromagnetic field in Eq. 3.111, the wave function of the total system becomes

$$\Phi = \sum_{\boldsymbol{k}} B_{\boldsymbol{k}}(t)\Psi_{\boldsymbol{k}}|N_{\boldsymbol{k}}\rangle \tag{3.114}$$

where $|N_{\boldsymbol{k}}\rangle$ describes the photon field with $N_{\boldsymbol{k}}$ as the number of photons associated with electron-system state $\boldsymbol{k}$. By solving the time-dependent Schrödinger equation

$$i\hbar\frac{\mathrm{d}}{\mathrm{d}t}\Phi = H(\boldsymbol{r}, t)\Phi \tag{3.115}$$

Following the generalized Golden rule presented in Chapter 2 we now consider the optical transition induced by the external excitation radiation

$$\begin{aligned} i\hbar\frac{\mathrm{d}B_{\boldsymbol{q}}(t)}{\mathrm{d}t} &= \frac{e}{m_0}\sqrt{\frac{\hbar}{2\omega\epsilon\Omega}}\sum_{\boldsymbol{k}}\langle\Psi_{\boldsymbol{q}}|\boldsymbol{a}\cdot\boldsymbol{p}|\Psi_{\boldsymbol{k}}\rangle B_{\boldsymbol{k}}(t) \\ &\quad \left\{\sqrt{N_{\hbar\omega}}\exp\left[\frac{i(E_{\boldsymbol{q}} - E_{\boldsymbol{k}} - \hbar\omega)t}{\hbar}\right]\right. \\ &\quad \left. + \sqrt{N_{\hbar\omega}+1}\exp\left[\frac{i(E_{\boldsymbol{q}} - E_{\boldsymbol{k}} + \hbar\omega)t}{\hbar}\right]\right\} \end{aligned} \tag{3.116}$$

Here we have set $\mathrm{e}^{i\boldsymbol{s}\cdot\boldsymbol{r}} = 1$ in Eq. 2.89 because of the small wave number of the photon and the nanoscale of the QD. For many bio-imaging applications, the optical power of the excitation light is rather high so that $N_{\hbar\omega} + 1 \approx N_{\hbar\omega} = N$. Within a time interval from t to $t + \Delta t$, the modification of the system-state coefficient becomes

$$i\hbar\Delta B_{\boldsymbol{q}}^{\mathrm{st}}(t) = \frac{e}{m_0\omega}\sqrt{\frac{2S^{\mathrm{st}}}{\epsilon c}}\sum_{\boldsymbol{k}}\langle\Psi_{\boldsymbol{q}}|\boldsymbol{a}\cdot\boldsymbol{p}|\Psi_{\boldsymbol{k}}\rangle B_{\boldsymbol{k}}(t)G_{\boldsymbol{qk}}(t) \tag{3.117}$$

$S^{\mathrm{st}} = N\hbar\omega c/\Omega$, which connects the quantization of the electromagnetic field with the macroscopic quantity of the optical excitation power; superscript "st" stands for "stimulated" and c is the light velocity.

$$
\begin{aligned}
G^{\rm st}_{\boldsymbol{qk}}(t) = & \frac{\hbar}{i(E_{\boldsymbol{q}} - E_{\boldsymbol{k}} - \hbar\omega)} \exp\left[\frac{i(E_{\boldsymbol{q}} - E_{\boldsymbol{k}} - \hbar\omega)t}{\hbar}\right] \\
& \times \left\{\exp\left[\frac{i(E_{\boldsymbol{q}} - E_{\boldsymbol{k}} - \hbar\omega)\Delta t}{\hbar}\right] - 1\right\} \\
& + \frac{\hbar}{i(E_{\boldsymbol{q}} - E_{\boldsymbol{k}} + \hbar\omega)} \exp\left[\frac{i(E_{\boldsymbol{q}} - E_{\boldsymbol{k}} + \hbar\omega)t}{\hbar}\right] \\
& \times \left\{\exp\left[\frac{i(E_{\boldsymbol{q}} - E_{\boldsymbol{k}} + \hbar\omega)\Delta t}{\hbar}\right] - 1\right\}
\end{aligned}
\tag{3.118}
$$

Let us express the steady-state solution of the system-state coefficient in Eq. 3.117 up to the second order in the form of

$$
B_{\boldsymbol{q}}(t) = B^{(0)}_{\boldsymbol{q}} + B^{(1)}_{\boldsymbol{q}}(t) + B^{(2)}_{\boldsymbol{q}}(t) \tag{3.119}
$$

By writing the external field as $\sum_s h^s \mathrm{e}^{-i\omega_s t}$ and letting $t \to \infty$, we have

$$
\begin{aligned}
B^{(1)}_{\boldsymbol{q}}(\infty) &= \sum_{\boldsymbol{sk}} \frac{h^{\boldsymbol{s}}_{\boldsymbol{q}k} B^{(0)}_{\boldsymbol{k}}}{E_{\boldsymbol{q}} - E_{\boldsymbol{k}} - \hbar\omega_{\boldsymbol{s}} + i\Gamma_{\boldsymbol{q}k}} \\
B^{(2)}_{\boldsymbol{q}}(\infty) &= \sum_{\boldsymbol{sitk}} \frac{h^{\boldsymbol{s}}_{\boldsymbol{q}i} h^{\boldsymbol{t}}_{\boldsymbol{i}k} B^{(0)}_{\boldsymbol{k}}}{E_{\boldsymbol{i}} - E_{\boldsymbol{k}} - \hbar\omega_{\boldsymbol{s}} + i\Gamma_{\boldsymbol{i}k}} \\
\times & \left(\frac{1}{E_{\boldsymbol{q}} - E_{\boldsymbol{i}} - \hbar\omega_{\boldsymbol{s}} + i\Gamma_{\boldsymbol{q}i}} - \frac{1}{E_{\boldsymbol{q}} - E_{\boldsymbol{k}} - \hbar\omega_{\boldsymbol{s}} - \hbar\omega_{\boldsymbol{t}} + i\Gamma_{\boldsymbol{q}k}} \right)
\end{aligned}
\tag{3.120}
$$

For single beam excitation, $\boldsymbol{s} = \boldsymbol{t}$ in $B^{(2)}_{\boldsymbol{q}}$ so that the two-photon process has been included implicitly in Eq. 3.117 which is the general nonperturbative solution of the time-dependent Schrödinger equation Eq. 3.115.

In the Monte Carlo scheme, the probability of an event of one optical transition from system state $\Psi_{\boldsymbol{k}}$ to $\Psi_{\boldsymbol{q}}$ induced by the external excitation light in the time interval is

$$
P^{\rm st}_{\boldsymbol{qk}}(t) = \left| \frac{e}{m_0\hbar\omega} \sqrt{\frac{2S^{\rm st}}{\epsilon c}} \langle \Psi_{\boldsymbol{q}} | \boldsymbol{a} \cdot \boldsymbol{p} | \Psi_{\boldsymbol{k}} \rangle B_{\boldsymbol{q}}(t) G^{\rm st}_{\boldsymbol{qk}}(t) \right|^2 \tag{3.121}
$$

For spontaneous radiative transition from system state $\Psi_{\boldsymbol{k}}$ to $\Psi_{\boldsymbol{q}}$, the photon energy is $\hbar\omega_{\boldsymbol{i}}$, which has a resonance value of $\hbar\omega_{\boldsymbol{i}} = E_{\boldsymbol{k}} - E_{\boldsymbol{q}}$. Since the spontaneously emitted photon does not accumulate in the QD, the number of such photons inside the QD is zero before the transition. It becomes one directly after the transition so that the corresponding optical power is

$$
S^{\rm sp}_{\boldsymbol{qki}} = \frac{\hbar\omega_{\boldsymbol{i}}\, c}{\Omega} \tag{3.122}
$$

where superscript "sp" denotes "spontaneous." Expressions about the probability, similar to Eqs. 3.118, 3.121, can be written down for the spontaneous photon emission event.

$\langle \Psi_{\boldsymbol{q}} | \boldsymbol{a} \cdot \boldsymbol{p} | \Psi_{\boldsymbol{k}} \rangle$ in Eq. 3.121 is calculated numerically by first expanding the system wavefunctions $\Psi_{\boldsymbol{q}}$ and $\Psi_{\boldsymbol{k}}$ in terms of single-electron state ψ_i using Eq. 3.113. $\langle \psi_j | \boldsymbol{a} \cdot$

$\boldsymbol{p}|\psi_i\rangle = p_{cv}\langle\psi_j|\psi_i\rangle$ for inter-band transitions, $\langle\psi_j|\boldsymbol{a}\cdot\boldsymbol{p}|\psi_i\rangle = \hbar\langle\psi_j|\nabla|\psi_i\rangle$ for intra-band transitions. p_{cv} is the momentum matrix element. ψ_i is numerically calculated by using the effective mass approximated described in Section 3.1.

We adopt Eq. 3.99 to describe the nonradiative electron–phonon interaction. As discussed before in sub-Section 3.5.2, for most commonly used semiconductors, $D_t K$ is in the order of $10^8\,\mathrm{eV/cm^{-1}}$ so that the electron-phonon scattering rate is in the order of $10^{12}\,\mathrm{s^{-1}}$. This is comparable with the probabilities of optical transitions, typical around $10^{13}\,\mathrm{s^{-1}}$ for the sum of all possible optical transition events (about 10^{12}–$10^2\,\mathrm{s^{-1}}$ for individual event) for an optical excitation power of $10^8\,\mathrm{W/cm^2}$.

The general framework of the Monte Carlo simulation scheme has been well established. It was used to simulate individual electron (or hole) transition trajectories in semiconductor electronic devices.[64] Wysin *et al.* reported an Monte Carlo calculation to study the carrier relaxation of the photoexcited GaAs based on relevant scattering rates of electrons.[83] We apply a similar Monte Carlo scheme here by incorporating the system-state transition probabilities obtained from the above equations. The time interval of Δt adopted in our Monte Carlo scheme is chosen in such a way that maximally only one transition event can occur. For the whole system containing n electrons and N electron states, there are $n \times (N-n)$ probable final states for a one-electron transition process. After calculating the probability for every possible transition within Δt, a random number is generated in the Monte Carlo scheme to determine the occurrence of transition events.

Before switching on the excitation light, all VB sublevels are occupied and the CB ones are unoccupied. After the switch on of an external continuous-wave (CW) optical radiation, electrons originally occupying VB sublevels will be photoexcited to CB sublevels, leaving holes in the VB band. There can be nonraditiate energy relaxations for the high-energy electrons in the CB and holes in the VB. Electrons in the CB and holes in the VB can also radiatively recombine. Schematic temporal development of the sublevel occupation in the QD is shown in Fig. 3.8. Note that all the sublevels are identified by an index number according to their relative energies. For example, the ground sublevel is tagged as number "0," and the first-excited sublevel as "1." The occupied sublevels in the CB and the unoccupied sublevels in the VB (i.e., holes) are marked by black dots. As schematically shown in Fig. 3.8, an electron originally staying at VB state 1 at time 0 is excited to CB state 3 at time interval 1 by absorbing one photon. Transitions in time interval 2 are nonradiative phonon scattering processes from CB3 to CB1 and from VB1 to VB0.

We now consider the CdS QD with a radius of 3.7 nm, where the QD band gap (energy separation between the ground states in the CB and VB) is 2.655 eV. Figure 3.9 shows the dynamic occupations of the CB and VB sublevels using the electron-hole picture after the CW optical excitation is switched on at $t = 0$. This Monte Carlo scheme includes all the probability calculations of possible transitions. And a random number is generated to pick up one transition when it is smaller than the sum of the probabilities of all possible transitions.

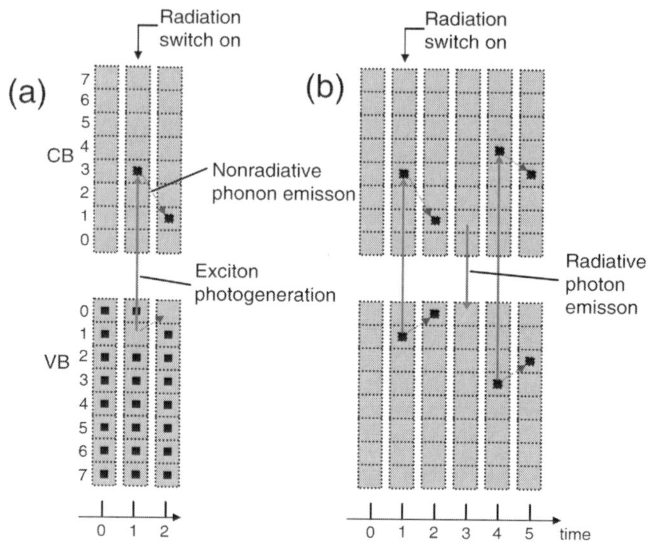

Fig. 3.8 (a) Schematic diagram of transition processes. (a) Electron picture. Black dots are electrons. (b) Electron-hole picture. Black dots in the conduction band (CB) are electrons. They are holes in the valence band (VB). T.-T. Han, Y. Fu, and H. Ågren, Dynamic photon emission from multiphoton-excited semiconductor quantum dot, J. Appl. Phys. vol. 103, p. 93703(6), 2008. Copyright (2008) by The American Institute of Physics. See also Color Insert.

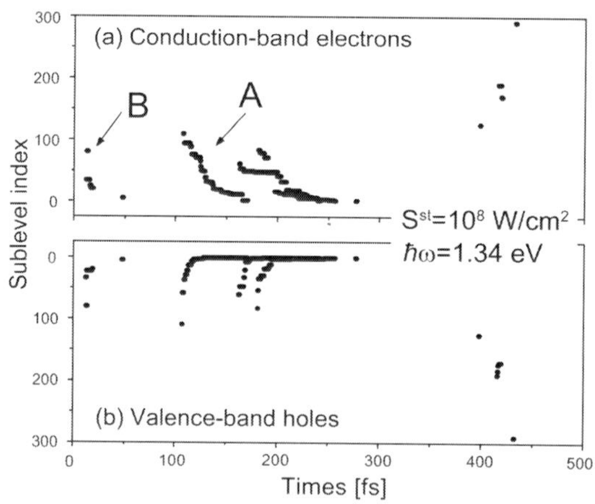

Fig. 3.9 Temporal development of occupations of sublevels in a CdS QD with a radius of 3.7 nm (its energy band gap is 2.67 eV). The excitation power $S^{st} = 10^8$ W/cm^2 and excitation photon energy $\hbar\omega = 1.34$ eV, corresponds to a two-photon excitation. T.-T. Han, Y. Fu, and H. Ågren, Dynamic photon emission from multiphoton-excited semiconductor quantum dot, J. Appl. Phys. vol. 103, p. 93703(6), 2008. Copyright (2008) by The American Institute of Physics.

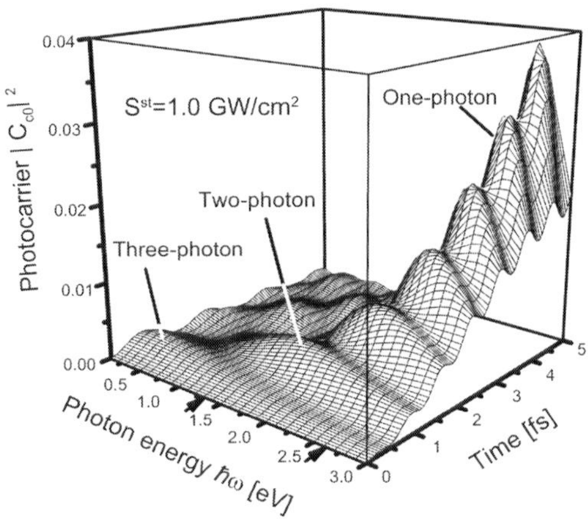

Fig. 3.10 Multiphoton excitation of QD exciton in an ultrashort period excited by a CW light source switched on at $t = 0$. S^{st} denotes the optical power of the excitation light source. Y. Fu, T.-T. Han, Y. Luo, and H. Ågren, Multi-photon excitation of quantum dots by ultra-short and ultra-intense laser pulse, Appl. Phys. Lett. vol. 88, p. 221114(3), 2006. Copyright (2006) by The American Institute of Physics.

Otherwise the system remains unmodified until the next time interval. For the excited electrons, stepwise tracks are observed in Fig. 3.9 (the points marked with arrow "A") due to the electron–phonon nonradiative relaxation processes from high-energy- to low-energy sublevels. Points marked by arrow "B" correspond to the radiative recombination processes. Event B in Fig. 3.9 occurs just a few fem to seconds after the turn-on of the external field. It is a single event in the time domain, which happens to be very fast.

Figure 3.10 shows the occupation of the ground CB sublevel, i.e., $|C_{c0}|^2$ in Eq. 3.113, after the CW optical excitation is switched on at $t = 0$. We observe that the excitation spectral range is very broad. Single-photon excitation (i.e., photon energy larger than 2.655 eV) is very fast and strong, whereas for the photon energy from about 1.3 to 2.6 eV, i.e., two-photon excitation between the two arrows along the photon energy axis in Fig. 3.10, the calculated excited exciton rate is also very significant, even when the photon energy is below 1.3 eV (more-than-two-photon excitation).

Table 3.5 lists the details about three types of core-multishell CdSe QDs, which are denoted as QD556, QD600, and QD622, respectively, that were chemically synthesized. Before optical characterization, the QD samples were diluted by toluene to concentrations of 2.0×10^{-6}, 1.3×10^{-6}, and 1.6×10^{-6} mol/L. PL spectroscopy measurements were performed using the fluorescence spectrometer (Cary Eclipse, VARIAN) at excitation wavelengths of 350, 400, and 450 nm (one-photon excitation), respectively, which shows that the one-photon PL spectra of these QDs are independent of the excitation wavelength. The results of the three QD assemblies

Table 3.5 Structure and optical properties of core-multishell CdSe QDs.

	QD556	QD600	QD622
CdSe diameter (nm)	3.6	4.5	5.4
CdS (ML)	1	2	3
$Cd_{0.5}Zn_{0.5}S$ (ML)	1	1	1
ZnS (ML)	1	1	1
Wavelength (nm)	556	600	622
FWHM (nm/meV)	32/121	29/104	24/77
Transition energy (nm)	557.7	597.5	623.2

are listed in Table 3.5 and presented in Fig. 3.11. Except for the difference in the peak wavelength, the three PL peaks are rather similar both qualitatively and quantitatively with a full width at half maximum (FWHM) of about 30 nm. We observe the decrease of the FWHM when the QD is coated, which is believed to be the result of reduced surface states in the CdSe core that deteriorate the optical properties of low-energy states confined in the CdSe core.

Two-photon-induced fluorescence experiments were performed using a femtosecond laser with a pulse width of 130 fs, peak wavelength of 800 nm, FWHM of 7 nm,

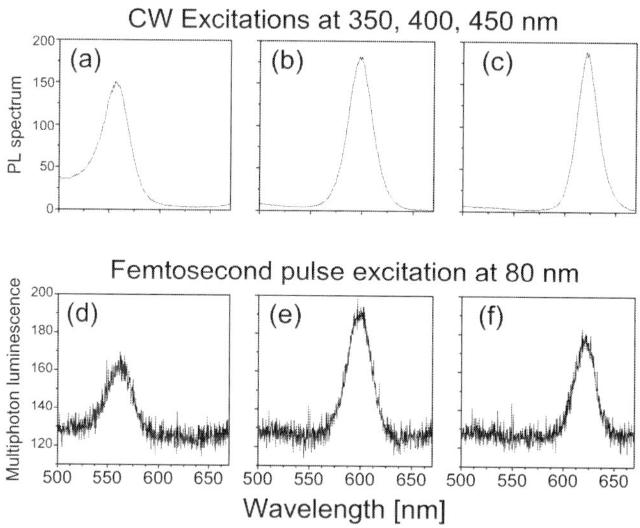

Fig. 3.11 PL spectra of the three QD assemblies. The excitation wavelength is 400 nm. (a) QD556. The PL peak wavelength is 556 nm and the FWHM is 32 nm. (b) QD600. PL peak at 600 nm and FWHM=29 nm. (c) QD622. PL peak at 622 nm and FWHM=24 nm. (d+e+f) Fluorescence spectra of the three QD assemblies multiphoton-excited by femtosecond laser pulses. Y. Fu, T.-T. Han, H. Ågren, L. Lin, P. Chen, Y. Liu, G.-Q. Tang, J. Wu, Y. Yue, and N. Dai, Design of semiconductor CdSe-core ZnS/CdS-multishell quantum dots for multiphoton applications, Appl. Phys. Lett. vol. 90, p. 173102(3), 2007. Copyright (2007) by The American Institute of Physics.

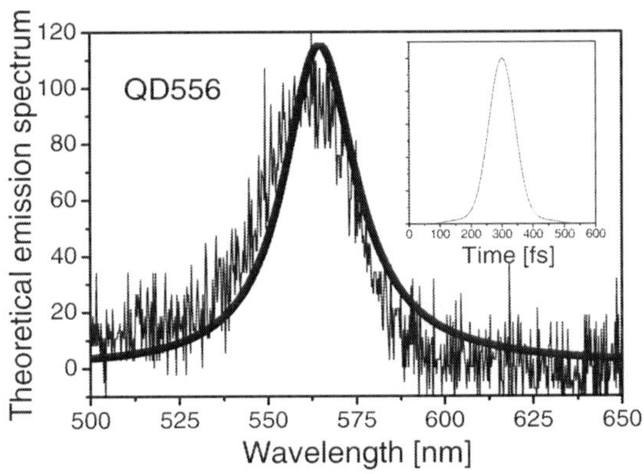

Fig. 3.12 Optical spectrum of spontaneously emitted photons from the QD556 assembly. The optical peak power inside the QD used in the calculation is 10^9 W/cm^2, and the excitation photon energy is 1.553 eV. The inset shows the laser pulse shape. Y. Fu, T.-T. Han, H. Ågren, L. Lin, P. Chen, Y. Liu, G.-Q. Tang, J. Wu, Y. Yue, and N. Dai, Design of semiconductor CdSe-core ZnS/CdS-multishell quantum dots for multiphoton applications, Appl. Phys. Lett. vol. 90, p. 173102(3), 2007. Copyright (2007) by The American Institute of Physics.

and average power output of about 630 mW at 800 nm at a pulse frequency of 76 MHz. The laser pulse beam was focused into the center of a 1.0 cm path length square sample cell using a 10 cm focal length lens, and the laser beam had a diameter of approximately 30 μm on the sample. The two-photon excitation fluorescence of the samples was led to a monochrometer; the results are presented in Figs. 3.11 d–f. By removing the base level count of about 125, the luminescence spectra of Figs. 3.11d–f are strikingly identical to the ones in Figs. 3.11a–c, both the shapes including the FWHMs and the relative intensities.

The theoretical result of the QD556 assembly is presented in Fig. 3.12 together with the experimental spectrum for direct comparison, where the inset shows the femtosecond laser pulse used in the numerical calculation. An almost perfect agreement is observed here. Similar conclusions are obtained for the QD600 and QD622 assemblies. The principal reason of such a robustness is the nonzero optical transition rate between the CB and VB states as well as the large numbers of confined electron and hole states in the QDs.

3.7 AUGER RECOMBINATION AND IMPACT IONIZATION

The Auger recombination is a process in which an electron and a hole recombine and the excess energy is transferred to either another electron or a hole, see Fig. 3.14. We first examine the case in semiconductor bulk materials, where we denote that state $\boldsymbol{k}_1$ is full (electron), $\boldsymbol{k}_2$ is empty (hole), and $\boldsymbol{k}_3$ is full (electron) (or empty when the third partner is a hole). The probability of such a configuration is

$$W(\boldsymbol{k}_1, \boldsymbol{k}_2, \boldsymbol{k}_3) = f(\boldsymbol{k}_1)\big[1 - f(\boldsymbol{k}_2)\big]f(\boldsymbol{k}_3) \tag{3.123}$$

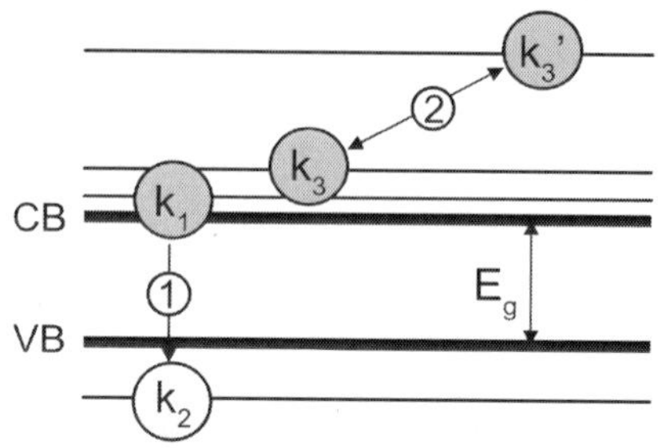

Fig. 3.13 Schematic depiction of Auger recombination process of an electron $\boldsymbol{k}_1$ and a hole $\boldsymbol{k}_2$ (process 1) and the excess energy is transferred (process 2) to another electron $\boldsymbol{k}_3$, which then transits to state $\boldsymbol{k}_3'$. CB/VB=conduction/valence band.

where

$$\begin{aligned} f(\boldsymbol{k}_1) &= \frac{n}{N_c}\exp\left(\frac{-E_{c\boldsymbol{k}_1}}{k_BT}\right) \\ 1-f(\boldsymbol{k}_2) &= \frac{p}{N_v}\exp\left(\frac{-E_{v\boldsymbol{k}_2}}{k_BT}\right) \\ f(\boldsymbol{k}_3) &= \frac{n}{N_c}\exp\left(\frac{-E_{c\boldsymbol{k}_3}}{k_BT}\right) \end{aligned} \tag{3.124}$$

for nondegenerate semiconductors. Here n and p are electron and hole concentrations and N_c and N_v are the CB and VB effective densities of states, respectively.

We further assume that the final state $\boldsymbol{k}_3'$ for the third electron is always available (i.e., empty), since it is a high-energy electron state in the CB. The total probability is thus

$$W(\boldsymbol{k}_1,\boldsymbol{k}_2,\boldsymbol{k}_3) \approx \frac{n}{N_c}\exp\left(-\frac{E_g+E_{c\boldsymbol{k}_1}+E_{v\boldsymbol{k}_2}+E_{c\boldsymbol{k}_3}}{k_BT}\right) \tag{3.125}$$

We now try to minimize the energy

$$\frac{\hbar^2k_1^2}{2m_c^*}+\frac{\hbar^2k_2^2}{2m_v^*}+\frac{\hbar^2k_3^2}{2m_c^*} \tag{3.126}$$

(in the parabolic band approximation) for optimal probability factor, which is consistent with energy and momentum conservations. It is found that the threshold is reached when the high-energy state $\boldsymbol{k}_3'$ lines opposite to the other three states. And the conditions of

$$\boldsymbol{k}_1 = \boldsymbol{k}_3 = \mu\boldsymbol{k}_2 \tag{3.127}$$

give us a maximum probability factor

$$W(\boldsymbol{k}_1,\boldsymbol{k}_2,\boldsymbol{k}_3) = \frac{n}{N_c}\exp\left[-\frac{(1+2\mu)E_g}{(1+\mu)k_BT}\right] \tag{3.128}$$

where $\mu = m_c^*/m_v^*$.

The final Auger rate is obtained by evaluating the multiple integral of the matrix element multiplied by the probability factor over all the available states. The interaction is Coulombic among the charged particles.

In general, the Auger rate increases exponentially as the bandgap is decreased. They also increase exponentially as the temperature increases. These are the direct results of the energy and momentum conservations and the carrier statistics. Auger processes are more or less unimportant in semiconductors with bandgaps larger than approximately 1.5 eV (e.g., GaAs, AlGaAs, InP). However, they become quite important in narrow bandgap materials such as InGaAs and HgCdTe ($E_g < 0.5\,\text{eV}$), and are thus a serious hindrance for the development of long-wavelength lasers. In general, Auger processes can be mediated by defects as well. Moreover, deep levels in the bandgap can be involved in the Auger processes.

We now consider the case of semiconductor nanoparticles. Nozik proposed that impact ionization (i.e., the reverse process of the Auger effect) might be enhanced in semiconductor QDs to increase the efficiency of solar cells up to about 66%.[84] This was later verified experimentally by Schaller and Klimov.[85] Impact ionization was reported to produce multiple exciton generation (MEG) per photon in one QD that results in a very high quantum yield (up to 300% when the photon energy reaches four times the QD bandgap) in QD solar cells.[86] Multiple carrier extraction (> 100%) was observed at photon energies greater than 2.8 times the PbSe QD bandgap with about 210% measured at 4.4 times the bandgap.[87] In a recent communication in *Nano Letters*, Trinh *et al.*[88] showed compelling support for carrier multiplication in PbSe QDs, while at the same time the obtained MEG efficiency was very close to the one in bulk material.

Figure 3.14 shows a highly photoexcited electron and hole pair that evolves into a multiple-exciton state through the impact ionization. The carrier–carrier interaction is expressed by the Coulomb potential in the form of

$$V = \frac{e^2}{4\pi\epsilon|\boldsymbol{r}_1 - \boldsymbol{r}_2|} \tag{3.129}$$

to account for the two-body interaction of two electrons from an initial state $\phi_j(\boldsymbol{r}_1)\psi_i(\boldsymbol{r}_2)\exp[-i(E_j + E_i)t/\hbar]$ to a final state

$$\frac{1}{\sqrt{2}}\begin{vmatrix} \psi_n(\boldsymbol{r}_1)\psi_m(\boldsymbol{r}_1) \\ \psi_n(\boldsymbol{r}_2)\psi_m(\boldsymbol{r}_2) \end{vmatrix} \exp[-i(E_n + E_m)t/\hbar] \tag{3.130}$$

Note that the two electrons in the final state are not distinguishable so that we use the Slater determinant.

Let us discuss the scattering matrix element of the Auger-type process

$$\int\int \Psi^*_{c\boldsymbol{k}_1}(\boldsymbol{r}_1)\Phi^*_{v\boldsymbol{k}_2}(\boldsymbol{r}_2)\,\frac{1}{|\boldsymbol{r}_1 - \boldsymbol{r}_2|}\,\Psi_{c\boldsymbol{k}'_1}(\boldsymbol{r}_1)\Psi_{c\boldsymbol{k}'_2}(\boldsymbol{r}_2)\,\mathrm{d}\boldsymbol{r}_1\mathrm{d}\boldsymbol{r}_2 \tag{3.131}$$

where an incident high-energy electron in the CB with a wave vector $\boldsymbol{k}_1$ collides with

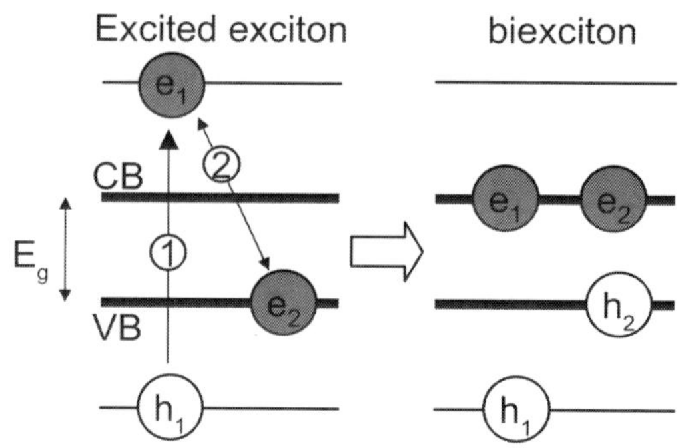

Fig. 3.14 Schematic depiction of impact ionization of a high-energy electron-hole pair. (1) Electron-hole pair (e_1 and h_1) is photogenerated. (2) e_2 gets excited from a valence-band state to a conduction-band state via Coulomb interaction with e_1, leaving hole h_2 behind. CB/VB=conduction/valence band. The reverse process is referred to be Auger recombination. Y. Fu, Y.-H. Zhou, H. Su, F. Y. C. Boey, and H. Ågren, Impact ionization and Auger recombination rates in semiconductor quantum dots, J. Physical Chemistry C, vol. 114, p. 3743–7, 2010. Copyright (2010) by American Chemical Society. See also Color Insert.

a second electron that occupies a VB state $\boldsymbol{k}_2$, resulting in two CB electrons $\boldsymbol{k}'_1$ and $\boldsymbol{k}'_2$. See, for example, the textbook of Landau and Lifshitz.[89] Here $\Psi_{c\boldsymbol{k}}(\boldsymbol{r})$ $[\Phi_{v\boldsymbol{k}}(\boldsymbol{r})]$ denotes the total wave function of CB electron (VB hole) state $\boldsymbol{k}$. For semiconductor systems and within the effective mass approximation, $\Psi_{c\boldsymbol{k}}(\boldsymbol{r}) = \psi_{c\boldsymbol{k}}(\boldsymbol{r})u_{c\boldsymbol{k}}(\boldsymbol{r})$ and $\Phi_{v\boldsymbol{k}}(\boldsymbol{r}) = \phi_{v\boldsymbol{k}}(\boldsymbol{r})u_{v\boldsymbol{k}}(\boldsymbol{r})$, where ψ and ϕ are envelop functions of conduction-band electron and valence-band hole, respectively, and u's are periodic Block functions, the above expression becomes

$$\begin{aligned}\int\!\!\int & \psi^*_{c\boldsymbol{k}_1}(\boldsymbol{r}_1)u^*_{c\boldsymbol{k}_1}(\boldsymbol{r}_1)\phi^*_{v\boldsymbol{k}_2}(\boldsymbol{r}_2)u^*_{v\boldsymbol{k}_2}(\boldsymbol{r}_2)\,\frac{1}{|\boldsymbol{r}_1-\boldsymbol{r}_2|}\,\psi_{c\boldsymbol{k}'_1}(\boldsymbol{r}_1)\\ &\times u_{c\boldsymbol{k}'_1}(\boldsymbol{r}_1)\psi_{c\boldsymbol{k}'_2}(\boldsymbol{r}_2)u_{c\boldsymbol{k}'_2}(\boldsymbol{r}_2)\,\mathrm{d}\boldsymbol{r}_1\mathrm{d}\boldsymbol{r}_2 \end{aligned} \tag{3.132}$$

See, for example, Eq. (6.128) in Ref.[65] Following overlap integrals are thus involved

$$\begin{aligned} I_1 &= \int_{\text{cell}} u^*_{c\boldsymbol{k}_1}(\boldsymbol{r}_1)u_{c\boldsymbol{k}'_1}(\boldsymbol{r}_1)\,\mathrm{d}\boldsymbol{r}_1\\ I_2 &= \int_{\text{cell}} u^*_{v\boldsymbol{k}_2}(\boldsymbol{r}_2)u_{c\boldsymbol{k}'_2}(\boldsymbol{r}_2)\,\mathrm{d}\boldsymbol{r}_2 \end{aligned} \tag{3.133}$$

in the evaluation of the Auger-type scattering processes. The first overlap integral can be approximated to unity. Because of the orthogonality of u functions for different bands but the same $\boldsymbol{k}$, the second overlap integral I_2 is, in crudest approximation, zero. This however is not correct since the periodic Bloch functions are functions of wave vectors.[90] For small $\boldsymbol{k}$ we can write

$$
\begin{aligned}
u_{c\boldsymbol{k}}(\boldsymbol{r}) &= u_c(\boldsymbol{r}) + \boldsymbol{k}\nabla_{\boldsymbol{k}} u_c(\boldsymbol{r}) + \cdots \\
u_{v\boldsymbol{k}}(\boldsymbol{r}) &= u_v(\boldsymbol{r}) + \boldsymbol{k}\nabla_{\boldsymbol{k}} u_v(\boldsymbol{r}) + \cdots
\end{aligned}
\tag{3.134}
$$

where $u_c(\boldsymbol{r})$ and $u_v(\boldsymbol{r})$ are periodic Bloch functions at the conduction- and valence-band edges, respectively. See also Eq. 3.61. $\int_{\text{cell}} u_v^*(\boldsymbol{r}_2) u_c(\boldsymbol{r}_2)\,\mathrm{d}\boldsymbol{r}_2 = 0$. Substituting these expressions into the overlap integrals we obtain the squared overlap integral in terms of the heavy-hole mass m_v [90, 91]

$$
|I_2|^2 = \frac{\hbar^2}{2E_g}\left(\frac{1}{m_0} + \frac{1}{m_v}\right)|\boldsymbol{k}_2 - \boldsymbol{k}_2'|^2 \tag{3.135}
$$

where E_g is the energy band gap of the bulk material. By using the inverse Bohr radius of shallow impurities as a measure about the $\boldsymbol{k}$ values, Landsberg and Adams obtained that $|I_2| = 0.223$ for shallow-impurity-assisted Auger-type processes in bulk CdS and $|I_2| = 0.265$ in GaAs.[91] Note that the inverse Bohr radius of shallow impurities in bulk semiconductors is small. For QDs under investigation, the effective Bohr radius of the electron and hole distribution in the QD is largely determined by the QD size, which is about 5 nm, i.e., very small compared with the Bohr radius of shallow impurities in bulk semiconductors (about 100 nm in CdS and GaAs[91]). This results in an large value of $|I_2|$ that may exceed unity depending on the quantum confinement of the carrier inside the QDs. Under this specific circumstance, we approximate $|I_2| = 1$. In other words, the electrons and holes in QDs, described by effective masses with the presence of the QD confinement potentials, interact with each other via the Coulomb force of Eq. (3.129), see also Ref.[92] The approach has been adopted for describing carrier interactions in many electronic devices such as tunnel junctions where the kinetic energies of relevant carriers are large, see for examples Refs.[93, 94]

By the scattering theory and the generalized Golden rule presented in Chapter 2 and denoting $\langle \phi_j \psi_i | \hat{T}(t) | \phi_j \psi_i \rangle \approx \mathrm{e}^{-w_{ji}t/2}$ as the temporal development $\hat{T}(t)$ of state ji, it is easy to obtain

$$
w_{ji} = \frac{2\pi}{\hbar} \sum_{nm \neq ji} \left| \langle \phi_j \psi_i | V | \psi_n \psi_m \rangle \right|^2 \delta(E_j + E_i - E_n - E_m) \tag{3.136}
$$

$\Gamma_{ji} = \hbar w_{ji}/2$ is the relaxation energy and $1/w_{ji}$ the decaying time. In numerical calculations the δ functions in the above equation are replaced by $\Gamma_{ji}/[\Gamma_{ji}^2 + (E_j + E_i - E_n - E_m)^2]$.

To estimate the Coulomb potential of Eq. 3.129, we make the following analysis. Consider the best configuration for Eq. 3.129 that the four wavefunctions involved in the impact ionization are all uniformly distributed within the QD volume. The corresponding Coulomb potential is

$$
\frac{e}{8\pi\epsilon}
\begin{cases}
\dfrac{2}{r} & , r > a \\
\dfrac{1}{a}\left(3 - \dfrac{r^2}{a^2}\right) & , r \leq a
\end{cases}
\tag{3.137}
$$

where a is the radius of the QD. When the hole is assumed to distribute also uniformly in the same QD volume, the Coulomb interaction energy between this electron–hole pair is $3e^2/10\pi\epsilon a$, which is about 28.8 meV when $a = 5$ nm and $\epsilon = 12$ (common semiconductors) when we neglect the energy conservation requirement, i.e., we neglect the δ function in Eq. 3.136. This is very large compared with the value in bulk because of the quantum confinements (electrons and holes are now confined within a spherical volume with a radius of a). Such an interaction energy is much large as compared with the light–matter interaction, which is normally less than 0.1 meV in similar QDs; see Eq. 3.98. Large impact ionization and also Auger recombination are thus expected in colloidal QDs.

To calculate the impact ionization interaction from an initial state of

$$R_{\ell_1}(r_1)Y_{\ell_1 m_1}(\theta_1,\phi_1)R_{\ell_2}(r_2)Y_{\ell_2 m_2}(\theta_2,\phi_2)$$

to a final state

$$R_{\ell_3}(r_1)Y_{\ell m}(\theta_2,\phi_2)R_{\ell_4}(r_2)Y_{\ell_4 m_4}(\theta_2,\phi_2)$$

(the first term one in Eq. 3.130), we notice

$$\frac{1}{4\pi|\boldsymbol{r}_1-\boldsymbol{r}_2|}=\sum_{\ell=0}^{\infty}\sum_{m=-\ell}^{\ell}\frac{1}{2\ell+1}\frac{r_<^{\ell}}{r_>^{\ell+1}}Y_{\ell m}^*(\theta_2,\phi_2)Y_{\ell m}(\theta_1,\phi_1) \tag{3.138}$$

where $r_< = \min\{|\boldsymbol{r}_1|,|\boldsymbol{r}_2|\}$ and $r_> = \max\{|\boldsymbol{r}_1|,|\boldsymbol{r}_2|\}$ so that the impact ionization energy consists of

$$\begin{aligned}&\frac{r_<^{\ell}}{r_>^{\ell+1}}R_{\ell_1}(r_1)R_{\ell_2}(r_2)R_{\ell_3}(r_1)R_{\ell_4}(r_2)\\&\times Y_{\ell_1 m_1}^*(\theta_1,\phi_1)Y_{\ell_2 m_2}^*(\theta_2,\phi_2)Y_{\ell m}^*(\theta_2,\phi_2)\\&\times Y_{\ell m}(\theta_1,\phi_1)Y_{\ell_3 m_3}(\theta_1,\phi_1)Y_{\ell_4 m_4}(\theta_2,\phi_2)\end{aligned} \tag{3.139}$$

for which we utilize the addition theorem for spherical harmonics of Eq. 3.109.

A little detailed analysis shows a few selection rules about ℓs and ms such as $m_4 - m_2 = m_1 - m_3$. However, these selection rules can be easily fulfilled in nanoscale QDs because of the large number of available states confined in the QDs. The most important qualitative selection rules are about the radial functions that

$$\begin{aligned}&\int\int\frac{r_<^{\ell}}{r_>^{\ell+1}}R_{\ell_1}(r_1)R_{\ell_2}(r_2)R_{\ell_3}(r_1)R_{\ell_4}(r_2)\\&=\int_0^a R_{\ell_1}(r_1)R_{\ell_3}(r_1)\left[\int_0^{r_1}\frac{r_2^{\ell}}{r_1^{\ell+1}}R_{\ell_2}(r_2)R_{\ell_4}(r_2)r_2^2\mathrm{d}r_2\right.\\&\left.+\int_{r_1}^{a}\frac{r_1^{\ell}}{r_2^{\ell+1}}R_{\ell_2}(r_2)R_{\ell_4}(r_2)r_2^2\mathrm{d}r_2\right]r_1^2\mathrm{d}r_1\end{aligned} \tag{3.140}$$

which shows that a direct spatial overlapping between $R_{\ell_2}(r_2)$ (VB hole state) and $R_{\ell_4}(r_4)$ (CB electron sate) will result in a large impact ionization. Impact ionization in type-II QDs is thus negligible since the electron–hole wavefunction overlapping

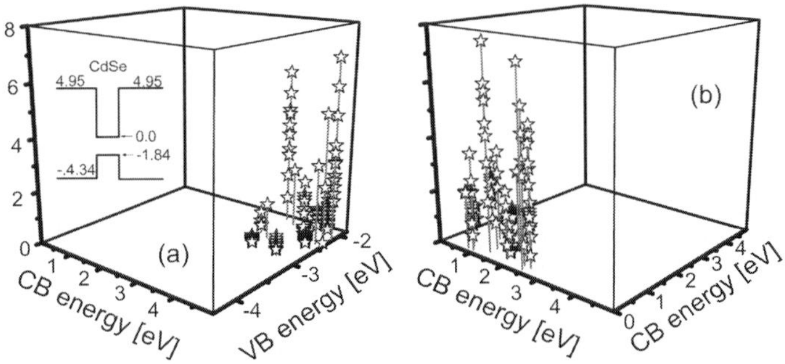

Fig. 3.15 (a) Impact ionization energy between initial CB state(s) and initial VB state(s). Inset shows the band structure of the CdSe QD in vacuum. (b) Auger recombination. The QD radius $a = 4$ nm. Y. Fu, Y.-H. Zhou, H. Su, F. Y. C. Boey, and H. Ågren, Impact ionization and Auger recombination rates in semiconductor quantum dots, J. Physical Chemistry C, vol. 114, p. 3743–7, 2010. Copyright (2010) by American Chemical Society. See also Color Insert.

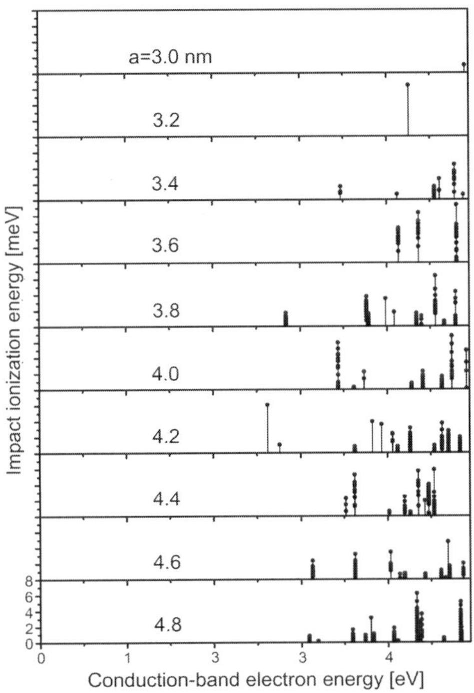

Fig. 3.16 Impact ionization energy vs. initial CB states as functions of the QD radius a. Y. Fu, Y.-H. Zhou, H. Su, F. Y. C. Boey, and H. Ågren, Impact ionization and Auger recombination rates in semiconductor quantum dots, J. Physical Chemistry C, vol. 114, p. 3743–7, 2010. Copyright (2010) by American Chemical Society.

is small. Furthermore, to ensure energy conservation, $R_{\ell_1}(r)$ is normally a high-energy CB state, while both $R_{\ell_3}(r)$ and $R_{\ell_4}(r)$ are low-energy CB states. Impact ionization in many core-shell-structured QDs will be small, since the high-energy wavefunctions are much more extended (extended into shells) than the ground-state wavefunction (deeply confined in the core region).

We pick up one of the CB state's from Fig. 3.4a and one VB state from Fig. 3.4b to calculate the total impact ionization energy to all possible combinations of two final CB states in Fig. 3.4a, which is shown in Fig. 3.15a.

Figure 3.15a agrees what can be expected intuitively that the initial CB state has to be high while the VB state low in order to fulfil the energy conservation requirement. Furthermore, as mentioned before, high-energy states can be highly degenerate because of large ℓ, so Γ can be also high. For the case of Fig. 3.15 where $a = 4\,\text{nm}$, we observe a maximal Γ of about $7\,\text{meV}$. We further observe 10 of Γ over $4\,\text{meV}$ while majorities are about $1\,\text{meV}$. These are very high as compared with the light–matter interaction in a similar CdSe QD; thus, we can be very optimistic about MEG processes in colloidal QDs, as has been very much reported as well as anticipated for significant solar cell applications. This is more clearly reflected in Fig. 3.16 about the relationship between the impact ionization and initial CB state(s). For the CdSe QDs under investigation, MEG effect can be expected when the excitation radiation energy exceeds a threshold of about $2E_g$ ($E_g = 1.84\,\text{eV}$ for CdSe) when the QD radius $a = 4.8\,\text{nm}$, while the threshold energy will be $3E_g$ when $a = 4\,\text{nm}$, and $4E_g$ for $a = 3\,\text{nm}$.

Auger recombination rates (reverse processes of the impact ionizations) shown in Fig. 3.15b can be similarly analyzed. Note that the energy range of the initial CB states in Auger recombination processes is much wider than the impact ionization, especially for small QDs. This is due to the large density of states at high energy so that there are more available final CB states for two initial CB states to interact via the Auger recombination process.

References

[1] *Excitons*, ed. M. D. Sturge and E. I. Rashba, North-Holland, Amsterdam, 1982.

[2] G. D. Scholes and G. Rumbles, "Excitons in nanoscale systems," Nat. Mater., **5**, 683–696, 2006.

[3] C. Weisbuch, H. Benisty, and R. Houdré, "Overview of fundamentals and applications of electrons, excitons and photons in confined structures," J. Lumin., **85**, 271–293, 2000.

[4] P. E. Lippens and M. Lannoo, "Comparison between calculated and experimental values of the lowest excited electronic state of small CdSe crystallites," Phys. Rev. B, **41**, 6079–6081, 1990.

[5] P. E. Lippens and M. Lannoo, "Calculation of the band gap for small CdS and ZnS crystallites," Phys. Rev. B, **39**, 10935–10942, 1989.

[6] S. Sapra and D. D. Sarma, "Evolution of the electronic structure with size in II-VI semiconductor nanocrystals," Phys. Rev. B, **69**, 125304(7), 2004.

[7] L. E. Brus, "Electron-electron and electron–hole interactions in small semiconductor crystallites: the size dependence of the lowest excited electronic state," J. Chem. Phys., **80**, 4403–4409, 1984.
[8] S. V. Nair, S. Sinha, and K. C. Rustagi, "Quantum size effects in spherical semiconductor microcrystals," Phys. Rev. B, **35**, 4098–4101, 1987.
[9] Y. Kayanuma and H. Momiji, "Incomplete confinement of electrons and holes in microcrystals," Phys. Rev. B, **41**, 10261–10263, 1990.
[10] *Data in Science and Technology: Semiconductors Other Than Group IV Elements and III–V Compounds*, ed. O. Madelung, Springer, Boston, 1992.
[11] P. Lawaetz, "Valence-band parameters in cubic semiconductors," Phys. Rev. B, **4**, 3460–3467, 1971.
[12] V. Ya. Grabovskis, Ya. Ya. Dzenis, A. I. Ekimov, I. A. Kudryavtsev, M. N. Tolstoi, and U. T. Rogulis, "Photoionization of semiconducting microcrystals in glass [luminescence studies]," Fizika Tverdogo Tela, vol. 31, 272–275, 1989 (Soviet Physics – Solid State, **31**, 149–151, 1989).
[13] R. K. Swank, "Surface properties of II–VI compounds," Phys. Rev., **153**, 844–849, 1967.
[14] M. Bujatti, "CdS-metal barriers from photovoltage measurements," Brit. J. Appl. Phys. (J. Phys. D), ser. 2, **1**, 581–584, 1968.
[15] Y. Fu, M. Willander, and E. L. Ivchenko, "Photonic dispersions of semiconductor-quantum-dot-array-based photonic crystals in primitive and face-centered cubic lattices," Superlatt. Microstruct., **27**, 255–264, 2000.
[16] *Semiconductors Group IV Elements and III–V Compounds*, ed. O. Madelung, Springer-Verlag, Berlin, 1991.
[17] *Landolt-Börnstein, New Series, Group III*, vol. 17a–b, Springer-Verlag, Berlin, 1982.
[18] W. Wilson, P. Szajowski, and L. Brus, "Quantum confinement in size-selected, surface-oxidized silicon nanocrystals," Science, **262**, 1242–1244, 1993.
[19] J. Perriere, E. Millon, M. Chammaro, M. Morcrette, and C. Andreazza, "Formation of GaAs nanocrystals by laser ablation," Appl. Phys. Lett., **78**, 2949(3), 2001.
[20] M. Hirasawa, N. Ichikawa, Y. Egashira, I. Honma, and H. Komiyama, "Synthesis of GaAs nanoparticles by digital radio frequency sputtering," Appl. Phys. Lett., **67**, 3483(3), 1995.
[21] H. X. Fu, L. W. Wang, and A. Zunger, "Applicability of the kp method to the electronic structure of quantum dots," Phys. Rev. B, **57**, 9971–9987, 1998.
[22] X. Q. Wang, S. J. Clark, and R. A. Abram, "Ab initio calculations of the structural and electronic properties of Hg_mTe_n clusters," Phys. Rev. B, **70**, 235328(6), 2004.
[23] G. L. Gutsev, E. Johnson, M. D. Mochena, and C. W. Bauschlicher Jr., J. Chem. Phys., **128**, 144707, (2008).
[24] P. P. Korambath and S. P. Karna, J. Phys. Chem. A, **104**, 4801, (2000).
[25] N. D. Drummond, A. J. Williamson, R. J. Needs, and G. Galli, "Electron emission from diamondoids: a diffusion quantum Monte Carlo study," Phys. Rev. Lett., **95**, 096801(4), 2005.
[26] J. Jiang, K. Liu, W. Lu, and Y. Luo, "An elongation method for first principle simulations of electronic structures and electron transport properties of finite nanostructures," J. Chem. Phys., **124**, 214711(9), 2006.
[27] J. Jiang, B. Gao, T.-T. Han, and Y. Fu, "Ab initio study of energy band structures of GaAs nanoclusters," Appl. Phys. Lett., **94**, 092110(3), 2009.
[28] J. Nayak, R. Mythili, M. Vijayalakshmi, and S. N. Sahu, "Size quantization effect in GaAs nanocrystals," Physica E, **24**, 227–233, 2004.

[29] M. A. Malik, P. O'Brien, S. Norager, and J. Smith, "Gallium arsenide nanoparticles: synthesis and characterisation," J. Mater. Chem., **13**, 2591–2595, 2003.

[30] G. T. Einevoll, "Confinement of excitons in quantum dots," Phys. Rev. B, **45**, 3410–3417, 1992.

[31] S. V. Nair, L. M. Ramaniah, and K. C. Rustagi, "Electron states in a quantum dot in an effective-bond-orbital model," Phys. Rev. B, **45**, 5969–5979, 1992.

[32] P. Vogl, H. P. Hjalmarson, and J. D. Dow, "A semi-empirical tight-binding theory of the electronic structure of semiconductor," J. Phys. Chem. Solids, **44**, 365–378, 1983.

[33] S. Sapra, N. Shanthi, and D. D. Sarma, "Realistic tight-binding model for the electronic structure of II-VI semiconductors," Phys. Rev. B, **66**, 205202(8), 2002.

[34] Al. L. Efros and A. L. Efros, "Interband absorption of light in a semiconductor sphere," Sov. Phys. Semicond., **16**, 772–775, 1982.

[35] Y. Kayanuma, "Wannier exciton in microcrystals," Solid state Commun., **59**, 405–408, 1986.

[36] H. M. Schmidt and H. Weller, "Quantum size effects in semiconductor crystallites: calculation of the energy spectrum for the confined exciton," Chem. Phys. Lett., **129**, 615–618, 1986.

[37] R. L. Liboff, *Introductory Quantum Mechanics*, 4th edn, Addison Wesley, San Francisco, 2003, p. 373.

[38] X.Peng, "Green chemical approaches toward high-quality semiconductor nanocrystals," Chem. Eur. J., **8**, 334–339, 2002.

[39] C. P. Raptopoulou, V. Tangoulis, and E. Devlin, "$[\{Fe(OMe)_2[O_2CC(OH)Ph_2]\}_{12}]$: synthesis and characterization of a new member in the family of molecular ferric wheels with the carboxylatobis(alkoxo) bridging unit," Angew. Chem. Int. Ed., **41**, 2386–2389, 2002.

[40] R. Xie, U. Kolb, J. Li, T. Basch, and A. Mews, "Synthesis and characterization of highly luminescent CdSe-Core $CdS/Zn_{0.5}Cd_{0.5}S/ZnS$ multishell nanocrystals," J. Am. Chem. Soc., **127**, 7480–7488, 2005.

[41] J. O. Dimmock, "Chapter 7 Introduction to the theory of exciton states in semiconductors," *Semiconductors and Semimetals*, vol. 3, ed. R. K. Willardson and A. C. Beer, Academic Press, New York, 1967, pp. 259–319.

[42] D. A. B. Miller, D. S. Chemla, D. J. Eilenberg, P. W. Smith, A. C. Gossard, and W. T. Tsang, "Large room-temperature optical nonlinearity in $GaAs/Ga_{1-x}Al_xAs$ multiple quantum well structures," Appl. Phys. Lett., **41**, 679–681, 1982.

[43] Y. Fu and K. A. Chao, "Exciton binding energy in GaAs/AlGaAs multiple quantum wells," Phys. Rev., **B43**, 12626–12629, 1991.

[44] H. D. Sun, T. Makino, Y. Segawa, M. Kawasaki, A. Ohtomo, K. Tamura, and H. Koinuma, "Enhancement of exciton binding energies in ZnO/ZnMgO multiquantum wells," J. Appl. Phys., **91**, 1993–1997. 2002.

[45] T. Yatsui, S. Sangu, T. Kawazoe, M. Ohtsu, S. J. An, J. Yoo, and G.-C. Yi, "Nanophotonic switch using ZnO nanorod double-quantum-well structures," Appl. Phys. Lett., **90**, 223110(3), 2007.

[46] T. Yatsui, G.-C. Yib, and M. Ohtsu, "Progress in developing nanophotonic integrated circuits," Proc. SPIE, **7007**, 700703(8), 2008.

[47] M Zamfirescu, A. Kavokin, B. Gil, G. Malpuech, and M. Kaliteevski, "ZnO as a material mostly adapted for the realization of room-temperature polariton lasers," Phys. Rev. B, **65**, 161205(4), 2002.

[48] J. L. Birman and N. Q. Huong, "Wannier-Frenkel hybrid exciton in organic-semiconductor quantum dot heterostructures," J. Lumin., **125**, 196–200, 2007.

[49] G. Bacher, R. Weigand, J. Seufert, V. D. Kulakovskii, N. A. Gippius, A. Forchel, K. Leonardi, and D. Hommel, "Biexciton versus exciton lifetime in a single semiconductor quantum dot," Phys. Rev. Lett., **83**, 4417(4), 1999.

[50] T. Kawazoe, K. Kobayashi, and M. Ohtsu, "Anti-parallel dipole coupling of quantum dots via an optical near-field interaction," IEICE Trans. Electron, **E88-C**, 1845–1849, 2005.

[51] S. A. Crooker, T. Barrick, J. A. Hollingsworth, and V. I. Klimov, "Multiple temperature regimes of radiative decay in CdSe nanocrystal quantum dots: intrinsic limits to the dark-exciton lifetime," Appl. Phys. Lett., **82**, 2793(3), 2003.

[52] T. Okada, K. Komori, K. Goshima, S. Yamauchi, I. Morohashi, T. Sugaya, O. Yamazaki, M. Ogura, and T. Hattori, "Coherent control of exciton in a single quantum dot using high-resolution Michelson interferometer," Jpn. J. Appl. Phys., **43**, 6093–6096, 2004.

[53] H. P. Wagner, H.-P. Tranitz, H. Preis, W. Langbein, and J. M. Hvam, "Dephasing and interaction of excitons in CdSe/ZnSe islands," J. Crys. Growth, **214/215**, 747–751, 2000.

[54] Y. Viale, P. Gilliot, O. Crégut, J.-P. Likforman, B. Hönerlage, R. Levy, L. Besombes, L. Marshal, K. Kheng, and H. Mariette, "Excitonic dynamics in CdTe/ZnTe quantum dots," Mater. Sc. Eng. B, **101**, 55–59, 2003.

[55] E. Dekel, D. V. Regelman, D. Gershoni, E. Ehrenfreund, W. V. Schoenfeld, and P.M. Petroff, "Radiative lifetimes of single excitons in semiconductor quantum dots - manifestation of the spatial coherence effect," Solid State Commun., **117**, 395–400, 2001.

[56] R. R. Cooney, S. L. Sewall, E. A. Dias, D. M. Sagar, K. E. H. Anderson, and P. Kambhampati, "Unified picture of electron and hole relaxation pathways in semiconductor quantum dots," Phys. Rev. B, **75**, 245311(14), 2007.

[57] V. M. Huxter and G. D. Scholes, "Nonlinear optical approach to multiexciton relaxation dynamics in quantum dots," J. Chem. Phys., **125**, 144716–14427, 2006.

[58] J. Suffczyński, T. Kazimierczuk, M. Goryca, B. Piechal, A. Trajnerowicz, K. Kowalik, P. Kossacki, A. Golnik, K. P. Korona, M. Nawrocki, and J. A. Gaj, "Excitation mechanisms of individual CdTe/ZnTe quantum dots studied by photon correlation spectroscopy," Phys. Rev. B, **74**, 085319(7), 2006.

[59] R. P. Prasankumar, R. S. Attaluri, R. D. Averitt, J. Urayama, N. Weisse-Bernstein, P. Rotella, A. D. Stintz, S. Krishna, and A. J. Taylor, "Ultrafast carrier dynamics in an InAs/InGaAs quantum dots-in-a-well heterostructure," Opt. Express, **16**, 1165–1173, 2008.

[60] P. Borri, W. Langbein, S. Schneider, U. Woggon, R. L. Sellin, D. Ouyang, and D. Bimberg, "Exciton relaxation and dephasing in quantum-dot amplifiers from room to cryogenic temperature," IEEE J. Sel. Top. Quantum Electron., **8**, 984–991, 2002.

[61] T.-T. Han, Y. Fu, and H. Ågren, "Dynamic photon emission from multiphoton-excited semiconductor quantum dot," J. Appl. Phys., **103**, 93703(6), 2008.

[62] J. Förstner, C. Weber, J. Danckwerts, and A. Knorr, "Phonon-assisted damping of Rabi oscillations in semiconductor quantum dots," Phys. Rev. Lett., **91**, 127401(4), 2003.

[63] Y. Fu, H. Ågren, J. M. Kowalewski, H. Brismar, J. Wu, Y. Yue, N. Dai, and L. Thylén, "Radiative and nonradiative recombination of photoexcited excitons in multi-shell-coated CdSe/CdS/ZnS quantum dots," EuroPhys. Lett., **86**, 37003(6), 2009.

[64] C. Jacobini and L. Reggiani, "The Monte Carlo method for the solution of charge transport in semiconductors with applications to covalent materials," Rev. Mod. Phys., **55**, 645–705 1983.

[65] B. K. Ridley, *Quantum Processes in Semiconductors*, 2nd edn., Clarendon Press, Oxford, 1988.

[66] D. C. Reynolds, C. W. Litton, and T. C. Collins, "Bound-phonon quasiparticle in CdS," Phys. Rev. B, **4**, 1868–1872, 1971.

[67] A. L. Pan, R. B. Liu, and B. S. Zou, "Phonon-assisted stimulated emission from single CdS nanoribbons at room temperature," Appl. Phys. Lett., **88**, 173102(3), 2006.

[68] K. C. Rustagi and W. Weber, "Adiabatic bond charge model for the phonons in A^3B^5 semiconductors," Solid State Commun., **18**, 673–675, 1976.

[69] M. Sugawara, T. Yamamoto, and H. Ebe, "Quantum-dot-based photonic devices," Fujitsu Sci. Tech. J., **43**, 495–501, 2007.

[70] B. N. Murdin, A. R. Hollingworth, M. Kamal-Saadi, R. T. Kotitschke, C. M. Ciesla, C. R. Pidgeon, P. C. Findlay, H. P. M. Pellemans, C. J. G. M. Langerak, A. C. Rowe, R. A. Stradling, and E. Gornik, "Suppression of LO phonon scattering in Landau quantized quantum dots," Phys. Rev. B, **59**, R7817–R7820, 1999.

[71] Y. Fu, T.-T. Han, H. Ågren, L. Lin, P. Chen, Y. Liu, G.-Q. Tang, J. Wu, Y. Yue, and N. Dai, "Design of semiconductor CdSe-core ZnS/CdS-multishell quantum dots for multiphoton applications," Appl. Phys. Lett., **90**, 173102(3), 2007.

[72] Y. Fu, F. Ferdos, M. Sadeghi, S. M. Wang, and A. Larsson, "Photoluminescence of an assembly of size-distributed self-assembled InAs quantum dots," J. Appl. Phys., **92**, 3089–3092, 2002.

[73] M. B. Mohamed, C. Burda, and M. A. El-Sayed, "Shape dependent ultrafast relaxation dynamics of CdSe nanocrystals: nanorods vs nanodots," Nano Lett., **1**, 589–593, 2001.

[74] M. Goeppert Mayer, "Elementary processes with two quantum jumps," Ann. Physik (Leipzig), **9**, 273–294, 1931.

[75] J.-F. Lami, P. Gilliot, and C. Hirlimann, "Observation of interband two-photon absorption saturation in CdS," Phys. Rev. Lett., **77**, 1632–1635, 1996.

[76] F. Helmchen, K. Svododa, W. Denk, D. Kleinfeld, and D. W. Tank, "In vivo dendritic calcium dynamics in deep-layer cortical pyramidal neurons," Nat. Neurosci., **2**, 989–996, 1996.

[77] H. Yokoyama, H. Guo, T. Yoda, K. Takashima, K.-I. Sato, H. Taniguchi, and H. Ito, "Two-photon bioimaging with picosecond optical pulses from a semiconductor laser," Opt. Express, **14**, 3467–3471, 2006.

[78] B. S. Wherrett, "Scaling rules for multiphoton interband absorption in semiconductors," J. Opt. Soc. Am. B, **1**, 67–72, 1984.

[79] M. E. Schmidt, S. A. Blanton, M. A. Hines, and P. Guyot-Sionnest, "Size-dependent two-photon excitation spectroscopy of CdSe nanocrystals," Phys. Rev. B, **53**, 12629–12632, 1996.

[80] Y. Fu, T.-T. Han, Y. Luo, and H. Ågren, "Multi-photon excitation of quantum dots by ultra-short and ultra-intense laser pulse," Appl. Phys. Lett., **88**, 221114(3), 2006.

[81] B. L. van der Waerden, *Die Gruppentheoretische Methode in der Quantenmechanik*, Springer, Berlin, 1932.

[82] G. Racah, "Theory of complex spectra. II," Phys. Rev., **62**, 438–462, 1942.

[83] G. M. Wysin, D. L. Smith, and A. Redondo, "Picosecond response of photoexcited GaAs in a uniform electric field by Monte Carlo dynamics," Phys. Rev. B, **38**, 12514–12524, 1988.

[84] A. J. Nozik, "Quantum dot solar cells," Physica E, **14**, 115–120, 2002.

[85] R. D. Schaller and V. I. Klimov, "High efficiency carrier multiplication in PbSe nanocrystals: implications for solar energy conversion," Phys. Rev. Lett., **92**, 186601(4), 2004.

[86] M. Hanna, R. J. Ellingson, M. Beard, P. Yu, and A. J. Nozik, "Quantum dot solar cells: high efficiency through impact ionization," 2004 DOE Solar Energy Technologies Program Review Meeting, October 25–28, 2004, Denver, USA.

[87] S. J. Kim, W. J. Kim, Y. Sahoo, A. N. Cartwright, and P. N. Prasad, "Multiple exciton generation and electrical extraction from a PbSe quantum dot photoconductor," Appl. Phys. Lett., **92**, 31107(3), 2008.

[88] M. T. Trinh, A. J. Houtepen, J. M. Schins, T. Hanrath, J. Piris, W. Knulst, A. P. L. M. Goossens, and L. D. A. Siebbeles, "In spite of recent doubts carrier multiplication does occur in PbSe nanocrystals," Nano Lett., **8**, 1713–1718, 2008.

[89] L. D. Landau and E. M. Lifshitz, *Quantum Mechanics*, 3rd Ed. Pergamon Press Oxford, New York, 1962, p. 278.

[90] B. K. Ridley, *Quantum Processes in Semiconductors*, Clarendon Press, Oxford 1988, pp. 269–278.

[91] P. T. Landsberg and M. J. Adams, "Theory of donor-acceptor radiative and Auger recombination in simple semiconductors," Proc. R. Soc. Lond. A, **334**, 523–539, 1973.

[92] E. Abrahams, "Electron-electron scattering in Alkali metals," Phys. Rev., **95**, 839–810, 1954.

[93] N. Takenaka, M. Inoue, and Y. Inuishi, "Influence of inter-carrier scattering on hot electron distribution function in GaAs," J. Phys. Soc. Jpn., **47**, 861–868, 1979.

[94] P. Rodina, U. Ebert, W. Hundsdorfer, and I. Grekhov, "Tunneling-assisted impact ionization fronts in semiconductors," J. Appl. Phys., **92**, 958–964, 2002.

Chapter 4

Exciton Polariton

Reference[1] presents a detailed discussion of energy dispersion of exciton–polaritons in microcavities (mostly in quantum wells, QWs). In this chapter we discuss mostly the exciton polariton in quantum dots.

4.1 EXCITON EXCITATION

In the absence of external fields, the exciton composed of a conduction-band electron $\boldsymbol{r}_e$ and a valence-band hole $\boldsymbol{r}_h$ is described by a Hamiltonian of

$$H_0 = -\frac{\hbar^2\nabla_e^2}{2m_0} + V_0(\boldsymbol{r}_e) - \frac{\hbar^2\nabla_h^2}{2m_0} + V_0(\boldsymbol{r}_h) - \frac{\mathrm{e}^2}{4\pi\epsilon_b|\boldsymbol{r}_e - \boldsymbol{r}_h|} \tag{4.1}$$

where $V_0(\boldsymbol{r}_e)$ and $V_0(\boldsymbol{r}_h)$ are the lattice potentials for the electron and hole. ϵ_b is the dielectric constant. Here we neglect the spatial variation of the dielectric constant ϵ_b in hetero materials.

Neglecting the exchange interaction between the electron and the hole, it was shown in Section 3.2 that the Schrödinger equation of the exciton can be written as

$$\left[E_c(-i\nabla_e) - E_v(-i\nabla_h) - \frac{\mathrm{e}^2}{4\pi\epsilon_b|\boldsymbol{r}_\mathrm{e} - \boldsymbol{r}_h|} - E_{n\boldsymbol{K}}\right]\psi_{n\boldsymbol{K}}(\boldsymbol{r}_e, \boldsymbol{r}_h) = 0 \tag{4.2}$$

where $E_c(-i\nabla_e)$ is the expression obtained by replacing $\boldsymbol{k}_e$ in the single-particle CB energy $E_c(\boldsymbol{k}_e)$ by $-i\nabla_e$.

By the Bloch theorem, the single-particle wave functions are expressed as

$$\begin{aligned}\psi_{c\boldsymbol{k}_e}(\boldsymbol{r}_e) &= \frac{1}{\sqrt{\Omega}}\,\mathrm{e}^{i\boldsymbol{k}_e\cdot\boldsymbol{r}_e}u_c(\boldsymbol{r}_e)\\ \psi_{v\boldsymbol{k}_h}(\boldsymbol{r}_h) &= \frac{1}{\sqrt{\Omega}}\,\mathrm{e}^{i\boldsymbol{k}_h\cdot\boldsymbol{r}_h}u_v(\boldsymbol{r}_h)\end{aligned} \tag{4.3}$$

for electron and hole in bulk material, where Ω is the volume of the system, and $u_c(\boldsymbol{r}_e)$ and $u_v(\boldsymbol{r}_h)$ are Bloch functions normalized in the unit cell. we can easily extend the above single-particle wave functions in bulk materials to nanostructures by using the envelop function theory for which $\boldsymbol{k}$ denotes the quantum number of the energy level in the nanostructures and $\mathrm{e}^{i\boldsymbol{k}_e\cdot\boldsymbol{r}_e}/\sqrt{\Omega}$ is to be replaced by the corresponding envelop wave function.

Alternatively we simply concentrate on the effective-mass expression Eq. 4.4 of the exciton, which also includes the confinement potentials $V_c(\boldsymbol{r}_e)$ and $V_v(\boldsymbol{r}_h)$ of electron and hole

$$\left[E_c(-i\nabla_e) + V_c(\boldsymbol{r}_e) - E_v(-i\nabla_h) - V_v(\boldsymbol{r}_h) - \frac{\mathrm{e}^2}{4\pi\epsilon_b|\boldsymbol{r}_e - \boldsymbol{r}_h|}\right]\psi_{n\boldsymbol{K}}(\boldsymbol{r}_e, \boldsymbol{r}_h)$$
$$= E_{n\boldsymbol{K}}\psi_{n\boldsymbol{K}}(\boldsymbol{r}_e, \boldsymbol{r}_h) \tag{4.4}$$

so that after the inclusion of the Coulomb interaction between the electron and the hole, the exciton eigenstate consists of single-particle electron and hole states multiplied by an envelop function (see Chapter 3)

$$\Psi_{n\boldsymbol{K}}(\boldsymbol{r}_e, \boldsymbol{r}_h) = \psi_{n\boldsymbol{K}}(\boldsymbol{r}_e, \boldsymbol{r}_h) u_{c\boldsymbol{k}_e}(\boldsymbol{r}_e) u_{v\boldsymbol{k}_h}(\boldsymbol{r}_h) \tag{4.5}$$

where $\boldsymbol{K} = \boldsymbol{k}_e + \boldsymbol{k}_h$ is the total wave vector of the exciton state n, and $u_{c\boldsymbol{k}_e}(\boldsymbol{r}_e)$ and $v_{v\boldsymbol{k}_h}(\boldsymbol{r}_h)$ are Bloch functions of the CB electron and VB hole, respectively. We have neglected the contributions from other remote bands.

We consider an optical absorption process in which an allowed electric dipole transition creates an exciton $\Psi_{n\boldsymbol{K}}$ from a filled valence band, and in this scheme, the wave function of the initial state is simply unity, which is denoted by Ψ_0 in the formalism of second quantization. Assume that our electron–hole system is initially in its ground state $\Psi_0(t)$ for time $t < 0$. We switch on an external radiation of $\boldsymbol{E}(\boldsymbol{r}, t)$ for $t \geq 0$, the first-order perturbative Hamiltonian is

$$V = \int \boldsymbol{d}(\boldsymbol{r}) \cdot \boldsymbol{E}(\boldsymbol{r}, t)\, \mathrm{d}\boldsymbol{r} \tag{4.6}$$

where $\boldsymbol{d}(\boldsymbol{r})$ is the dipole-moment operator

$$\boldsymbol{d}(\boldsymbol{r}) = -e\boldsymbol{r}_e\delta(\boldsymbol{r} - \boldsymbol{r}_e) + e\boldsymbol{r}_h\delta(\boldsymbol{r} - \boldsymbol{r}_h) \tag{4.7}$$

Denote the wave function of the electron–hole system as $|\boldsymbol{r}_e, \boldsymbol{r}_h, t\rangle$; its time-dependent Schrödinger equation is

$$i\hbar\frac{\partial}{\partial t}|\boldsymbol{r}_e, \boldsymbol{r}_h, t\rangle = (H_0 + V)|\boldsymbol{r}_e, \boldsymbol{r}_h, t\rangle \tag{4.8}$$

where H_0 is given by Eq. 4.1 or its effective mass expression of Eq. 4.4.

By the first-order perturbative approximation where we consider only the exciton ground state $\Psi_0(t)$ to the first excited state $\Psi_{n\boldsymbol{K}}(\boldsymbol{r}_e, \boldsymbol{r}_h, t)$ the time-dependent wave function

$$|\boldsymbol{r}_e, \boldsymbol{r}_h, t\rangle = |\Psi_0(t)\rangle + c_{n\boldsymbol{K}}(t)|\Psi_{n\boldsymbol{K}}(\boldsymbol{r}_e, \boldsymbol{r}_h, t)\rangle \tag{4.9}$$

Here we simply assume that $|c_{n\boldsymbol{K}}(t)| \ll 1$. We will discuss more closely in the coming sections.

We now formally denote that both $\Psi_0(t)$ and $\Psi_{n\boldsymbol{K}}(\boldsymbol{r}_e, \boldsymbol{r}_h, t)$ as eigen functions of H_0 such that

$$\begin{aligned}
&H_0\Psi_0(t) = E_0\Psi_0(t) = \hbar\omega_0\Psi_0(t)\\
&H_0\Psi_{n\boldsymbol{K}}(\boldsymbol{r}_e, \boldsymbol{r}_h, t) = E_{n\boldsymbol{K}}\Psi_{n\boldsymbol{K}}(\boldsymbol{r}_e, \boldsymbol{r}_h, t) = \hbar\omega_{n\boldsymbol{K}}\Psi_{n\boldsymbol{K}}(\boldsymbol{r}_e, \boldsymbol{r}_h, t)\\
&\Psi_0(t) = \Psi_0\mathrm{e}^{-i\omega_0 t}\\
&\Psi_{n\boldsymbol{K}}(\boldsymbol{r}_e, \boldsymbol{r}_h, t) = \Psi_{n\boldsymbol{K}}(\boldsymbol{r}_e, \boldsymbol{r}_h)e^{-i\omega_{n\boldsymbol{K}}t}
\end{aligned} \tag{4.10}$$

Insert Eq. 4.9 into Eq. 4.8 becomes

$$\begin{aligned}
&i\hbar\frac{\partial\Psi_0(t)}{\partial t} + i\hbar\frac{\partial c_{n\boldsymbol{K}}(t)}{\partial t}\Psi_{n\boldsymbol{K}}(\boldsymbol{r}_e, \boldsymbol{r}_h, t) + i\hbar c_{n\boldsymbol{K}}(t)\frac{\partial\Psi_{n\boldsymbol{K}}(\boldsymbol{r}_e, \boldsymbol{r}_h, t)}{\partial t}\\
&= H_0\Psi_0(t) + c_{n\boldsymbol{K}}(t)H_0\Psi_{n\boldsymbol{K}}(\boldsymbol{r}_e, \boldsymbol{r}_h, t) + V\Psi_0(t) + c_{n\boldsymbol{K}}(t)V\Psi_{n\boldsymbol{K}}(\boldsymbol{r}_e, \boldsymbol{r}_h, t)
\end{aligned}$$

By Eq. 4.10,

$$\begin{aligned}
&\hbar\omega_0\Psi_0(t) + i\hbar\frac{\partial c_{n\boldsymbol{K}}(t)}{\partial t}\Psi_{n\boldsymbol{K}}(\boldsymbol{r}_e, \boldsymbol{r}_h, t) + \hbar\omega_{n\boldsymbol{K}}\, c_{n\boldsymbol{K}}(t)\Psi_{n\boldsymbol{K}}(\boldsymbol{r}_e, \boldsymbol{r}_h, t)\\
&= E_0\Psi_0(t) + c_{n\boldsymbol{K}}(t)E_{n\boldsymbol{K}}\Psi_{n\boldsymbol{K}}(\boldsymbol{r}_e, \boldsymbol{r}_h, t) + V\Psi_0(t) + c_{n\boldsymbol{K}}(t)V\Psi_{n\boldsymbol{K}}(\boldsymbol{r}_e, \boldsymbol{r}_h, t)
\end{aligned}$$

which is

$$i\hbar\frac{\partial c_{n\boldsymbol{K}}(t)}{\partial t}\Psi_{n\boldsymbol{K}}(\boldsymbol{r}_e, \boldsymbol{r}_h, t) = V\Psi_0(t) + c_{n\boldsymbol{K}}(t)V\Psi_{n\boldsymbol{K}}(\boldsymbol{r}_e, \boldsymbol{r}_h, t) \tag{4.11}$$

Multiplying the above equation by $\langle\Psi_{n\boldsymbol{K}}(\boldsymbol{r}_e, \boldsymbol{r}_h, t)|$ we come to

$$\frac{\partial c_{n\boldsymbol{K}}(t)}{\partial t} = \frac{1}{i\hbar}\langle\Psi_{n\boldsymbol{K}}(\boldsymbol{r}_e, \boldsymbol{r}_h, t)|\int \boldsymbol{d}(\boldsymbol{r})\cdot\boldsymbol{E}(\boldsymbol{r}, t)\,\mathrm{d}\boldsymbol{r}|\Psi_0(t)\rangle \tag{4.12}$$

Note that due to the symmetry consideration,

$$\langle\Psi_{n\boldsymbol{K}}(\boldsymbol{r}_e, \boldsymbol{r}_h, t)|\int \boldsymbol{d}(\boldsymbol{r})\cdot\boldsymbol{E}(\boldsymbol{r}, t)\,\mathrm{d}\boldsymbol{r}|\Psi_{n\boldsymbol{K}}(\boldsymbol{r}_e, \boldsymbol{r}_h, t)\rangle = 0$$

For $\boldsymbol{E}(\boldsymbol{r}, t) = \boldsymbol{E}(\boldsymbol{r})\mathrm{e}^{-i\omega t} + \text{c.c.}$, it is easy to obtain the solution of Eq. 4.12

$$\begin{aligned}
c_{n\boldsymbol{K}}(t) = &\frac{\mathrm{e}^{i(\omega_{n\boldsymbol{K}}-\omega_0-\omega)t}}{\hbar(\omega_{n\boldsymbol{K}}-\omega_0-\omega)}\langle\Psi_{n\boldsymbol{K}}(\boldsymbol{r}_e, \boldsymbol{r}_h)|\int \boldsymbol{d}(\boldsymbol{r})\cdot\boldsymbol{E}(\boldsymbol{r})\,\mathrm{d}\boldsymbol{r}|\Psi_0\rangle\\
&+\frac{e^{i(\omega_{n\boldsymbol{K}}-\omega_0+\omega)t}}{\hbar(\omega_{n\boldsymbol{K}}-\omega_0+\omega)}\langle\Psi_{n\boldsymbol{K}}(\boldsymbol{r}_e, \boldsymbol{r}_h)|\int \boldsymbol{d}(\boldsymbol{r})\cdot\boldsymbol{E}(\boldsymbol{r})\,\mathrm{d}\boldsymbol{r}|\Psi_0\rangle
\end{aligned} \tag{4.13}$$

Note that $\Psi_0(t)$ is the ground exciton state, i.e., without any electron and hole in the CB and VB so that $\omega_0 = 0$. Furthermore, it is noticed that the term with $\mathrm{e}^{i(\omega_{n\boldsymbol{K}}+\omega)t}$ oscillates much fast, which corresponds, by Chapter 2, to the emission of a photon $\hbar\omega$ during the transition of an electron from a VB state to a CB state. Its probability is small so that

$$c_{n\boldsymbol{K}}(t) = \frac{\mathrm{e}^{i(\omega_{n\boldsymbol{K}}-\omega)t}}{\hbar(\omega_{n\boldsymbol{K}}-\omega)}\langle\Psi_{n\boldsymbol{K}}(\boldsymbol{r}_e, \boldsymbol{r}_h)|\int \boldsymbol{d}(\boldsymbol{r})\cdot\boldsymbol{E}(\boldsymbol{r})\,\mathrm{d}\boldsymbol{r}|\Psi_0\rangle \tag{4.14}$$

Now let us calculate Eq. 4.14 as much as possible. We first calculate the matrix element of the electron momentum $\boldsymbol{p}_e$

$$\begin{aligned}\langle\Psi_{n\boldsymbol{K}}(\boldsymbol{r}_e,\boldsymbol{r}_h)|\boldsymbol{p}_e|\Psi_0\rangle &= m_0\left\langle\Psi_{n\boldsymbol{K}}(\boldsymbol{r}_e,\boldsymbol{r}_h)\left|\frac{\partial\boldsymbol{r}_e}{\partial t}\right|\Psi_0\right\rangle \\ &= \frac{im_0}{\hbar}\left\langle\Psi_{n\boldsymbol{K}}(\boldsymbol{r}_e,\boldsymbol{r}_h)\right|[H_0,\boldsymbol{r}_e]\left|\Psi_0\right\rangle \\ &= im_0(\omega_{n\boldsymbol{K}}-\omega_0)\langle\Psi_{n\boldsymbol{K}}(\boldsymbol{r}_e,\boldsymbol{r}_h)|\boldsymbol{r}_e|\Psi_0\rangle \end{aligned} \tag{4.15}$$

Here

$$\begin{aligned}[H_0,\boldsymbol{r}_e] &= H_0\boldsymbol{r}_e-\boldsymbol{r}_eH_0 \\ H_0\Psi_{n\boldsymbol{K}}(\boldsymbol{r}_e,\boldsymbol{r}_h) &= \hbar\omega_{n\boldsymbol{K}}\Psi_{n\boldsymbol{K}}(\boldsymbol{r}_e,\boldsymbol{r}_h) \\ H_0\Psi_0 &= \hbar\omega_0\Psi_0 \\ \hbar\omega_0 &= 0\end{aligned}$$

By Eq. 4.5,

$$\begin{aligned}\langle\Psi_{n\boldsymbol{K}}(\boldsymbol{r}_e,\boldsymbol{r}_h)|\boldsymbol{p}_e|\Psi_0\rangle &= \iint\psi^*_{n\boldsymbol{K}}(\boldsymbol{r}_e,\boldsymbol{r}_h)u_{c\boldsymbol{k}_e}(\boldsymbol{r}_e)\boldsymbol{p}_eu_{v\boldsymbol{k}_h}(\boldsymbol{r}_h)\,\mathrm{d}\boldsymbol{r}_e\mathrm{d}\boldsymbol{r}_h \\ &= \int\psi^*_{n\boldsymbol{K}}(\boldsymbol{r}_e,\boldsymbol{r}_e)u_{c\boldsymbol{k}_e}(\boldsymbol{r}_e)\boldsymbol{p}_eu_{v\boldsymbol{k}_h}(\boldsymbol{r}_e)\,\mathrm{d}\boldsymbol{r}_e\end{aligned} \tag{4.16}$$

Since $u_{c\boldsymbol{k}_e}(\boldsymbol{r}_e)$ and $u_{v\boldsymbol{k}_h}(\boldsymbol{r}_e)$ are periodic within a unit cell while $\psi^*_{n\boldsymbol{K}}(\boldsymbol{r}_e,\boldsymbol{r}_h)$ varies in a space (nano scale for nanostructures) much larger than the unit cell (Å scale),

$$\langle\Psi_{n\boldsymbol{K}}(\boldsymbol{r}_e,\boldsymbol{r}_h)|\boldsymbol{p}_e|\Psi_0\rangle = \boldsymbol{p}_{cv}\int\psi^*_{n\boldsymbol{K}}(\boldsymbol{r}_e,\boldsymbol{r}_e)\,\mathrm{d}\boldsymbol{r}_e \tag{4.17}$$

where again $\boldsymbol{p}_{cv}=\langle u_{c\boldsymbol{k}}(\boldsymbol{r})|\boldsymbol{p}_{cv}|u_{v\boldsymbol{k}}(\boldsymbol{r})\rangle$.

By Eq. 4.7 it is then easy to calculate the optical matrix element from the followings

$$\begin{aligned}&\langle\Psi_{n\boldsymbol{K}}(\boldsymbol{r}_e,\boldsymbol{r}_h)|\int(-e)\boldsymbol{r}_e\delta(\boldsymbol{r}-\boldsymbol{r}_e)\cdot\boldsymbol{E}(\boldsymbol{r})\,\mathrm{d}\boldsymbol{r}|\Psi_0\rangle \\ &= \frac{-e}{im_0\omega_{n\boldsymbol{K}}}\langle\Psi_{n\boldsymbol{K}}(\boldsymbol{r}_e,\boldsymbol{r}_h)|\int\boldsymbol{p}_e\delta(\boldsymbol{r}-\boldsymbol{r}_e)\cdot\boldsymbol{E}(\boldsymbol{r})\,\mathrm{d}\boldsymbol{r}|\Psi_0\rangle \\ &= \frac{-e}{im_0\omega_{n\boldsymbol{K}}}\int\boldsymbol{p}_{cv}\left[\int\delta(\boldsymbol{r}-\boldsymbol{r}_e)\psi^*_{n\boldsymbol{K}}(\boldsymbol{r}_e,\boldsymbol{r}_e)\,\mathrm{d}\boldsymbol{r}_e\right]\cdot\boldsymbol{E}(\boldsymbol{r})\,\mathrm{d}\boldsymbol{r} \\ &= \frac{-e}{im_0\omega_{n\boldsymbol{K}}}\int\psi^*_{n\boldsymbol{K}}(\boldsymbol{r},\boldsymbol{r})\boldsymbol{p}_{cv}\cdot\boldsymbol{E}(\boldsymbol{r})\,\mathrm{d}\boldsymbol{r}\end{aligned} \tag{4.18}$$

Inserting the above relationship into Eq. 4.14, we finally obtain the occupation of the excited exciton state due to the external field

$$c_{n\boldsymbol{K}}(t) = \frac{-e\,e^{i(\omega_{n\boldsymbol{K}}-\omega)t}}{im_0\hbar\omega_{n\boldsymbol{K}}(\omega_{n\boldsymbol{K}}-\omega)}\int\psi^*_{n\boldsymbol{K}}(\boldsymbol{r},\boldsymbol{r})\boldsymbol{p}_{cv}\cdot\boldsymbol{E}(\boldsymbol{r})\,\mathrm{d}\boldsymbol{r} \tag{4.19}$$

4.2 EXCITONIC POLARIZATION

The excitation of the excited exciton state, i.e., the appearance of the electron and hole, results in an electric dipole; thus, the exciton contribution to the dielectric polarization is given by

$$\boldsymbol{P}_{n\boldsymbol{K}}(\boldsymbol{r},t) = \langle \boldsymbol{r}_e, \boldsymbol{r}_h, t|\boldsymbol{d}(\boldsymbol{r})|\boldsymbol{r}_e, \boldsymbol{r}_h, t\rangle = \langle \Psi_0|\boldsymbol{d}(\boldsymbol{r})|\Psi_{n\boldsymbol{K}}(\boldsymbol{r}_e, \boldsymbol{r}_h, t)\rangle c_{n\boldsymbol{K}}(t) + \text{c.c.} \tag{4.20}$$

It is easy to calculate $\langle \Psi_0|\boldsymbol{d}(\boldsymbol{r})|\Psi_{n\boldsymbol{K}}(\boldsymbol{r}_e, \boldsymbol{r}_h)\rangle$ by following similar steps in the previous section and the result is

$$\langle \Psi_0|\boldsymbol{d}(\boldsymbol{r})|\Psi_{n\boldsymbol{K}}(\boldsymbol{r}_e, \boldsymbol{r}_h)\rangle = \frac{-e}{im_0\omega_{n\boldsymbol{K}}}\,\psi_{n\boldsymbol{K}}(\boldsymbol{r},\boldsymbol{r})\boldsymbol{p}_{cv} \tag{4.21}$$

Inserting Eqs. 4.19 and 4.21 into Eq. 4.20, we get

$$\boldsymbol{P}_{n\boldsymbol{K}}(\boldsymbol{r},t) = \frac{\mathrm{e}^2\boldsymbol{p}_{cv}}{\hbar(\omega_{n\boldsymbol{K}}-\omega)m_0^2\omega_{n\boldsymbol{K}}^2}\,\psi_{n\boldsymbol{K}}(\boldsymbol{r},\boldsymbol{r})\int \psi^*_{n\boldsymbol{K}}(\boldsymbol{r}',\boldsymbol{r}')\boldsymbol{p}_{cv}\cdot\boldsymbol{E}(\boldsymbol{r}',t)\,\mathrm{d}\boldsymbol{r}' \tag{4.22}$$

Coupling the nonradiative damping rate Γ ($=\hbar\gamma$) of the exciton state (see Chapter 2 and more discussions late in this chapter), it can be shown that

$$\boldsymbol{P}_{n\boldsymbol{K}}(\boldsymbol{r},t) = \frac{\mathrm{e}^2\boldsymbol{p}_{cv}}{\hbar(\omega_{n\boldsymbol{K}}-\omega-i\gamma)m_0^2\omega_{n\boldsymbol{K}}^2}\,\psi_{n\boldsymbol{K}}(\boldsymbol{r},\boldsymbol{r})\int \psi^*_{n\boldsymbol{K}}(\boldsymbol{r}',\boldsymbol{r}')\boldsymbol{p}_{cv}\cdot\boldsymbol{E}(\boldsymbol{r}',t)\,\mathrm{d}\boldsymbol{r}' \tag{4.23}$$

We shall discuss the damping more closely shortly.

Define

$$T_{n\boldsymbol{K}}(\omega) = \frac{\epsilon_b\omega_{\mathrm{LT}}\pi a_{\mathrm{Br}}^3}{\omega_{n\boldsymbol{K}}-\omega-i\gamma} \tag{4.24}$$

where

$$\epsilon_b\omega_{\mathrm{LT}}a_{\mathrm{Br}}^3 = \frac{4\mathrm{e}^2p_{cv}^2}{\hbar\omega_{n\boldsymbol{K}}^2m_0^2}$$

where $p_{cv} = |\boldsymbol{p}_{cv}|$, and ω_{LT} and a_{Br} are the exciton longitudinal-transverse splitting and Bohr radius in the corresponding bulk semiconductor; the exciton contribution to the dielectric polarization is

$$\boldsymbol{P}_{n\boldsymbol{K}}(\boldsymbol{r},t) = T_{n\boldsymbol{K}}(\omega)\,\psi_{n\boldsymbol{K}}(\boldsymbol{r},\boldsymbol{r})\int \psi^*_{n\boldsymbol{K}}(\boldsymbol{r}',\boldsymbol{r}')\boldsymbol{E}(\boldsymbol{r}',t)\,\mathrm{d}\boldsymbol{r}' \tag{4.25}$$

where we have assumed $\boldsymbol{p}_{cv}(\boldsymbol{p}_{cv}\cdot\boldsymbol{E}) = p_{cv}^2\boldsymbol{E}$.

We now discuss briefly the significance of such a dielectric polarization in a QD. As expressed by Eq. 3.92, the first-excited exciton state in a spherical QD with radius R is described as

$$\psi_{n\boldsymbol{K}}(\boldsymbol{r}_e,\boldsymbol{r}_h) = \frac{1}{|\boldsymbol{r}-\boldsymbol{a}|\sqrt{2\pi R}}\sin\left(\frac{\pi|\boldsymbol{r}-\boldsymbol{a}|}{R}\right)\frac{1}{\sqrt{\pi a_{\mathrm{Br}}^3}}e^{-\frac{|\boldsymbol{r}_e-\boldsymbol{r}_h|}{a_{\mathrm{Br}}}} \tag{4.26}$$

which is confined and real. Here $\boldsymbol{a}$ is the center of the QD. Note that the above expression is valid when R is in the same order as a_{Br} so that one may neglect the internal motion of the electron–hole pair in the QD. And

$$\boldsymbol{r} = \frac{m_e \boldsymbol{r}_e + m_h \boldsymbol{r}_h}{m_e + m_h} \tag{4.27}$$

is the exciton center of mass.

For $\boldsymbol{r}_e = \boldsymbol{r}_h = \boldsymbol{r}$,

$$\psi_{n\boldsymbol{K}}(\boldsymbol{r}, \boldsymbol{r}) = \frac{1}{|\boldsymbol{r} - \boldsymbol{a}|\sqrt{2\pi R}} \sin\left(\frac{\pi|\boldsymbol{r} - \boldsymbol{a}|}{R}\right) \frac{1}{\sqrt{\pi a_{\mathrm{Br}}^3}} \tag{4.28}$$

Assuming small spatial variation of the excitation field within the QD, i.e., $\boldsymbol{E}(\boldsymbol{r}) \approx \boldsymbol{E}(\boldsymbol{a})$,

$$\int \psi_{n\boldsymbol{K}}^*(\boldsymbol{r}', \boldsymbol{r}')\boldsymbol{E}(\boldsymbol{r}')\, \mathrm{d}\boldsymbol{r}' = \boldsymbol{E}(\boldsymbol{a}) \int \psi_{n\boldsymbol{K}}^*(\boldsymbol{r}', \boldsymbol{r}')\, \mathrm{d}\boldsymbol{r}' = \frac{4R^2}{\pi\sqrt{2Ra_{\mathrm{Br}}^3}} \boldsymbol{E}(\boldsymbol{a}) \tag{4.29}$$

In other words, it is assumed that the electric field inside the QD is uniform since the wavelength of interest is about 1 μm whereas the QD diameter is about 20 nm or less. As before, ϵ_b is the background dielectric constant of the QD material.

Let $\alpha_{n\boldsymbol{K}} = \pi|\boldsymbol{r} - \boldsymbol{a}|/R$ and by Eqs. 4.24 and 4.29

$$\boldsymbol{P}_{n\boldsymbol{K}}(\boldsymbol{r}) = \frac{2\epsilon_b \omega_{\mathrm{LT}}}{\omega_{n\boldsymbol{K}} - \omega - i\gamma} \frac{\sin \alpha_{n\boldsymbol{K}}}{\alpha_{n\boldsymbol{K}}} \boldsymbol{E}(\boldsymbol{a})$$

By approximating $\boldsymbol{E}(\boldsymbol{a}) \approx \boldsymbol{E}(\boldsymbol{r})$, we obtain the dielectric polarization of the exciton due to its optical generation by the external optical field $\boldsymbol{E}(\boldsymbol{r})$

$$\boldsymbol{P}_{n\boldsymbol{K}}(\boldsymbol{r}) = \frac{2\epsilon_b \omega_{\mathrm{LT}}}{\omega_{n\boldsymbol{K}} - \omega - i\gamma} \frac{\sin \alpha_{n\boldsymbol{K}}}{\alpha_{n\boldsymbol{K}}} \boldsymbol{E}(\boldsymbol{r}) \tag{4.30}$$

Due to such an exciton polarization, the nonlocal material equation relating the electric field $\boldsymbol{E}(\boldsymbol{r})$ and the displacement vector $\boldsymbol{D}(\boldsymbol{r})$ is

$$\boldsymbol{D}(\boldsymbol{r}) = \epsilon_b \boldsymbol{E}(\boldsymbol{r}) + \boldsymbol{P}(\boldsymbol{r}) = \epsilon_{\mathrm{QD}}(\boldsymbol{r})\boldsymbol{E}(\boldsymbol{r}) \tag{4.31}$$

where an effective permittivity can be defined for the QD exciton–polariton

$$\epsilon_{\mathrm{QD}}(\boldsymbol{r}) = \epsilon_b \left(1 + \frac{2\omega_{\mathrm{LT}}}{\omega_{n\boldsymbol{K}} - \omega + i\gamma} \frac{\sin \alpha_{n\boldsymbol{K}}}{\alpha_{n\boldsymbol{K}}}\right) \tag{4.32}$$

By averaging $\epsilon_{\mathrm{QD}}(\boldsymbol{r}, \omega)$ over the QD, we obtain the expression for the effective permittivity of the QD due to the ground-state exciton excited by an external electromagnetic field

$$\epsilon_{\mathrm{QD}} = \frac{1}{\Omega} \int \epsilon_{\mathrm{QD}}(\boldsymbol{r})\, \mathrm{d}\boldsymbol{r} = \epsilon_b \left[1 + \frac{6\omega_{\mathrm{LT}}}{\pi^2 \left(\omega_{n\boldsymbol{K}} - \omega + i\gamma\right)}\right] \tag{4.33}$$

where $\Omega = 4\pi R^3/3$ is the volume of the QD with a radius of R.

There were experimental reports such as Rayleigh scattering at photon frequencies near an exciton resonance[2] that confirm the real and imaginary parts of the dielectric polarizability of a single QD, i.e., Eq. 4.32. Similar results were obtained in CdSe QDs of controlled radius (1–2.5 nm) and the electronic response was instantaneous even for THz frequencies at room temperature.[3]

Equation 4.32 shows that resonant exciton states in QDs provide a source for both the high and negative dielectric constant. In the following sections, we discuss the photonic effect of such exciton polarization in various nanostructures.

4.3 DAMPING RATE

In the previous section we have assumed that the QD stays largely at its ground state and the probability of optical excitation is very small. Let us examine this assumption closely in this section. Due to the external radiation, the time-dependent wave function

$$|\boldsymbol{r}_e, \boldsymbol{r}_h, t\rangle = c_0(t)|\Psi_0(\boldsymbol{r}_e, \boldsymbol{r}_h)\rangle + \sum_n c_n(t)|\Psi_n(\boldsymbol{r}_e, \boldsymbol{r}_h, t)\rangle \tag{4.34}$$

from which

$$\begin{aligned} i\hbar\frac{\mathrm{d}c_n(t)}{\mathrm{d}t} &= c_0(t)\langle\Psi_n(\boldsymbol{r}_e, \boldsymbol{r}_h)|V(t)|\Psi_0(\boldsymbol{r}_e, \boldsymbol{r}_h)\rangle\, \mathrm{e}^{i\omega_n t} \\ i\hbar\frac{\mathrm{d}c_0(t)}{\mathrm{d}t} &= \sum_{n(\neq 0)} c_n(t)\langle\Psi_0(\boldsymbol{r}_e, \boldsymbol{r}_h)|V(t)|\Psi_n(\boldsymbol{r}_e, \boldsymbol{r}_h)\rangle\, \mathrm{e}^{-i\omega_n t} \end{aligned} \tag{4.35}$$

where $V(t)$ is the perturbation due to the optical field. Here we have ignored the exciton wave vector $\boldsymbol{K}$, since it is zero for exciton states confined in a QD. Note that optical couplings among exciton states are zero because of the selection rules. Furthermore, only a limited number of exciton states have nonzero optical couplings with the ground state.

Let $V(t) = V\mathrm{e}^{i\omega t}$. We consider a two-state system

$$\begin{aligned} i\hbar\frac{\mathrm{d}c_0(t)}{\mathrm{d}t} &= c_1(t)\mathrm{e}^{-i(\omega_1-\omega)t}\langle 0|V|1\rangle \equiv \hbar\gamma c_1(t)\mathrm{e}^{-i(\omega_1-\omega)t} \\ i\hbar\frac{\mathrm{d}c_1(t)}{\mathrm{d}t} &= \hbar\gamma\, c_0(t)\mathrm{e}^{i(\omega_1-\omega)t} \end{aligned} \tag{4.36}$$

where $\langle 0|V|1\rangle = \hbar\gamma/2$ simplifies the optical transition matrix, $\hbar\omega_1$ is the eigenvalue of $|1\rangle$, and the eigenvalue of $|0\rangle$ is set to be zero. In terms of the detuning parameter $\delta = \omega - \omega_1$, the above equations read

$$\begin{aligned} i\hbar\frac{\mathrm{d}c_0(t)}{\mathrm{d}t} &= \frac{\hbar\gamma}{2}\, c_1(t)\mathrm{e}^{i\delta t} \\ i\hbar\frac{\mathrm{d}c_1(t)}{\mathrm{d}t} &= \frac{\hbar\gamma}{2}\, c_0(t)\mathrm{e}^{-i\delta t} \end{aligned} \tag{4.37}$$

Differentiating Eq. 4.37 with respect to time yields

$$\frac{\mathrm{d}^2 c_0}{\mathrm{d}t^2} = \frac{\gamma\delta}{2}\, c_1\, \mathrm{e}^{i\delta t} - \frac{i\gamma}{2}\frac{\mathrm{d}c_1}{\mathrm{d}t}\mathrm{e}^{i\delta t} = i\delta\frac{\mathrm{d}c_0}{\mathrm{d}t} - \frac{\gamma^2}{4}\, c_0 \tag{4.38}$$

i.e.,

$$\frac{\mathrm{d}^2 c_0}{\mathrm{d}t^2} = i\delta\frac{\mathrm{d}c_0}{\mathrm{d}t} - \frac{\gamma^2}{4}\, c_0 \tag{4.39}$$

A trial solution of the form

$$c_0 = \mathrm{e}^{-i\chi t}$$

leads to

$$\chi^2 + \delta\chi - \gamma^2 = 0 \quad , \quad \chi = \chi_\pm = -\frac{\delta}{2} \pm \frac{1}{2}\sqrt{\delta^2 + \gamma^2} \tag{4.40}$$

Thus a general solution is

$$c_0 = \mathrm{e}^{i\delta t/2}\left[A\cos\left(\frac{t}{2}\sqrt{\delta^2+\gamma^2}\right) + B\sin\left(\frac{t}{2}\sqrt{\delta^2+\gamma^2}\right)\right] \tag{4.41}$$

By the first equation of Eq. 4.37,

$$\begin{aligned} c_1 &= \frac{2i}{\gamma}\mathrm{e}^{-i\delta t}\frac{\mathrm{d}c_0}{\mathrm{d}t} \\ &= -\mathrm{e}^{-i\delta t/2}\left\{\frac{\delta}{\gamma}\left[A\cos\left(\frac{t}{2}\sqrt{\delta^2+\gamma^2}\right) + B\sin\left(\frac{t}{2}\sqrt{\delta^2+\gamma^2}\right)\right]\right. \\ &\quad \left. +i\frac{\sqrt{\delta^2+\gamma^2}}{\gamma}\left[A\sin\left(\frac{t}{2}\sqrt{\delta^2+\gamma^2}\right) - B\cos\left(\frac{t}{2}\sqrt{\delta^2+\gamma^2}\right)\right]\right\} \end{aligned} \tag{4.42}$$

If the system is in the $|0\rangle$ state at time $t = 0$, i.e., $c_0(0) = 1$ and $c_1(0) = 0$, $A = 1$ and $B = -i\delta/\sqrt{\delta^2+\gamma^2}$. Thus at a later time

$$\begin{aligned} c_0(t) &= \mathrm{e}^{i\delta t/2}\left[\cos\left(\frac{t}{2}\sqrt{\delta^2+\gamma^2}\right) - \frac{i\delta}{\sqrt{\delta^2+\gamma^2}}\sin\left(\frac{t}{2}\sqrt{\delta^2+\gamma^2}\right)\right] \\ c_1(t) &= \mathrm{e}^{-i\delta t/2}\,\sin\left(\frac{t}{2}\sqrt{\delta^2+\gamma^2}\right)\frac{2i\gamma}{\sqrt{\delta^2+\gamma^2}} \end{aligned} \tag{4.43}$$

We can then discuss the probability of an electron transiting from its initial state $|0\rangle$ to $|1\rangle$ under the influence of $V(t)$. Expressing the total wave function as $c_0(t)|0\rangle + c_1(t)|1\rangle$, the probability of electron occupying state $|1\rangle$ at time t is simply

$$|c_1|^2 = \mathrm{e}^{-i\delta t}\left[1 - \cos\left(t\sqrt{\delta^2+\gamma^2}\right)\right]\frac{\gamma^2}{\delta^2+\gamma^2} \tag{4.44}$$

Note that δ is the detuning parameter and $\delta = \omega - \omega_1$; this is much alike the results we obtained in Chapter 2, (see Eq. 2.41 in sub-Section 2.1.2).

We now calculate the matrix element of momentum $\boldsymbol{p}$ (can be any physical parameter)

$$\Big(\langle 1|c_1^*(t) + \langle 0|c_0^*(t)\Big)\boldsymbol{p}\Big(c_0(t)|0\rangle + c_1(t)|1\rangle\Big) = c_0(t)c_1^*(t)\langle 1|\boldsymbol{p}|0\rangle + c_0^*(t)c_1(t)\langle 0|\boldsymbol{p}|1\rangle$$

Note that $\langle 0|\boldsymbol{p}|0\rangle = \langle 1|\boldsymbol{p}|1\rangle = 0$. The above equation shows that the physical variable is to be evaluated by $c_0(t)c_1^*(t)$, which is

$$c_0(t)c_1^*(t) = \mathrm{e}^{i\delta t}\left[\left(\sqrt{\delta^2+\gamma^2} - i\delta\right)\cos\left(t\sqrt{\delta^2+\gamma^2}\right) - i\delta\right]\frac{-i\gamma}{\delta^2+\gamma^2} \tag{4.45}$$

by Eq. 4.43. Including the term $e^{i\omega_1}$ in $\langle 1|\boldsymbol{p}|0\rangle$ (remember that the total wave function of an eigensolution contains such a term, see e.g., Eq. 2.2) and making a time average

$$\left\langle c_0(t)c_1^*(t)\langle 1|\boldsymbol{p}|0\rangle\right\rangle_t = |\langle 1|\boldsymbol{p}|0\rangle|\,\frac{(\omega_1-\omega)\gamma}{(\omega_1-\omega)^2+\gamma^2} \tag{4.46}$$

$\langle\,\rangle_t$ denotes the time average. Note that $\hbar\gamma/2 = \langle 0|V|1\rangle$ is the transition matrix element. It is easy to see that the above equation is equivalent to the real part of the dielectric polarization in Eq. 4.23. Repeating Eq. 4.23 here

$$\boldsymbol{P}_{n\boldsymbol{K}}(\boldsymbol{r},t) = \frac{e^2\boldsymbol{p}_{cv}}{\hbar(\omega_{n\boldsymbol{K}}-\omega-i\gamma)m_0^2\omega_{n\boldsymbol{K}}^2}\,\psi_{n\boldsymbol{K}}(\boldsymbol{r},\boldsymbol{r})\int\psi_{n\boldsymbol{K}}^*(\boldsymbol{r}',\boldsymbol{r}')\boldsymbol{p}_{cv}\cdot\boldsymbol{E}(\boldsymbol{r}',t)\,\mathrm{d}\boldsymbol{r}'$$

where apart from physical constants

$$\frac{1}{\omega_{n\boldsymbol{K}}-\omega-i\gamma} = \frac{\omega_{n\boldsymbol{K}}-\omega}{(\omega_{n\boldsymbol{K}}-\omega)^2+\gamma^2} + i\,\frac{\gamma}{(\omega_{n\boldsymbol{K}}-\omega)^2+\gamma^2}$$

Furthermore,

$$\gamma \propto \langle 0|V|1\rangle \propto \int\psi_{n\boldsymbol{K}}^*(\boldsymbol{r}',\boldsymbol{r}')\boldsymbol{p}_{cv}\cdot\boldsymbol{E}(\boldsymbol{r}',t)\,\mathrm{d}\boldsymbol{r}'$$

Here we see the damping rate of the initial state due to the excitation of the excited states. A more advanced treatment of multiple states (we considered only two states in the above derivations) can be done by the generalized Golden rule, which will result in a damping rate due to the interactions among all the states; see Chapter 2 and Ref.[4]

We now calculate the energy dissipation. First we consider Poynting's theorem describing conservation of energy

$$\oint \boldsymbol{E}\times\boldsymbol{H}\cdot\mathrm{d}\boldsymbol{S} = -\int\left[\boldsymbol{E}\cdot\frac{\partial\boldsymbol{D}}{\partial t}+\boldsymbol{H}\cdot\frac{\partial\boldsymbol{B}}{\partial t}\right]\mathrm{d}\Omega \tag{4.47}$$

where $\boldsymbol{S}$ denotes the surface and Ω the volume. Denote $\epsilon = \epsilon' + i\epsilon''$ and $\mu = \mu' + i\mu''$. For monochromatic light with a frequency ω, an amplitude E_0 for its $\boldsymbol{E}$ field and H_0 for $\boldsymbol{H}$ field, the time average of Poynting's theorem in Eq. 4.47 gives

$$\oint\langle\boldsymbol{E}\times\boldsymbol{H}\rangle_t\cdot\mathrm{d}\boldsymbol{S} = -\int\left[\omega\epsilon''\frac{E_0^2}{2}+\omega\mu''\frac{H_0^2}{2}\right]\mathrm{d}\Omega \tag{4.48}$$

Loss requirements are satisfied through the signs of ϵ'' and μ''. Assuming $\mu''=0$ and a small volume $\Delta\Omega$, the power dissipated is thus

$$-\oint\langle\boldsymbol{E}\times\boldsymbol{H}\rangle_t\cdot\mathrm{d}\boldsymbol{S} = \frac{E_0^2}{2}\omega\epsilon''\Delta\Omega \tag{4.49}$$

On the other hand, consider a thin slab of material with thickness Δz and normal surface area A, and a plane wave travelling in the z-direction with complex wave number $k = k' + ik''$. The time average Poynting vector, assuming a single travelling wave, is

$$\boldsymbol{S} = \boldsymbol{z}_0\frac{E_0^2}{2}\Re\left\{\frac{1}{\eta^*}e^{-2k''z}\right\} \tag{4.50}$$

where $\eta = \sqrt{\mu/\epsilon}$ is the complex wave impedance and $\boldsymbol{z}_0$ is the unit vector of the z direction. $\Re A$ denotes the real part of a complex variable A. The closed surface integral of the Poynting vector, i.e., the power dissipated in $\Delta\Omega = A\Delta z$, is thus

$$P_d = \frac{AE_0^2}{2}\Re\left\{\frac{1}{\eta^*}\left(1 - \mathrm{e}^{-2k''\Delta z}\right)\right\} \tag{4.51}$$

Writing $\eta = \sqrt{\mu_0/\epsilon}$ and assuming Δz very small so that $\mathrm{e}^{-2k''\Delta z} \ll 1$,

$$P_d = \frac{AE_0^2}{2\sqrt{\mu_0}}\Re\left\{2k''\Delta z(\sqrt{\epsilon})^*\right\} \tag{4.52}$$

Since $k = \omega\sqrt{\epsilon}/c$, where c is the speed of light,

$$\Re\left\{k''(\sqrt{\epsilon})^*\right\} = \frac{\omega\epsilon''}{2c} \tag{4.53}$$

so that

$$P_d = \frac{AE_0^2}{2}\omega\epsilon''\Delta z = \frac{E_0^2}{2}\omega\epsilon''\Delta\Omega \tag{4.54}$$

as Eq. 4.49. Therefore, the energy and Poynting vector descriptions for dissipation are in agreement, as they must be from Eq. 4.47.

We now calculate the optical loss of the external electromagnetic field due to the QD exciton excitation by writing the continuity equation for the photon density

$$\frac{\mathrm{d}n_{\hbar\omega}(z)}{\mathrm{d}t} = \frac{\partial n_{\hbar\omega}(z)}{\partial t} - \frac{\partial\left[c\, n_{\hbar\omega}(z)\right]}{\partial z} \tag{4.55}$$

Here we consider a beam of photons travelling along the z-axis. The first term on the right side of the above equation represents the absorption rate of photons and the second term the photons leaving due to the photon flux. c is the speed of light in the medium. We have used the above equation in Section 2.5 when discussing the optical spectrum of the materials. At steady state we have, in general

$$n_{\hbar\omega}(z) = n_{\hbar\omega}(0)e^{-\epsilon''(\omega)z} \tag{4.56}$$

which defines the optical loss $\epsilon''(\omega)$ as in Eqs. 4.49 and 4.54. Since

$$\frac{\partial n_{\hbar\omega}(z)}{\partial t} = p_{n\boldsymbol{K}\leftarrow 0}(1 - f_{n\boldsymbol{K}})f_0 - p_{0\leftarrow n\boldsymbol{K}}(1 - f_0)f_{n\boldsymbol{K}} \tag{4.57}$$

where f_0 and $f_{n\boldsymbol{K}}$ are the occupations of state Ψ_0 and $\Psi_{n\boldsymbol{K}}$, respectively. $f_0 = |c_0|^2$ and $f_{n\boldsymbol{K}} = |c_{n\boldsymbol{K}}(\infty)|^2$. $p_{n\boldsymbol{K}\leftarrow 0}$ is the transition probability per unit time

$$\begin{aligned} p_{n\boldsymbol{K}\leftarrow 0} &= |c_{n\boldsymbol{K}}(\infty)|^2\gamma \\ &= \frac{\mathrm{e}^2\gamma}{\hbar^2\left[(\omega_{n\boldsymbol{K}} - \omega)^2 + \gamma^2\right]\omega_{n\boldsymbol{K}}^2 m_0^2}\left[\int \psi_{n\boldsymbol{K}}(\boldsymbol{r}', \boldsymbol{r}')\boldsymbol{p}_{cv} \cdot \boldsymbol{E}(\boldsymbol{r}')\,\mathrm{d}\boldsymbol{r}'\right]^2 \end{aligned} \tag{4.58}$$

At the steady state, $\mathrm{d}n_{\hbar\omega}(z)/\mathrm{d}t = 0$,

$$\epsilon'' = \frac{\mathrm{e}^2\gamma\left(f_0 - f_{n\boldsymbol{K}}\right)}{\hbar^2\left[(\omega_{n\boldsymbol{K}} - \omega)^2 + \gamma^2\right]c\, n_{\hbar\omega}\omega_{n\boldsymbol{K}}^2 m_0^2}\left[\int \psi_{n\boldsymbol{K}}(\boldsymbol{r}', \boldsymbol{r}')\boldsymbol{p}_{cv} \cdot \boldsymbol{E}(\boldsymbol{r}')\,\mathrm{d}\boldsymbol{r}'\right]^2 \tag{4.59}$$

We assume again that the electric field inside the QD is uniform and note that the energy density of the electromagnetic field equals $\epsilon_\infty E^2 = \hbar\omega n_{\hbar\omega}/\Omega$, where $\Omega = 4\pi R^3/3$ is the volume of the QD,

$$\epsilon'' = \frac{e^2 p_{cv}^2 \gamma\omega\Big(f_0 - f_{n\boldsymbol{K}}\Big)}{\Omega\hbar\Big[(\omega_{n\boldsymbol{K}} - \omega)^2 + \gamma^2\Big] c\, \epsilon_\infty \omega_{n\boldsymbol{K}}^2 m_0^2} \left[\int \psi_{n\boldsymbol{K}}(\boldsymbol{r}', \boldsymbol{r}')\, \mathrm{d}\boldsymbol{r}'\right]^2 \tag{4.60}$$

By inserting Eq. 3.92 into the above equation, we finally have

$$\epsilon'' = \frac{6\gamma\omega\omega_{LT}\Big(f_0 - f_{n\boldsymbol{K}}\Big)}{\pi^2 c\Big[(\omega_{n\boldsymbol{K}} - \omega)^2 + \gamma^2\Big]} = \frac{6\gamma\omega\omega_{LT}}{\pi^2 c\Big[(\omega_{n\boldsymbol{K}} - \omega)^2 + \gamma^2\Big]} \tag{4.61}$$

so that by Eq. 4.53

$$\epsilon''(\omega) = \frac{6\gamma\omega_{LT}}{\pi^2\Big[(\omega_{n\boldsymbol{K}} - \omega)^2 + \gamma^2\Big]} \tag{4.62}$$

per QD. The last equality in Eq. 4.61 is obtained, since $f_{n\boldsymbol{K}} \ll 1$ and $f_0 \approx 1$. Note that Eq. 4.62 is exactly the imaginary part of the complex effective permittivity of Eq. 4.33.

4.4 MICROCAVITY AND FOUR-WAVE MIXING

By putting a QD in a photonic crystal (PhC) microcavity, Gevaux *et al.* showed that by manipulating the density of optical states in the environment of the QD, the rate of spontaneous recombination was altered by the Purcell effect. In their work, temperature was used to spectrally tune an InAs self-assembled QD exciton and the mode of a PhC defect cavity.[5] Strong coupling was also manifested in PL data that display anti-crossings between the QD exciton and cavity-mode dispersion relations.[6] Radiative recombination of excitons in single QDs embedded in microcavities was experimentally probed and a fivefold enhancement of spontaneous emission outside the microcavity was found for QDs in resonance with the cavity mode, relative to detuned QDs.[7] The QD interacting with a PhC nanocavity was quantified theoretically to be in a strong coupling regime, which makes the QD–PhC nanocavity very promising for quantum-electrodynamics applications at the nanoscale level.[8] The exciton–polariton modes at finite in-plane momentum of a QD lattice embedded in a planar optical cavity were shown to be capable of being guided with long lifetimes.[9] Quite interestingly, Kessler *et al.* further calculated and showed that the effective mass for the exciton-polariton in a 2D QD lattice at various points in the Brillouin zone can be as low as $10^{-8}m_0$, where m_0 is the electron rest mass.

Over the past 10 years a new family of optoelectronic devices has emerged whose performance is enhanced by placing the active device structure inside a Fabry-Perot resonant microcavity. In such structures, the device functions largely as before, but is subject to the effects of the cavity, mainly wavelength selectivity and a large enhancement of the resonant optical field.[10] Photon confinement in microcavities with embedded QWs has considerably enhanced the exciton–photon coupling.[11–17] Quantum microvacities can thus be particularly interesting for nonlinear optics, since the nonlinear response is characterized by a stronger dependence upon the coupling constant.

The increased optical field allows photodetectors to be made thinner and therefore faster, while simultaneously increasing the quantum efficiency at the resonant wavelength. Since off-resonance wavelengths are rejected by the cavity, the photodetectors have both wavelength selectivity and high-speed response, making them ideal for wavelength division multiplexing applications. Optical modulators situated in a resonant cavity require fewer QWs to absorb the same fraction of the incident light, and can therefore operate at lower voltages. In the case of emitters, the cavity modifies the spontaneous emission of light-emitting diode, improving their spectral purity and directivity. For these reasons and others, devices of this genre are referred to as Resonant Cavity Enhanced.

Here, we present a theory of degenerate four-wave mixing in a semiconductor planar microcavity containing a QW between the cavity mirrors. A sequence of two coherent pulses with the wave vectors $\boldsymbol{k}_1$ and $\boldsymbol{k}_2$ creates a third-order dielectric polarization, which serves as a source for a new light wave propagating in the direction $(2\boldsymbol{k}_2 - \boldsymbol{k}_1)$. Note that first experiments on four-wave mixing response of QWs embedded in GaAs/AlGaAs Fabry-Perot microvacities have been recently reported.[18, 19]

We consider a multilayered heterostructure of Fig. 4.1 grown by using five compositional materials labeled as A, B, C, D, and F and consisting of N_l pairs of the C/B distributed Bragg reflector (DBR), an active region A with a QW D in its center, and N_r pairs of the B/C DBR grown on the substrate F.

The optical properties of the DBRs are characterized by the following amplitude reflection and transmission coefficients: r_{mj}, r'_{mj} (reflection from the left, $j = l$, and right, $j = r$, DBR for the light incident, respectively, from the active layer and from the external medium, vacuum or substrate) and t_{mj}, t'_{mj} (the same but for transmission through the DBR j). The reflection coefficient r_{mj} referred to the

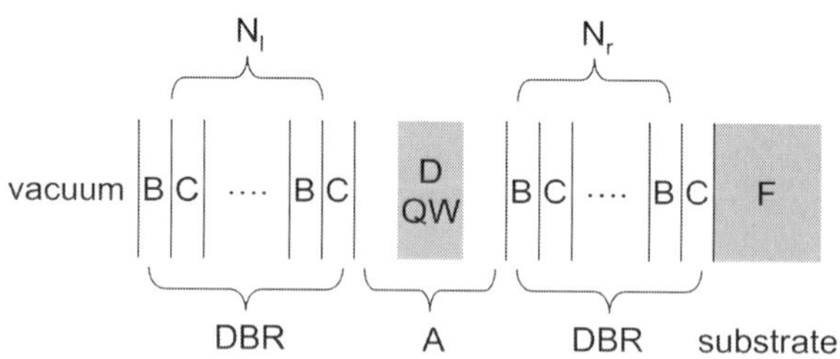

Fig. 4.1 Schematic structure of the QW cavity. See also Color Insert.

center of the cavity active layer can be approximately written in the form

$$\begin{aligned} r_{mj}(\omega) &\equiv |r_{mj}|e^{i\psi} \\ \psi &\approx N_c\pi + n_c\frac{\tilde{L}+L_c}{c}(\omega-\omega_c) \end{aligned} \tag{4.63}$$

where N_c is an integer (in the following an even integer in order to have an antinode in the cavity center), L_c is the thickness of the central layer,

$$\tilde{L} = \frac{\pi c}{\omega_c n_c^2}\frac{n_1 n_2}{n_2 - n_1}$$

the frequency ω_c is chosen to satisfy the conditions

$$\begin{aligned} n_c\frac{\omega_c}{c}L_c &= N_c\pi \\ n_1\frac{\omega_c}{c}a_1 &= n_2\frac{\omega_c}{c}a_2 = \frac{\pi}{2} \end{aligned}$$

where n_c is the refractive index of the active layer, n_1, n_2 and a_1, a_2 are the refractive indices and thicknesses of the layers in the DBRs, respectively. We assume $n_1 < n_2$ so that $\tilde{L}$ is positive. Since the DBRs are supposed to be transparent, the reflection and transmission coefficients are related by

$$\frac{n_{ext,j}}{n_c}|t_{mj}|^2 = \frac{n_c}{n_{ext,j}}|t'_{mj}|^2 = 1 - |r_{mj}|^2 = 1 - |r'_{mj}|^2 \tag{4.64}$$

where $n_{ext,j}$ is the refractive index of the outside medium close to DBR j.

The basic equations describing dynamics of the photonic mode in the microcavity and the exciton in the QW's can be written as

$$\begin{aligned} \left[\frac{\partial}{\partial t} + i(\omega_0 - \bar{\omega}) + \Gamma\right]\bar{P}(t) &= i\eta\Gamma_0\bar{E}(t) + \bar{F}_{NL}(t) \\ \left[\frac{\partial}{\partial t} + i(\omega_c - \bar{\omega}) + \gamma_c\right]\bar{E}(t) &= \tilde{\Gamma}\left[\frac{i}{\eta}\bar{P}(t) + \frac{t'_{ml}}{2}\bar{E}_0(t)\right] \end{aligned} \tag{4.65}$$

Here $\bar{\omega}$ is the current frequency of the light pulses, ω_0 and ω_c are the exciton and bare-photon-mode resonance frequencies, and Γ and Γ_0 are the exciton nonradiative and radiative damping rates in a single QW. γ_c is the photon-mode damping rate determining the cavity line width; it is a sum of the escape rates through the left and right DBRs

$$\gamma_c = \gamma_{cl} + \gamma_{cr} \tag{4.66}$$

where

$$\begin{aligned} \gamma_{cj} &\approx \frac{(1-|r_{mj}|)\tilde{\Gamma}}{4} \approx \frac{(1-|r_{mj}|^2)\tilde{\Gamma}}{8} \\ \tilde{\Gamma} &\equiv \frac{2c}{n_c(\tilde{L}+L_c)} \end{aligned} \tag{4.67}$$

$\bar{P}(t)$ is the slowly varying amplitude of the exciton polarization averaged over the QW width a, $\bar{E}(t)$ is the electric field in the center of the QW, $\bar{E}_0(t)$ is the slowly varying amplitude of the initial radiation. It consists of two pulses and, say, for Lorentzian pulses is given by

$$E_{0,1} \exp\left(ik_{1,x}x - \frac{|t-t_1|}{\tau_p}\right) + E_{0,2} \exp\left(ik_{2,x}x - \frac{|t-t_2|}{\tau_p}\right)$$

if (x, z) is the plane of incidence and $\boldsymbol{k}_j = (k_{jx}, 0, k_{jz})$. Usually in four-wave mixing experiments the $\boldsymbol{k}_1$ and $\boldsymbol{k}_2$ vectors make small angles with the sample normal z direction, and one may ignore an effect of the z component of the electric field as well as neglect angle dependences of the reflection, transmission, and nonlinear mixing coefficients. In the following we omit everywhere the exponential functions $\exp(ik_{j,x}x)$ bearing in mind that E_1, E_2, and E_3 describe the amplitudes of the three waves $\boldsymbol{k}_1$, $\boldsymbol{k}_2$, and $\boldsymbol{k}_3 = 2\boldsymbol{k}_2 - \boldsymbol{k}_1$.

Other notations introduced above are as follows

$$\eta = \frac{\varepsilon_b}{2\pi} \frac{1}{ka}, \; k = \frac{\omega_0}{c} n_c \tag{4.68}$$

Below we neglect the difference between n_c^2 and the background dielectric constant ε_b of the QW material. While deriving Eq. 4.65 we considered thin enough QWs satisfying the condition of $ka \ll 1$.

The term $\bar{F}_{NL}(t)$ in Eq. 4.65 is a nonlinear part giving rise to the third-order polarization. We first consider the so-called anharmonic-oscillator-like, P^3, and two-level-like, P^2E, contributions

$$\bar{F}_{NL}(t) = i|\bar{P}(t)|^2 \left[\beta_1 \bar{P}(t) + \beta_2 \bar{E}(t)\right] \tag{4.69}$$

where β_1 and β_2 are constant coefficients. According to Eq. 4.69 the polarization, $\bar{P}_3$, of the wave propagating in the $(2\boldsymbol{k}_2 - \boldsymbol{k}_1)$ direction is induced by the term

$$\bar{F}_{NL}(t) = i\bar{P}_1^*(t)\bar{P}_2(t) \left[\beta_1 \bar{P}_2(t) + \beta_2 \bar{E}_2(t)\right] \tag{4.70}$$

where the indices 1, 2 label the first and second laser pulses propagating in the directions of $\boldsymbol{k}_1$ and $\boldsymbol{k}_2$.

From the homogeneous equations corresponding to

$$\bar{E}_0(t) = 0$$

and

$$\bar{F}_{NL}(t) = 0$$

one can derive the equation for eigenfrequencies of the coupled modes, or exciton–polaritons:

$$(\omega - \omega_0 + i\Gamma)(\omega - \omega_c + i\gamma_c) = \tilde{\Gamma}\Gamma_0 \tag{4.71}$$

Turning now to the calculation of the four-wave mixing signal, it is worth to mention that the calculation procedure consists of solving

- Eq. 4.65 for the polarization $\bar{P}_1$ and the field $\bar{E}_1$ of the wave $\boldsymbol{k}_1$ induced by the initial pulse $E_{0,1}$,

- Eq. 4.65 for the response $\bar{P}_2$, $\bar{E}_2$ to the pulse $E_{0,2}$, and
- the linear equations for $\bar{P}_3$, $\bar{E}_3$ of the diffracted wave with the inhomogeneous term in the form of Eq. 4.70.

The outgoing electric field is expressed in terms of the field in the QW center as

$$\bar{E}_l(t) = \frac{1}{2} t_{ml} \bar{E}_3(t)$$
$$\bar{E}_r(t) = \frac{1}{2} t_{mr} \bar{E}_3(t) \tag{4.72}$$

where the indices l, r refer to the left- and right-hand-side external media. Let

$$\tau = t_2 - t_1$$

be the inter-pulse delay and put $t_2 = 0$; the time-integrated signal has the form

$$S_3(\tau) \equiv \int_{-\infty}^{+\infty} |\bar{E}_3(t)|^2 \, dt \tag{4.73}$$

Figure 4.2 illustrates the calculated time-dependent signal $|E_3(t)|^2$, while Fig. 4.3 is the nonlinear signal $S_3(\tau)$. For numerical computation, we used the $Al_{0.2}Ga_{0.8}As$ λ-cavity parameters of Ref.[11] with the DBRs comprising 24 $Al_{0.4}Ga_{0.6}As/AlAs$ stacks on the cavity front side (air interface) and 33 stacks on the substrate side. The refractive indices are: $n_{ext,l} = 1$, $n_{ext,r} = 3.63$, $n_1 = 3.17$, $n_2 = 3.45$, $n_c = 3.54$. The calculated values of γ_c and $\tilde{\Gamma}$ are 0.51 and 117 ps^{-1}, respectively. A single QW was assumed to be embedded in the center of the active layer, and values of $\Gamma = 1.0\,\text{ps}^{-1}$ and $\Gamma_0 = 0.05\,\text{ps}^{-1}$ were chosen for the exciton nonradiative and radiative damping rates. In this case the Rabi splitting 2Ω is $4.8\,\text{ps}^{-1}$. The signals in Figs. 4.2 and 4.3 exhibit quantum beats between the exciton and cavity modes, and a significant beat-like modulation is exposed in the dependences on both t and τ. The real time

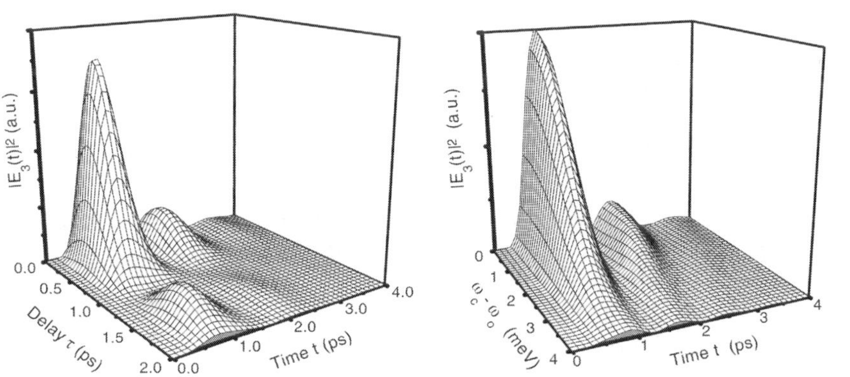

Fig. 4.2 (a) Time-resolved $2\boldsymbol{k}_2 - \boldsymbol{k}_1$ signal for different time delays between light pulses. (b) Effect of bare-photon-mode detuning from QW-exciton transition on time-resolved four-wave-mixing signal. $\tau = 0.5$ ps. Y. Fu, M. Willander, E. I. Ivchenko, and A. A. Kiselev, Four-wave mixing in microcavities with embedded quantum wells, Phys. Rev. B, vol. 55, p. 9872–9, 1997. Copyright (1997) by The American Physical Society.

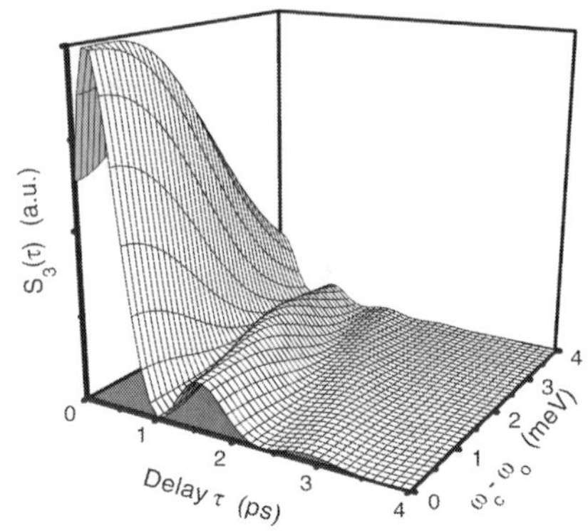

Fig. 4.3 Time-integrated nonlinear signal. Y. Fu, M. Willander, E. I. Ivchenko, and A. A. Kiselev, Four-wave mixing in microcavities with embedded quantum wells, Phys. Rev. B, vol. 55, p. 9872–9, 1997. Copyright (1997) by The American Physical Society.

t is referred to as the peak of pulse 2. The pattern reveals the damped oscillatory behavior in both τ- and t-dependences.

4.5 EXCITONIC PHOTONIC GAP

At present the study of photonic crystals,[20] i.e., dielectric materials with a periodicity matching optical wavelengths, is a subject of active research. Electron waves travelling in the periodic potential of a crystal are arranged into energy bands separated by gaps in which propagating states are prohibited. It is interesting to see whether analogous bands exist when electromagnetic waves propagate in a periodic dielectric structure. If such a bandgap exists, electromagnetic waves with frequencies inside the gap cannot propagate in any direction inside the material. These frequency gaps are referred as "photonic bandgaps."

Photonic bandgaps can have a profound impact on many areas in pure and applied physics.

1. Due to the absence of optical modes in the gap, spontaneous emission is suppressed for photons with frequencies in the forbidden region.
2. By tuning the photonic bandgap to overlap with the electronic band edge, the electron–hole recombination process can be controlled in a photonic bandgap material, leading to enhanced efficiency and reduced noise in the operation of semiconductor lasers and other solid state devices.
3. The suppression of spontaneous emission can also be used to prolong the lifetime of selected chemical species in catalytic processes.
4. Besides technical applications in various areas, scientists are interested in the possibility of observing the localization of electromagnetic waves by introducing defects and disorder in a photonic bandgap material.

5. Electromagnetic interaction governs many properties of atoms, molecules, and solids. The absence of electromagnetic modes and zero point fluctuations inside the photonic gap can lead to unusual physics phenomena; e.g., atoms or molecules embedded in such a material can be locked in excited states if the photons to be emitted have frequency within the forbidden gap.

Excitons in a QD lattice tuned into resonance by an external electromagnetic field can form coherent exciton–polariton states, which show clear photonic band structure. Equation 4.32 was coupled to an electromagnetic field to study the optical dispersion of exciton–polariton crystals[21, 22] stimulated by the experimental works of face-centered cubic (fcc) structured silica opals in Ref.[23] Maxwell equations describing the incident electromagnetic field in the QD photonic crystal, free of charges and free of drift-diffusion current, are written as

$$\begin{aligned} \nabla \times [\nabla \times \boldsymbol{E}(\boldsymbol{r})] &= \mu_0 \omega^2 \boldsymbol{D}(\boldsymbol{r}) \\ \nabla \cdot \boldsymbol{D}(\boldsymbol{r}) &= 0 \end{aligned} \tag{4.74}$$

in the MKS unit system, where $\boldsymbol{E}(\boldsymbol{r})$ is the electric field and $\boldsymbol{D}(\boldsymbol{r})$ is the displacement vector, ω is the angular frequency of the incident electromagnetic field. Here we have assumed that the QD consists of a uniform isotropic linear medium, for which $\boldsymbol{D} = \epsilon \boldsymbol{E}$ and $\boldsymbol{B} = \mu_0 \boldsymbol{H}$. $\boldsymbol{H}$ is the magnetizing field and $\boldsymbol{B}$ is the magnetic field. μ_0 is the magnetic constant (permeability of free space). The nonlocal material equation relating $\boldsymbol{D}(\boldsymbol{r})$ and $\boldsymbol{E}(\boldsymbol{r})$ is[17]

$$\boldsymbol{D}(\boldsymbol{r}) = \epsilon_b \boldsymbol{E}(\boldsymbol{r}) + \boldsymbol{P}(\boldsymbol{r}) \equiv \epsilon_b \boldsymbol{E}(\boldsymbol{r}) + \sum_{\boldsymbol{a}} \boldsymbol{P}_{\boldsymbol{a}}(\boldsymbol{r}) \tag{4.75}$$

where ϵ_b is the background dielectric constant, and $\boldsymbol{P}_{\boldsymbol{a}}(\boldsymbol{r})$ denotes the exciton contribution from the QD centered at $\boldsymbol{a}$ to the dielectric polarization expressed by Eq. 4.25. Note that the overlap of exciton envelope functions centered at different QDs has been neglected so that excitons excited in different QDs are assumed to be coupled only via the electromagnetic field, and the total polarization $\boldsymbol{P}(\boldsymbol{r})$ is a simple sum of all QDs' contributions $\boldsymbol{P}_{\boldsymbol{a}}(\boldsymbol{r})$.

It follows from the second Eq. 4.74 and Eq. 4.75 that

$$\nabla \cdot \boldsymbol{E}(\boldsymbol{r}) = -\frac{1}{\epsilon_b} \nabla \cdot \boldsymbol{P}(\boldsymbol{r}) \tag{4.76}$$

Furthermore,

$$\nabla \times (\nabla \times \boldsymbol{A}) \equiv \nabla(\nabla \cdot \boldsymbol{A}) - \nabla^2 \boldsymbol{A} \tag{4.77}$$

for any vector $\boldsymbol{A}$ so that the first Eq. 4.74 can be rewritten as

$$\nabla^2 \boldsymbol{E}(\boldsymbol{r}) + k^2 \boldsymbol{E}(\boldsymbol{r}) = -\frac{k_0^2}{\epsilon_0} \left\{ \boldsymbol{P}(\boldsymbol{r}) + \frac{1}{k^2} \nabla \Big[\nabla \cdot \boldsymbol{P}(\boldsymbol{r}) \Big] \right\} \tag{4.78}$$

where $k_0 = \omega/c$, $k = k_0 n_b = \omega n_b / c$, and $n_b = \sqrt{\epsilon_b/\epsilon_0}$.

When we order the QDs in a periodic lattice denoted by lattice vector $\boldsymbol{a}$, a reciprocal lattice $\boldsymbol{b}$ can be defined, as we discussed in Section 1.2. For such a periodic QD lattice, we seek for Bloch-like solutions of Eq. 4.78 satisfying

$$\boldsymbol{E}_{\boldsymbol{q}}(\boldsymbol{r}+\boldsymbol{a}) = \mathrm{e}^{i\boldsymbol{q}\cdot\boldsymbol{a}}\boldsymbol{E}_{\boldsymbol{q}}(\boldsymbol{r}) \tag{4.79}$$

for which we expand the vector function $\boldsymbol{E}_{\boldsymbol{q}}(\boldsymbol{r})$ in the Fourier series

$$\boldsymbol{E}_{\boldsymbol{q}}(\boldsymbol{r}) = \sum_{\boldsymbol{b}} \mathrm{e}^{i(\boldsymbol{q}+\boldsymbol{b})\cdot\boldsymbol{r}}\boldsymbol{E}_{\boldsymbol{q}+\boldsymbol{b}} \tag{4.80}$$

where $\boldsymbol{b}$ are the reciprocal lattice vectors. It is easy to see that

$$\begin{aligned}\boldsymbol{E}_{\boldsymbol{q}}(\boldsymbol{r}+\boldsymbol{a}) &= \sum_{\boldsymbol{b}} \mathrm{e}^{i(\boldsymbol{q}+\boldsymbol{b})\cdot(\boldsymbol{r}+\boldsymbol{a})}\boldsymbol{E}_{\boldsymbol{q}+\boldsymbol{b}}\\ &= \mathrm{e}^{i\boldsymbol{q}\cdot\boldsymbol{a}}\sum_{\boldsymbol{b}} \mathrm{e}^{i(\boldsymbol{q}+\boldsymbol{b})\cdot\boldsymbol{r}}\boldsymbol{E}_{\boldsymbol{q}+\boldsymbol{b}} = \mathrm{e}^{i\boldsymbol{q}\cdot\boldsymbol{a}}\boldsymbol{E}_{\boldsymbol{q}}(\boldsymbol{r})\end{aligned} \tag{4.81}$$

as required by Eq. 4.79. Here we have utilized the relationship of $\mathrm{e}^{i\boldsymbol{b}\cdot\boldsymbol{a}} \equiv 1$ between real- and reciprocal-space vectors.

4.5.1 *Atomic Lattice*

It is interesting to compare the QD exciton lattice with an atomic lattice in vacuum, which has

$$\begin{aligned}\epsilon_b &= 1\\ \boldsymbol{P} &= 4\pi\sum_{\boldsymbol{a}}\delta(\boldsymbol{r}-\boldsymbol{a})\alpha(\omega)\boldsymbol{E}(\boldsymbol{r})\\ \alpha(\omega) &= \frac{3c_0^3\Delta}{\omega_0^2}\frac{1}{\omega_0^2-\omega^2}\end{aligned} \tag{4.82}$$

The dispersion relation of such an atomic lattice can be written as

$$\mathrm{Det}||\delta_{\alpha\beta} - R_{\alpha\beta}|| = 0 \tag{4.83}$$

where $\alpha, \beta = x, y, z$, and

$$R_{\alpha\beta} = \frac{\alpha(\omega)k_0^2}{v_0}\sum_{\boldsymbol{b}}\left[\delta_{\alpha\beta} - \frac{(q_\alpha+b_\alpha)(q_\beta+b_\beta)}{k_0^2}\right]\frac{1}{|\boldsymbol{q}+\boldsymbol{b}|^2-k_0^2} \tag{4.84}$$

where v_0 is the lattice unit cell volume. For an fcc lattice $v_0 = a^3/4$ when a is the lattice constant.

Let us derive Eqs. 4.83 and 4.84. For this purpose we substitute Eq. 4.80 into the second Eq. 4.82 and obtain

$$\begin{aligned}\boldsymbol{P}(\boldsymbol{r}) &= \sum_{\boldsymbol{a}}\delta(\boldsymbol{r}-\boldsymbol{a})\alpha(\omega)\sum_{\boldsymbol{b}}\mathrm{e}^{i(\boldsymbol{q}+\boldsymbol{b})\cdot\boldsymbol{r}}\boldsymbol{E}_{\boldsymbol{q}+\boldsymbol{b}}\\ &= \alpha(\omega)\sum_{\boldsymbol{b}}\boldsymbol{E}_{\boldsymbol{q}+\boldsymbol{b}}\sum_{\boldsymbol{a}}\mathrm{e}^{i\boldsymbol{q}\cdot\boldsymbol{a}}\delta(\boldsymbol{r}-\boldsymbol{a})\end{aligned} \tag{4.85}$$

where the identity $e^{i\boldsymbol{b}\cdot\boldsymbol{a}} = 1$ is used. We further substitute Eq. 4.80 and 4.85 into Eq. 4.78, multiply by $e^{-i(\boldsymbol{q}+\boldsymbol{b}')\cdot\boldsymbol{r}}$ with a particular $\boldsymbol{b}'$, and integrate over the sample volume. The result reads

$$\left(k_0^2 - |\boldsymbol{q}+\boldsymbol{b}|^2\right) E_{\boldsymbol{q}+\boldsymbol{b},\alpha} = -\frac{\alpha(\omega)k_0^2}{v_0\epsilon_0}\left[\delta_{\alpha\beta} - \frac{(q_\alpha + b_\alpha)(q_\beta + b_\beta)}{k_0^2}\right]\sum_{\boldsymbol{b}'} E_{\boldsymbol{q}+\boldsymbol{b}',\beta} \quad (4.86)$$

Note that for the atomic lattice in vacuum, $\epsilon_b = 1$ so that $k = k_0$.

In the following we use a notation $S_{\alpha\beta}(\boldsymbol{q}+\boldsymbol{b})$ for

$$\delta_{\alpha\beta} - \frac{(q_\alpha + b_\alpha)(q_\beta + b_\beta)}{k_0^2}$$

Then Eq. 4.86 can be rewritten in the form

$$E_{\boldsymbol{q}+\boldsymbol{b},\alpha} = -\frac{\alpha(\omega)k_0^2}{v_0\epsilon_0}\frac{S_{\alpha\beta}(\boldsymbol{q}+\boldsymbol{b})}{k_0^2 - |\boldsymbol{q}+\boldsymbol{b}|^2}\sum_{\boldsymbol{b}'} E_{\boldsymbol{q}+\boldsymbol{b}',\beta} \quad (4.87)$$

Summing Eq. 4.87 over $\boldsymbol{b}$ and denoting

$$\sum_{\boldsymbol{b}'} E_{\boldsymbol{q}+\boldsymbol{b}',\beta} \equiv E_\beta \quad (4.88)$$

we obtain

$$E_\alpha = -\frac{\alpha(\omega)k_0^2}{v_0\epsilon_0}\sum_{\boldsymbol{b}}\frac{S_{\alpha\beta}(\boldsymbol{q}+\boldsymbol{b})}{k_0^2 - |\boldsymbol{q}+\boldsymbol{b}|^2} E_\beta \quad (4.89)$$

which is nothing more than the equation $E_\alpha = \sum_\beta R_{\alpha\beta}E_\beta$ with the matrix $R_{\alpha\beta}$ introduced by Eq. 4.84. Let us introduce the notations

$$\Omega = \frac{\omega}{\omega_0}\ ,\ Q = \frac{2\Delta}{\omega_0}\ ,\ P = \left(\frac{\pi\sqrt{3}c_0}{a\omega_0}\right)^3 \quad (4.90)$$

then Eq. 4.84 can be transformed into

$$R_{\alpha\beta} = \frac{8}{\pi^2\sqrt{3}}\frac{P}{Q}\frac{\Omega^2}{\Omega^2 - 1}\sum_{\boldsymbol{b}}\frac{S_{\alpha\beta}(\boldsymbol{q}+\boldsymbol{b})}{\Omega^2 - \Omega^2(\boldsymbol{q}+\boldsymbol{b})} \quad (4.91)$$

where

$$\Omega(\boldsymbol{q}+\boldsymbol{b}) = \frac{c_0|\boldsymbol{q}+\boldsymbol{b}|}{\omega_0} \quad (4.92)$$

While deriving Eq. 4.91 we used, for fcc lattices, $v_0 = a^3/4$ and

$$\frac{\alpha(\omega)k_0^2}{v_0}\frac{1}{k_0^2 - |\boldsymbol{q}+\boldsymbol{b}|^2} = \frac{8}{\pi^2\sqrt{3}}\frac{P}{Q}\frac{\Omega^2}{\Omega^2 - 1}\frac{1}{\Omega^2 - \Omega^2(\boldsymbol{q}+\boldsymbol{b})} \quad (4.93)$$

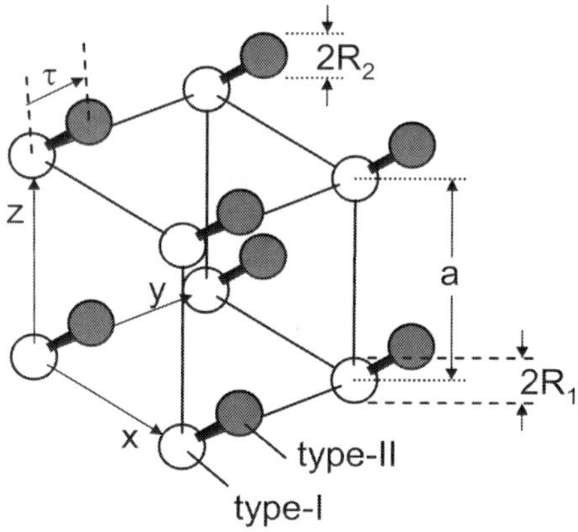

Fig. 4.4 Schematic QD binary compound lattice in a primitive cubic lattice structure. R_1 and R_2 denote the radii of the QDs and a the lattice constant of the primitive cubic lattice.

4.5.2 *Photonic Band in QD Lattice*

As we discussed in Section 4.2, the dielectric polarization due to the exciton photogeneration in a QD can be expressed by Eq. 4.25, which is repeated as

$$\boldsymbol{P}_{\boldsymbol{a}}(\boldsymbol{r}) = T(\omega)\psi_{\boldsymbol{a}}(\boldsymbol{r},\boldsymbol{r}) \int \psi^*_{\boldsymbol{a}}(\boldsymbol{r}',\boldsymbol{r}')\boldsymbol{E}(\boldsymbol{r}')\,\mathrm{d}\boldsymbol{r}' \tag{4.94}$$

where we have neglected contributions from high-energy exciton states and denoted the dielectric polarization of individual QD by the center $\boldsymbol{a}$ of the QD. We now study a binary compound lattice composed of two types of QDs denoted by exciton wave functions ψ_1 and ψ_2 and exciton energies ω_1 and ω_2. Let $t = \epsilon\,\omega_{LT}\pi a_B^3$; the total exciton polarization becomes

$$\boldsymbol{P}(\boldsymbol{r}) = \sum_{\boldsymbol{a}i} \frac{t}{\omega_i - \omega + i\gamma}\,\psi_{\boldsymbol{a}i}(\boldsymbol{r},\boldsymbol{r}) \int \psi_{\boldsymbol{a}i}(\boldsymbol{r}',\boldsymbol{r}')\boldsymbol{E}(\boldsymbol{r}')\mathrm{d}\boldsymbol{r}' \tag{4.95}$$

where $\boldsymbol{a}$ denotes the lattice and $i = 1, 2$. As schematically shown in Fig. 4.4, type-I QDs occupy the regular compound lattice sites $\boldsymbol{a}$, while type-II QDs are displaced from the type-I QDs by $\boldsymbol{\tau}$.

Since the exciton wave function is real, we have dropped off the complex conjugation of the wave function in the integrals. The two integrals in the above equation can be transformed into

$$\begin{aligned}\int \psi_{1,\boldsymbol{a}}(\boldsymbol{r})\boldsymbol{E}(\boldsymbol{r})\mathrm{d}\boldsymbol{r} &= \mathrm{e}^{i\boldsymbol{q}\cdot\boldsymbol{a}} \sum_{\boldsymbol{b}} I_{1,\boldsymbol{q}+\boldsymbol{b}}\boldsymbol{E}_{\boldsymbol{q}+\boldsymbol{b}} \equiv \mathrm{e}^{i\boldsymbol{q}\cdot\boldsymbol{a}}\boldsymbol{\Lambda}_1 \\ \int \psi_{2,\boldsymbol{a}+\boldsymbol{\tau}}(\boldsymbol{r})\boldsymbol{E}(\boldsymbol{r})\mathrm{d}\boldsymbol{r} &= \mathrm{e}^{i\boldsymbol{q}\cdot\boldsymbol{a}} \sum_{\boldsymbol{b}} I_{2,\boldsymbol{q}+\boldsymbol{b}}\boldsymbol{E}_{\boldsymbol{q}+\boldsymbol{b}} \equiv \mathrm{e}^{i\boldsymbol{q}\cdot\boldsymbol{a}}\boldsymbol{\Lambda}_2\end{aligned} \tag{4.96}$$

where

$$I_{1,\boldsymbol{q}+\boldsymbol{b}} = \int \psi_{1,0}(\boldsymbol{r})\mathrm{e}^{i(\boldsymbol{q}+\boldsymbol{b})\cdot\boldsymbol{r}}\mathrm{d}\boldsymbol{r}$$
$$I_{2,\boldsymbol{q}+\boldsymbol{b}} = \int \psi_{2,0}(\boldsymbol{r})\mathrm{e}^{i(\boldsymbol{q}+\boldsymbol{b})\cdot(\boldsymbol{r}+\boldsymbol{\tau})}\mathrm{d}\boldsymbol{r} \tag{4.97}$$

The sums $\sum_{\boldsymbol{a}} \psi_{1,\boldsymbol{a}}(\boldsymbol{r})\mathrm{e}^{i\boldsymbol{q}\cdot\boldsymbol{a}}$ and $\sum_{\boldsymbol{a}+\boldsymbol{\tau}} \psi_{2,\boldsymbol{a}+\boldsymbol{\tau}}(\boldsymbol{r})\mathrm{e}^{i\boldsymbol{q}\cdot\boldsymbol{a}}$ satisfy the translational symmetry and can be presented as

$$\sum_{\boldsymbol{a}} \psi_{1,\boldsymbol{a}}(\boldsymbol{r})\mathrm{e}^{i\boldsymbol{q}\cdot\boldsymbol{a}} = \sum_{\boldsymbol{b}} \mathrm{e}^{i(\boldsymbol{q}+\boldsymbol{b})\cdot\boldsymbol{r}} \frac{I^*_{1,\boldsymbol{q}+\boldsymbol{b}}}{v_0}$$
$$\sum_{\boldsymbol{a}+\boldsymbol{\tau}} \psi_{2,\boldsymbol{a}+\boldsymbol{\tau}}(\boldsymbol{r})\mathrm{e}^{i\boldsymbol{q}\cdot\boldsymbol{a}} = \sum_{\boldsymbol{b}} \mathrm{e}^{i(\boldsymbol{q}+\boldsymbol{b})\cdot\boldsymbol{r}} \frac{I^*_{2,\boldsymbol{q}+\boldsymbol{b}}}{v_0} \tag{4.98}$$

The linear equations for $\boldsymbol{E}_{\boldsymbol{q}+\boldsymbol{b}}$ can thus be written

$$\left(|\boldsymbol{q}+\boldsymbol{b}|^2 - k^2\right)\boldsymbol{E}_{\boldsymbol{q}+\boldsymbol{b}}$$
$$= \frac{k_0^2 t}{v_0}\left[1 - \frac{1}{k^2}(\boldsymbol{q}+\boldsymbol{b})^2\right]\left[\frac{I^*_{1,\boldsymbol{q}+\boldsymbol{b}}}{\omega_1 - \omega + i\gamma}\boldsymbol{\Lambda}_1 + \frac{I^*_{2,\boldsymbol{q}+\boldsymbol{b}}}{\omega_2 - \omega + i\gamma}\boldsymbol{\Lambda}_2\right] \tag{4.99}$$

Let $S(\boldsymbol{Q})_{\alpha\beta} = \delta_{\alpha\beta} - Q_\alpha Q_\beta / k^2$, $\alpha, \beta = x, y, z$; $\delta_{\alpha\beta}$ is the Kronecker symbol. Dividing Eq. 4.99 by $(|\boldsymbol{q}+\boldsymbol{b}|^2 - k^2)$, multiplying it by $I_{1,\boldsymbol{q}+\boldsymbol{b}}$ and summing over $\boldsymbol{b}$

$$\sum_{\boldsymbol{b}} I_{1,\boldsymbol{q}+\boldsymbol{b}}\boldsymbol{E}_{\boldsymbol{q}+\boldsymbol{b}} \equiv \boldsymbol{\Lambda}_1$$
$$= \frac{k_0^2 t}{v_0}\sum_{\boldsymbol{b}} \frac{\widehat{S}(\boldsymbol{q}+\boldsymbol{b})}{|\boldsymbol{q}+\boldsymbol{b}|^2 - k^2}\left[\frac{|I_{1,\boldsymbol{q}+\boldsymbol{b}}|^2}{\omega_1 - \omega + i\gamma}\boldsymbol{\Lambda}_1 + \frac{I_{1,\boldsymbol{q}+\boldsymbol{b}} I^*_{2,\boldsymbol{q}+\boldsymbol{b}}}{\omega_2 - \omega + i\gamma}\boldsymbol{\Lambda}_2\right]$$
$$= \widehat{M_1}(\omega, \boldsymbol{q})\boldsymbol{\Lambda}_1 + \widehat{M_2}(\omega, \boldsymbol{q})\boldsymbol{\Lambda}_2 \tag{4.100}$$

where

$$\widehat{M_1}(\omega, \boldsymbol{q}) = \frac{k_0^2 t}{(\omega_1 - \omega + i\gamma)v_0}\sum_{\boldsymbol{b}} \frac{\widehat{S}(\boldsymbol{q}+\boldsymbol{b})}{|\boldsymbol{q}+\boldsymbol{b}|^2 - k^2}|I_{1,\boldsymbol{q}+\boldsymbol{b}}|^2$$
$$\widehat{M_2}(\omega, \boldsymbol{q}) = \frac{k_0^2 t}{(\omega_2 - \omega + i\gamma)v_0}\sum_{\boldsymbol{b}} \frac{\widehat{S}(\boldsymbol{q}+\boldsymbol{b})}{|\boldsymbol{q}+\boldsymbol{b}|^2 - k^2} I_{1,\boldsymbol{q}+\boldsymbol{b}} I^*_{2,\boldsymbol{q}+\boldsymbol{b}} \tag{4.101}$$

Similarly, dividing Eq. 4.99 by $\left(|\boldsymbol{q}+\boldsymbol{b}|^2 - k^2\right)$, multiplying it by $I_{2,\boldsymbol{q}+\boldsymbol{b}}$, and summing over $\boldsymbol{b}$

$$\sum_{\boldsymbol{b}} I_{2,\boldsymbol{q}+\boldsymbol{b}}\boldsymbol{E}_{\boldsymbol{q}+\boldsymbol{b}} \equiv \boldsymbol{\Lambda}_2$$
$$= \frac{k_0^2 t}{v_0}\sum_{\boldsymbol{b}} \frac{\widehat{S}(\boldsymbol{q}+\boldsymbol{b})}{|\boldsymbol{q}+\boldsymbol{b}|^2 - k^2}\left[\frac{I_{2,\boldsymbol{q}+\boldsymbol{b}} I^*_{1,\boldsymbol{q}+\boldsymbol{b}}}{\omega_1 - \omega + i\gamma}\boldsymbol{\Lambda}_1 + = \frac{|I_{2,\boldsymbol{q}+\boldsymbol{b}}|^2}{\omega_2 - \omega + i\gamma}\boldsymbol{\Lambda}_2\right]$$
$$= \widehat{M_3}(\omega, \boldsymbol{q})\boldsymbol{\Lambda}_1 + \widehat{M_4}(\omega, \boldsymbol{q})\boldsymbol{\Lambda}_2 \tag{4.102}$$

where

$$\widehat{M}_3(\omega, \boldsymbol{q}) = \frac{k_0^2 t}{(\omega_1 - \omega + i\gamma)v_0} \sum_{\boldsymbol{b}} \frac{\widehat{S}(\boldsymbol{q}+\boldsymbol{b})}{|\boldsymbol{q}+\boldsymbol{b}|^2 - k^2} I_{2,\boldsymbol{q}+\boldsymbol{b}} I^*_{1,\boldsymbol{q}+\boldsymbol{b}}$$

$$\widehat{M}_4(\omega, \boldsymbol{q}) = \frac{k_0^2 t}{(\omega_2 - \omega + i\gamma)v_0} \sum_{\boldsymbol{b}} \frac{\widehat{S}(\boldsymbol{q}+\boldsymbol{b})}{|\boldsymbol{q}+\boldsymbol{b}|^2 - k^2} |I_{2,\boldsymbol{q}+\boldsymbol{b}}|^2 \tag{4.103}$$

We arrive at the vector equations

$$\left[\mathrm{I} - \widehat{M}_1(\omega, \boldsymbol{q})\right] \boldsymbol{\Lambda}_1 = \widehat{M}_2(\omega, \boldsymbol{q}) \boldsymbol{\Lambda}_2$$

$$\left[\mathrm{I} - \widehat{M}_4(\omega, \boldsymbol{q})\right] \boldsymbol{\Lambda}_2 = \widehat{M}_3(\omega, \boldsymbol{q}) \boldsymbol{\Lambda}_1 \tag{4.104}$$

where I is a 3×3 unit matrix. We can rewrite the above equations as

$$D = \begin{pmatrix} 1 - M_{1,11} & -M_{1,12} & -M_{1,13} & -M_{2,11} & -M_{2,12} & -M_{2,13} \\ -M_{1,21} & 1 - M_{1,22} & -M_{1,23} & -M_{2,21} & -M_{2,22} & -M_{2,23} \\ -M_{1,31} & -M_{1,32} & 1 - M_{1,33} & -M_{2,31} & -M_{2,32} & -M_{2,33} \\ M_{3,11} & M_{3,12} & M_{3,13} & M_{4,11} - 1 & M_{4,12} & M_{4,13} \\ M_{3,21} & M_{3,22} & M_{3,23} & M_{4,21} & M_{4,22} - 1 & M_{4,23} \\ M_{3,31} & M_{3,32} & M_{3,33} & M_{4,31} & M_{4,32} & M_{4,33} - 1 \end{pmatrix} \tag{4.105}$$

$$M_{1,\alpha\beta}(\Omega, \boldsymbol{q}) = \frac{8}{\pi v_0} \frac{\omega_{\mathrm{LT}} R_1^3}{(\omega_1 - \omega + i\gamma)} \sigma_{11,\alpha\beta}(\Omega, \boldsymbol{q})$$

$$M_{2,\alpha\beta}(\Omega, \boldsymbol{q}) = \frac{8}{\pi v_0} \frac{\omega_{\mathrm{LT}} (R_1 R_2)^{3/2}}{(\omega_2 - \omega + i\gamma)} \sigma_{12,\alpha\beta}(\Omega, \boldsymbol{q})$$

$$M_{3,\alpha\beta}(\Omega, \boldsymbol{q}) = \frac{8}{\pi v_0} \frac{\omega_{\mathrm{LT}} (R_1 R_2)^{3/2}}{(\omega_1 - \omega + i\gamma)} \sigma_{21,\alpha\beta}(\Omega, \boldsymbol{q})$$

$$M_{4,\alpha\beta}(\Omega, \boldsymbol{q}) = \frac{8}{\pi v_0} \frac{\omega_{\mathrm{LT}} R_2^3}{(\omega_2 - \omega + i\gamma)} \sigma_{22,\alpha\beta}(\Omega, \boldsymbol{q})$$

$$\sigma_{st,\alpha\beta}(\Omega, \boldsymbol{q}) = \sum_{\boldsymbol{b}} \frac{f(|\boldsymbol{q}+\boldsymbol{b}|R_s)\, f(|\boldsymbol{q}+\boldsymbol{b}|R_t)\, S_{\alpha\beta}(\boldsymbol{q}+\boldsymbol{b})}{\Omega^2(\boldsymbol{q}+\boldsymbol{b}) - \Omega^2} e^{i(s-t)(\boldsymbol{q}+\boldsymbol{b})\cdot\boldsymbol{\tau}}$$

$$f(x) = \frac{\pi^2 \sin x}{x(\pi^2 - x^2)}$$

$$\Omega = \frac{\omega}{\omega_1}$$

$$\Omega(\boldsymbol{Q}) = \frac{c|\boldsymbol{Q}|}{\omega_1 n_b} \tag{4.106}$$

where $s, t = 1, 2$. The exciton–polariton dispersion $\omega(\boldsymbol{q})$ satisfies the equation

$$\mathrm{Det}||D|| = 0 \tag{4.107}$$

We can now consider practical three-dimensional QD arrays with periods comparable with the light wavelength and with sizes exceeding the bulk-exciton Bohr radius, which can be grown artificially or by embedding semiconductor nanocrystals into the pores of porous materials such as the synthetic opal.[23–25] It should be mentioned that the thus developed theory takes into account a contribution of only one exciton resonance, which is valid when the separation between the exciton size-quantization levels is much larger than the bulk value of the exciton longitudinal-transverse splitting, ω_{LT}. In the opposite limit of extremely large bulk-exciton translational effective mass one can use the local material relation

$$\boldsymbol{D}(\boldsymbol{r}) = \varepsilon(\boldsymbol{r})\boldsymbol{E}(\boldsymbol{r})$$

as it was done by Sigalas *et al.*[26] for phonon–polaritons in a two-dimensional lattice consisting of semiconductor cylinders.

For the fcc structures, $v_0 = a^3/4$. And for the fcc Bravais lattice with conventional cubic cell of side a, its reciprocal lattice is body-centered cubic,

$$\begin{aligned}\boldsymbol{b}_1 &= \frac{2\pi}{a}(\boldsymbol{y}_0 + \boldsymbol{z}_0 - \boldsymbol{x}_0)\\ \boldsymbol{b}_2 &= \frac{2\pi}{a}(\boldsymbol{z}_0 + \boldsymbol{x}_0 - \boldsymbol{y}_0)\\ \boldsymbol{b}_3 &= \frac{2\pi}{a}(\boldsymbol{x}_0 + \boldsymbol{y}_0 - \boldsymbol{z}_0)\end{aligned} \tag{4.108}$$

and we consider $XU - UL - L\Gamma - \Gamma X - XW - WK$ in the reciprocal $\boldsymbol{q}$-space (in unit of $2\pi/a$),

$$\begin{gathered}\Gamma = (0,0,0)\ ,\ X = (0,0,1)\ ,\ L = \left(\frac{1}{2},\frac{1}{2},\frac{1}{2}\right)\\ W = \left(\frac{1}{2},0,1\right)\ ,\ U = \left(\frac{1}{4},\frac{1}{4},1\right)\ ,\ K = \left(\frac{3}{4},0,\frac{3}{4}\right)\end{gathered} \tag{4.109}$$

For primitive cubic, $v_0 = a^3$,

$$\boldsymbol{b}_1 = \frac{2\pi}{a}\boldsymbol{x}_0\ ,\ \boldsymbol{b}_2 = \frac{2\pi}{a}\boldsymbol{y}_0\ ,\ \boldsymbol{b}_3 = \frac{2\pi}{a}\boldsymbol{z}_0 \tag{4.110}$$

and we consider $XR - R\Gamma - \Gamma X - XM$ in the reciprocal $\boldsymbol{q}$-space (in unit of $2\pi/a$)

$$\Gamma = (0,0,0)\ ,\ X = \left(0,0,\frac{1}{2}\right)\ ,\ R = \left(\frac{1}{2},\frac{1}{2},\frac{1}{2}\right)\ ,\ M = \left(\frac{1}{2},0,\frac{1}{2}\right) \tag{4.111}$$

Figure 4.5 shows the dispersion relationship of a photonic crystal that has a primitive cubic lattice ($a = 0.5a_{\mathrm{Br}}$). We have investigated two cases, $\omega_{\mathrm{LT}}/\omega_0 = 5.0\times 10^{-3}$ and $\omega_{\mathrm{LT}}/\omega_0 = 5.0\times 10^{-4}$, showing that the dispersion relationship depends weakly on the ratio of $\omega_{\mathrm{LT}}/\omega_0$. The result is to be expected theoretically that 5.0×10^{-3} is already rather small so that the effect can be treated as perturbation. Simple as the primitive cubic lattice structure, as can be realized experimentally (compared with the fcc structure), the corresponding dispersion relation of Fig. 4.5 is, however, rather complicated and we do not observe an overall bandgap in Fig. 4.5.

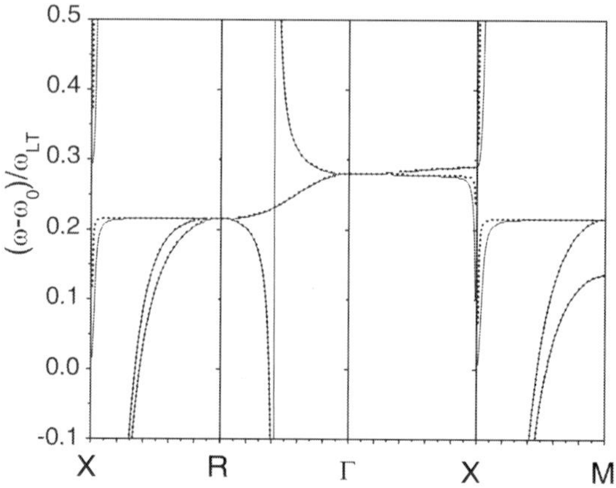

Fig. 4.5 Energy dispersion relation of a primitive cubic photonic crystal. $a = 0.5a_{\rm Br}$. Solid line: $\omega_{\rm LT}/\omega_0 = 5 \times 10^{-3}$; Dotted line: $\omega_{\rm LT}/\omega_0 = 5 \times 10^{-4}$. Y. Fu, M. Willander, and E. L. Ivchenko, Photonic dispersions of semiconductor-quantum-dot-array-based photonic crystals in primitive and face-centered cubic lattices, Superlattices and Microstructures, vol. 27, p. 255–64, 2000. Copyright (2000) by Elsevier.

The dispersion relationship for the fcc lattice is presented in Fig. 4.6 where we observe clearly an overall bandgap, most prominent along the Λ (Γ-L) line ([111] direction) characterized by a giant anti-crossing between the branches of bare transverse photon and exciton modes. At the X point, the gap is determined by the separation between the longitudinal and lower transverse branches; it is still remarkable

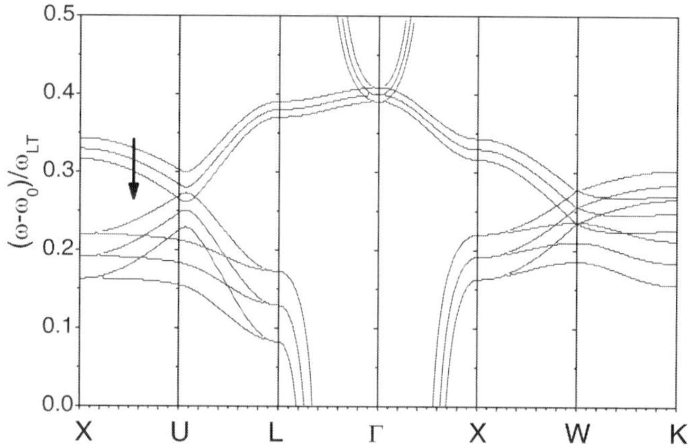

Fig. 4.6 Energy dispersion relation of a face-centred cubic photonic crystal. $\omega_{\rm LT}/\omega_0 = 5 \times 10^{-4}$. The arrow in the figure indicates the increase of the lattice constant a from 0.95, 1.01 and $1.1a_{\rm Br}$. Y. Fu, M. Willander, and E. L. Ivchenko, Photonic dispersions of semiconductor-quantum-dot-array-based photonic crystals in primitive and face-centered cubic lattices, Superlattices and Microstructures, vol. 27, p. 255–64, 2000. Copyright (2000) by Elsevier.

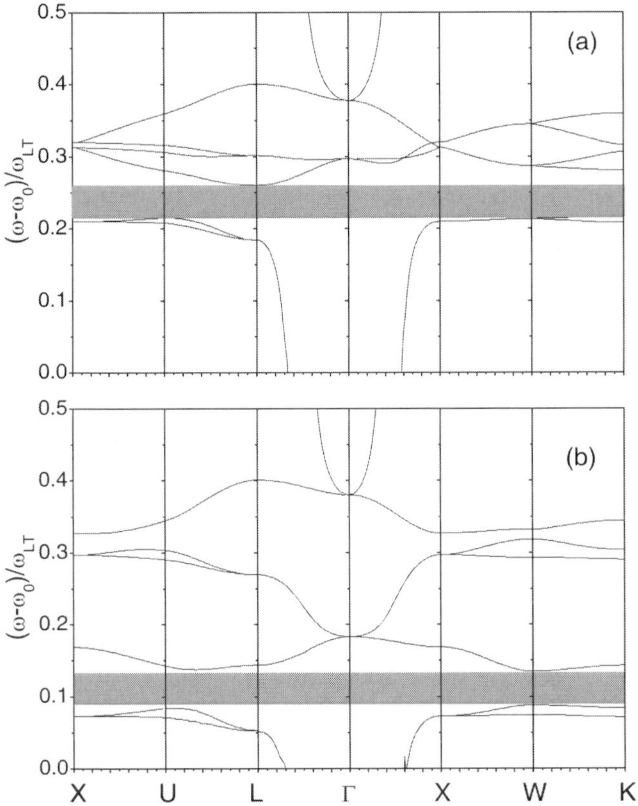

Fig. 4.7 (a) Energy dispersion relation of a QD diamond lattice. $a = 115$ nm, $\hbar\omega_{LT} = 0.5$ meV and the radius of QDs $R = 20$ nm. (b) Zinkblende lattice consisting of two types of QDs whose radii are different. $a = 100$ nm, $R_1 = 20$ nm, $R_2 = 20.05$ nm, and $\hbar\omega_{LT} = 0.5$ meV. Y. Zeng, Y. Fu, X. Chen, W. Lu, and H. Ågren, Complete band gaps in three-dimensional quantum-dot photonic crystals, Phys. Rev. B, vol. 74, p. 115325(5), 2006. Copyright (2006) by The American Physical Society.

and exceeds $0.5\omega_{LT}$. However, near the points U and W the exciton–polariton branches converge and the gap almost disappears. Note that the anti-crossing can be described with a high accuracy by retaining in the sum over $\boldsymbol{b}$ in Eq. 4.106 the two terms due to

$$\boldsymbol{b} = 0, -\frac{4\pi}{a}(0, 0, 1)$$

for the Δ points and

$$\boldsymbol{b} = 0, -\frac{2\pi}{a}(1, 1, 1)$$

for the Λ points. We have tried to tune the structure parameter a. It is found from Fig. 4.6 that the dispersion relationship vertically shifts down along the $(\omega-\omega_0)/\omega_{LT}$ axis following the increase of a/a_{Br} ratio.

A complete and wide bandgap is observed in a zincblende lattice, as shown in Fig. 4.7. Light propagation along these directions is thus suppressed for photon energies inside the bandgap. The complete bandgap prohibits propagation in all directions.

4.6 QD DIMER SYSTEM

Achieving a negative dielectric constant at an (optical) frequency ω where the loss goes to zero, i.e., $\epsilon = \epsilon' + i\epsilon''$ with $\epsilon' < 0$ and $\epsilon'' = 0$, is of fundamental importance in small-scale optics, which has come to be known as nanophotonics. A negative refractive index requires both negative dielectric constant and permeability.[27] Imaging with a negative index slab has a resolution that is limited only by the loss, with matched real parts of the constitutive parameters, to the background (vacuum, for instance).[28, 29] A plasmon–polariton surface wave exists at a metal–insulator interface because of the negative dielectric constant,[30] for a mode with magnetic field parallel to the interface.[31, 32] With two interacting surfaces, a metal-insulator-metal waveguide mode exists, and this propagates for arbitrarily small gaps. As the gap reduces, the wavelength reduces, for a fixed vacuum wavelength. It is thus possible in principle to shrink the size of optical components using such mechanisms. Unfortunately, the loss increases as the waveguide mode wavelength reduces, as measured per unit length. A means to offset this loss would offer the prospect of achieving arbitrarily small mode volumes and hence high-density optical elements and circuits.

Resonant states in semiconductors provide a source for negative dielectric constant, provided that the dipole strength and oscillator density is adequate to offset the background. Furthermore, semiconductors offer the prospect of pumping, either optically or electrically, to achieve a gain mechanism that can offset the loss. As we have thus far discussed in this chapter that QDs offer polarization insensitivity, i.e., an isotropic material, which would be particularly useful in optical components such as lenses, waveguides, and resonators. We address the issue of the dielectric constant from QD exciton–polaritons, and then present the influence of gain on the homogenized dielectric constant. By means of these examples, we describe the required QD density and pumping to achieve the desirable operation condition of $\epsilon' < 0$ and $\epsilon'' = 0$.

By the complex effective permittivity of Eq. 4.33 for QDs immersed into a background material, either a solid-state material or a polymer matrix, the effective ϵ of the composite system (QDs plus the background matrix) will be estimated using the dielectric theory of Maxwell-Garnet.[33, 34]

4.6.1 *Optical Excitation of QD*

For a normal system without any extra external modifications, expect the incident electromagnetic field of $\boldsymbol{E}(\boldsymbol{r}, t)$ under investigation, $f_{n\boldsymbol{K}}(t) = |c_{n\boldsymbol{K}}(t)|^2$, which is small (the first-order perturbation condition). High pumping rate can be achieved by increasing the optical power. The occupation of the excited state is normally smaller than the ground state so that $f_0 > f_{n\boldsymbol{K}}$ and optical absorption dominates. This means that the optical field is absorbed during the exciton excitation following its propagation through the QD. The situation is normally referred to as optical loss, for example, in photodetectors. Optical gain is achieved in light-emitting and laser devices by population inversion of $f_0 < f_{n\boldsymbol{K}}$.

To achieve optical gain in the QD, we increase the occupation of the exciton state. Here we consider such a pumping process that the exciton population $f_{n\boldsymbol{K}}$ is finite so the time-dependent wave function becomes

$$|\boldsymbol{r}_e, \boldsymbol{r}_h, t\rangle = \left[c_0^{(0)} + c_0(t)\right] |\Psi_0(\boldsymbol{r}_e, \boldsymbol{r}_h)\rangle + \sum_{n\boldsymbol{K}} \left[c_{n\boldsymbol{K}}^{(0)} + c_{n\boldsymbol{K}}(t)\right] |\Psi_{n\boldsymbol{K}}(\boldsymbol{r}_e, \boldsymbol{r}_h, t)\rangle \quad (4.112)$$

due to the external electromagnetic field $\boldsymbol{E}(\boldsymbol{r}, t)$. By following the same mathematical procedures of last section, we can obtain

$$\epsilon_{\mathrm{QD}}(\omega) = \epsilon_b \left[1 + \sum_{n\boldsymbol{K}} \frac{6\left(2|c_0^{(0)}|^2 - 1\right)\omega_{LT}}{\pi^2\left(\omega_{n\boldsymbol{K}} - \omega + i\gamma\right)}\right] \quad (4.113)$$

The QD becomes lossless when $|c_0^{(0)}|^2 = 1/2$. At the same time, the polarization changes sign.

Figure 4.8 shows the calculated dielectric constant for one QD before and after some fraction of the exciton states become populated. In this example, II-VI PbSe/ZnSe QD in another CdSe background is assumed. $\hbar\gamma = 1.0\,\mathrm{meV}$. In Fig. 4.8 and throughout this chapter we write $\epsilon = \epsilon' + i\epsilon''$; ϵ' and ϵ'' are expressed in units of ϵ_0. Here we observe the possibility of finding low-loss negative dielectric constant.

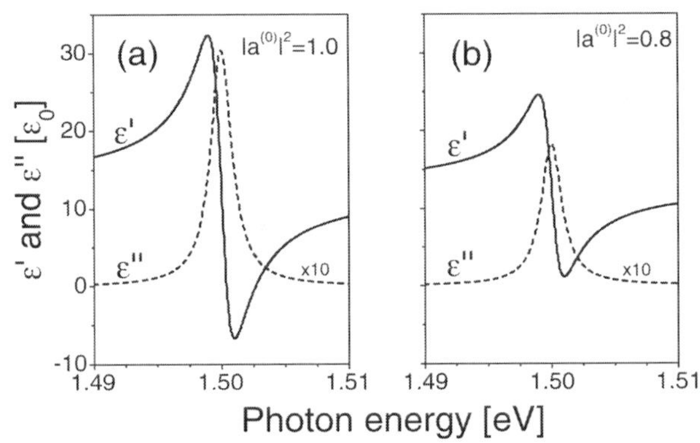

Fig. 4.8 Dielectric coefficients of the QD before and after the QD becomes populated. $\epsilon_\infty = 12.8$, $\hbar\omega_{n\boldsymbol{K}} = 1.5$ eV, $\hbar\omega_{\mathrm{LT}} = 5$ meV, and $\hbar\gamma = 1$ meV for typical II-VI semiconductor QDs, e.g., colloidal PbSe/ZnSe QD. Y. Fu, L. Thylén, and H. Ågren, A lossless negative dielectric constant from quantum dot exciton polaritons, Nano Letters, vol. 8, p. 1551–5, 2008. Copyright (2008) by American Chemical Society.

4.6.2 *Low-Loss Negative Dielectric Constant*

We require $\epsilon'' = 0$ at some frequency in order to achieve lossless waveguide modes. To investigate prospects for obtaining isotropic material, we consider distributions of two types of QDs, one providing an absorptive resonance and the other gain through either optical or electrical pumping. Consider a PbSe/ZnSe QD assembly immersed in a medium of dielectric constant ϵ_i. The dielectric constant ϵ for the ensemble of the QDs can be described by the dielectric theory of Maxwell-Garnet (MG),[33,34] which for two QD species can be written as

$$\frac{\epsilon - \epsilon_i}{\epsilon + 2\epsilon_i} = x_1 \frac{\epsilon_{\mathrm{QD1}} - \epsilon_i}{\epsilon_{\mathrm{QD1}} + 2\epsilon_i} + x_2 \frac{\epsilon_{\mathrm{QD2}} - \epsilon_i}{\epsilon_{\mathrm{QD2}} + 2\epsilon_i} \tag{4.114}$$

where x_i is the volume fraction of the ith QD species. It is assumed that the QDs are spherical and that macroscopic dielectric constants can be applied. While Eq. 4.114 works best for small x_i (less than 0.4), there is evidence that useful information can be available with higher concentrations.[35]

We first consider II–VI material as the background medium so that ϵ_i is about 10. We then consider conducting polymers with ϵ_i around 2. Furthermore, we assume a QD radius of 10 nm. Consider first the distribution of one type of QDs (i.e., $x_1 \neq 0$ and $x_2 = 0$); the spectra of ϵ for two QD densities and two background dielectric constants are shown in Fig. 4.9 by Eq. 4.114. Both ϵ' and ϵ'' increase with increasing QD density and decreasing ϵ_i. Achieving a low background dielectric constant is thus important in reducing the QD load to achieve $\epsilon' < 0$.

A mixture of two types of semiconductor QDs –, one type is lossy and the other has been pumped for optical gain, immersed in $\epsilon_i = 1.8$, (Fig. 4.10a) –, can

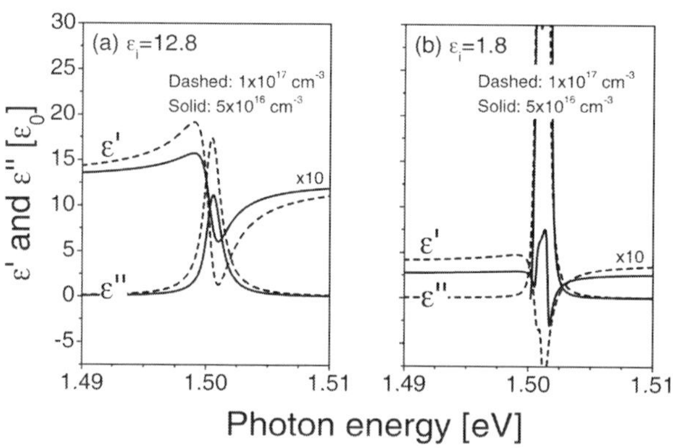

Fig. 4.9 Effective dielectric constant $\epsilon(\omega) = \epsilon'(\omega) + i\epsilon''(\omega)$ for an ensemble of QDs immersed in a medium of dielectric constant $\epsilon_i(\omega)$. The QD density are 5×10^{16} and 1×10^{17} cm^{-3}, respectively, correspond to volume percentages of 20.5% and 41%. (a) QDs embedded in semiconductor substrate $\epsilon_i = 12.8$. (b) QDs immersed in a medium with $\epsilon_i = 1.8$, e.g., a conducting polymer. Y. Fu, L. Thylén, and H. Ågren, A lossless negative dielectric constant from quantum dot exciton polaritons, Nano Letters, vol. 8, p. 1551–5, 2008. Copyright (2008) by American Chemical Society.

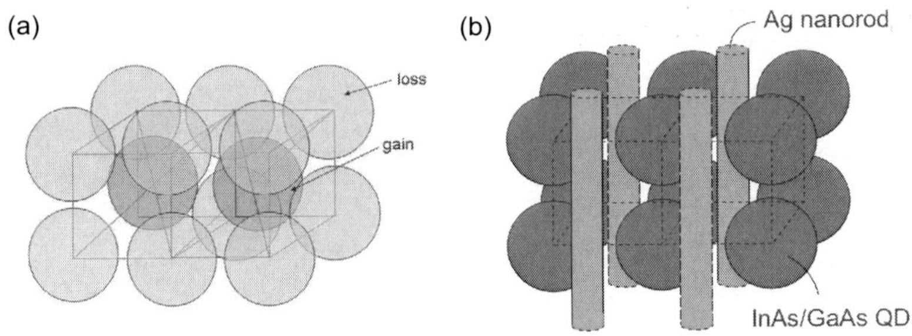

Fig. 4.10 (a) A periodic binary structure comprising gain and lossy QDs (pumped and lossy QDs, respectively) in a primitive cubic lattice. (b) A primitive cubic QD lattice interspersed with silver nanorods. See also Color Insert.

produce an effective dielectric constant that is lossless and negative, thus permitting, in concept, arbitrarily small scaling of the optical mode volume in the field of nanophotonics.[36] Figure 4.11a depicts the spectra of $\epsilon'_{\rm QD}$ and $\epsilon''_{\rm QD}$ of the two QDs. The combination of two types of QDs has one type lossy, while the other is pumped for optical gain; both are immersed in a conducting polymer with an effective dielectric constant $\epsilon_i = 1.8$. The MG mixing result is given in Fig. 4.11b showing an effective lossless negative dielectric constant of $\epsilon' = -8.5$ and $\epsilon'' = 0$ at 1.5015 eV with a total QD density of $14 \times 10^{16}\,\text{cm}^{-3}$ (58% volume fraction). Note that gain is at the higher energy, a condition we have found necessary in order to collectively achieve $\epsilon' < 0$ and $\epsilon'' = 0$.

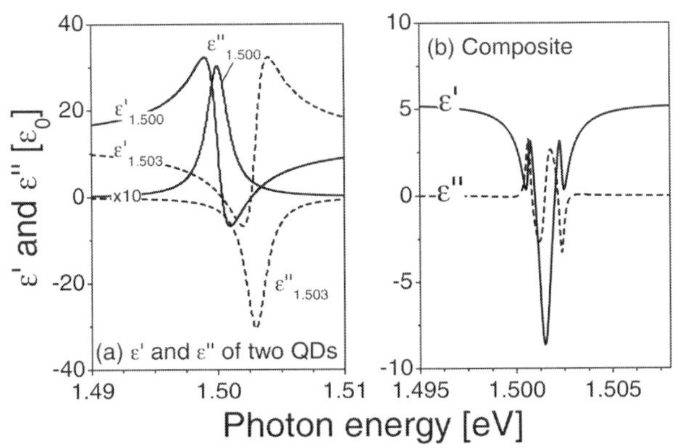

Fig. 4.11 (a) One type of QDs at 1.50 eV and $|a^{(0)}|^2 = 1.0$ (solid black lines), the other type of QDs at 1.503 eV and $|a^{(0)}|^2 = 0.0$ (dashed lines). (b) The two types of QDs are immersed in $\epsilon_i = 1.8$. The densities are 7×10^{16} cm^{-3}. Y. Fu, L. Thylén, and H. Ågren, A lossless negative dielectric constant from quantum dot exciton polaritons, Nano Letters, vol. 8, p. 1551–5, 2008. Copyright (2008) by American Chemical Society.

A binary mixture of QDs with silver nanorods, (Fig. 4.10b), was also shown to have a negative ϵ and compensated optical loss by using finite-difference time-domain (FDTD) calculations.[37]

Note that by combining lossy QDs with optically pumped QDs, the macroscopic permittivity of our structure is approximately in the form of

$$\frac{i}{\omega-\omega_r-i\gamma}=\frac{-\gamma}{(\omega-\omega_r)^2+\gamma^2}+i\,\frac{\omega-\omega_r}{(\omega-\omega_r)^2+\gamma^2} \tag{4.115}$$

where ω_r denotes the macroscopic resonance frequency of the dimer system and γ is the damping rate. The loss is minimal when ω approaches ω_r. The expression follows exactly the Kramers-Kronig relations as required by the stability of the system. Such a spectrum is fundamentally different from

$$\frac{1}{\omega_r-\omega-i\gamma}=\frac{\omega_r-\omega}{(\omega_r-\omega)^2+\gamma^2}+i\,\frac{\gamma}{(\omega_r-\omega)^2+\gamma^2} \tag{4.116}$$

for which γ is to be reduced in the form of $\gamma=0$, $\partial\gamma/\partial\omega=0$, and $\partial^2\gamma/\partial\omega^2>0$ in order to become lossless, which results in a pole in the upper half-plane and hence violates causality.[38]

4.6.3 *Photonic Band of QD-based Negative Dielectric Medium*

The Maxwell-Garnett formalism Eq. 4.114 has been used in the previous subsection, which could amount to a mean-field approximation; numerical FDTD calculations[37] based on calculating transmission and reflection coefficients and using a homogenization procedure[39] to extract the effective dielectric constant data confirmed the general concept that the electromagnetic field propagates normally through the QD dimer system.

In order to further elucidate the nature of propagation of the electromagnetic field in the QD dimer beyond the mean-field approximation as well as to avoid possible numerical difficulties (such as difficulties to resolve the issue of long-distance transport properties, since the simulation space is limited) in the FDTD simulations, here we discuss the photonic dispersion of the QD dimer system by considering perfect photonic crystal lattice structure so that periodic boundary conditions apply in order to study long-distance transport properties.

We construct a QD dimer photonic crystal by positioning two types of spherical QDs, having radii R_1 and R_2, respectively, in an fcc lattice having a lattice constant a, similarly shown in Fig. 4.4 (note that Fig. 4.4 shows a primitive cubic lattice). The lossy QDs, type-I QDs (exciton energies are $\hbar\omega_1$), occupy the normal fcc lattice sites, while the already excited QDs, type-II QDs ($\hbar\omega_2$) are displaced from the type-I QDs by $\boldsymbol{\tau}$. Let $c_{\boldsymbol{a}i}$ be the occupation of the ground state of QD i positioned at lattice site $\boldsymbol{a}$, i.e., the VB is completely filled and the CB is completely empty, the contribution of an excited exciton state in a QD $\boldsymbol{a}i$ to the dielectric polarization is given by

$$\boldsymbol{P}_{ai}(\boldsymbol{r})=\frac{e^2\boldsymbol{p}_{cv}\left(2|c_{\boldsymbol{a}i}|^2-1\right)}{\hbar(\omega_{\boldsymbol{a}i}-\omega+i\gamma)\omega_{\boldsymbol{a}i}^2m_0^2}\,\psi_{\boldsymbol{a}i}(\boldsymbol{r},\boldsymbol{r})\int\psi^*_{\boldsymbol{a}i}(\boldsymbol{r}',\boldsymbol{r}')\boldsymbol{p}_{cv}\cdot\boldsymbol{E}(\boldsymbol{r}')\,d\boldsymbol{r}' \tag{4.117}$$

where $\psi_{\boldsymbol{a}i}(\boldsymbol{r},\boldsymbol{r}) = \psi_{\boldsymbol{a}i}(\boldsymbol{r}_e,\boldsymbol{r}_h)|_{\boldsymbol{r}_e=\boldsymbol{r}_h=\boldsymbol{r}}$, $\psi_{\boldsymbol{a}i}(\boldsymbol{r}_e,\boldsymbol{r}_h)$ is the exciton wave function, $\boldsymbol{r}_e$ and $\boldsymbol{r}_h$ denote the electron and hole, respectively. $\hbar\omega_{\boldsymbol{a}i}$ is the exciton energy, and γ denotes the damping rate of the exciton state.

To find the dispersion properties of the QD dimer system, we denote $a_1 = \sqrt{2|c_1|^2-1}$, $a_2 = \sqrt{2|c_2|^2-1}$; $v_0 = a^3/4$ is the volume of the primitive unit cell of the fcc lattice; $\phi_i = a_i\psi_i$ and $t = \epsilon\, \omega_{\mathrm{LT}}\pi a_{\mathrm{Br}}^3$, so that the exciton polarization becomes

$$\boldsymbol{P}(\boldsymbol{r}) = \sum_{\boldsymbol{a}i} \frac{t}{\omega_i - \omega + i\gamma}\, \phi_{\boldsymbol{a}i}(\boldsymbol{r},\boldsymbol{r}) \int \phi^*_{\boldsymbol{a}i}(\boldsymbol{r}',\boldsymbol{r}')\boldsymbol{E}(\boldsymbol{r}')\mathrm{d}\boldsymbol{r}' \tag{4.118}$$

by which we can follow the steps of sub-section 4.5.2 and eventually reach the exciton–polariton dispersion $\omega(\boldsymbol{q})$ satifying

$$\mathrm{Det}|D| = 0 \tag{4.119}$$

where

$$D = \begin{pmatrix} 1-M_{1,11} & -M_{1,12} & -M_{1,13} & -M_{2,11} & -M_{2,12} & -M_{2,13} \\ -M_{1,21} & 1-M_{1,22} & -M_{1,23} & -M_{2,21} & -M_{2,22} & -M_{2,23} \\ -M_{1,31} & -M_{1,32} & 1-M_{1,33} & -M_{2,31} & -M_{2,32} & -M_{2,33} \\ M_{3,11} & M_{3,12} & M_{3,13} & M_{4,11}-1 & M_{4,12} & M_{4,13} \\ M_{3,21} & M_{3,22} & M_{3,23} & M_{4,21} & M_{4,22}-1 & M_{4,23} \\ M_{3,31} & M_{3,32} & M_{3,33} & M_{4,31} & M_{4,32} & M_{4,33}-1 \end{pmatrix} \tag{4.120}$$

$$\begin{aligned} M_{1,\alpha\beta}(\Omega,\boldsymbol{q}) &= \frac{8}{\pi v_0}\frac{\omega_{\mathrm{LT}}R_1^3|a_1|^2}{(\omega_1-\omega+i\gamma)}\sigma_{11,\alpha\beta}(\Omega,\boldsymbol{q}) \\ M_{2,\alpha\beta}(\Omega,\boldsymbol{q}) &= \frac{8}{\pi v_0}\frac{\omega_{\mathrm{LT}}(R_1R_2)^{3/2}a_1a_2^*}{(\omega_2-\omega+i\gamma)}\sigma_{12,\alpha\beta}(\Omega,\boldsymbol{q}) \\ M_{3,\alpha\beta}(\Omega,\boldsymbol{q}) &= \frac{8}{\pi v_0}\frac{\omega_{\mathrm{LT}}(R_1R_2)^{3/2}a_1^*a_2}{(\omega_1-\omega+i\gamma)}\sigma_{21,\alpha\beta}(\Omega,\boldsymbol{q}) \\ M_{4,\alpha\beta}(\Omega,\boldsymbol{q}) &= \frac{8}{\pi v_0}\frac{\omega_{\mathrm{LT}}R_2^3|a_2|^2}{(\omega_2-\omega+i\gamma)}\sigma_{22,\alpha\beta}(\Omega,\boldsymbol{q}) \\ \sigma_{st,\alpha\beta}(\Omega,\boldsymbol{q}) &= \sum_{\boldsymbol{g}} \frac{f(|\boldsymbol{g}+\boldsymbol{q}|R_s)\, f(|\boldsymbol{g}+\boldsymbol{q}|R_t)\, S_{\alpha\beta}(\boldsymbol{g}+\boldsymbol{q})}{\Omega^2(\boldsymbol{g}+\boldsymbol{q})-\Omega^2}\,\mathrm{e}^{i(s-t)(\boldsymbol{g}+\boldsymbol{q})\cdot\boldsymbol{\tau}} \\ f(x) &= \frac{\pi^2\sin x}{x(\pi^2-x^2)} \\ \Omega &= \frac{\omega}{\omega_1} \\ \Omega(\boldsymbol{Q}) &= \frac{c|\boldsymbol{Q}|}{\omega_1 n_b} \end{aligned} \tag{4.121}$$

$s,t = 1,2$. Note that by setting $|c_1|^2 = 0$ and $|c_2|^2 = 0.5$ (type-II QDs are totally transparent) we retrieve the case of the QD photonic crystal composed of only type-I QDs. By setting $|c_1|^2 = |c_2|^2 = 0$ and $\omega_1 = \omega_2$ we retrieve the case of compound QD photonic crystal, i.e., Eqs. 4.104, 4.105, 4.106 in sub-Section 4.5.2.

The dispersion relationship $\omega(\boldsymbol{q})$ for a QD dimer fcc lattice is presented in Fig. 4.12. The lattice is denoted by its lattice constant a, which is set to be $0.95a_{\text{Bragg}}$ (which is 116 nm for GaAs), where $a_{\text{Bragg}} = c\pi/\omega_1 n$. The lossy type-I

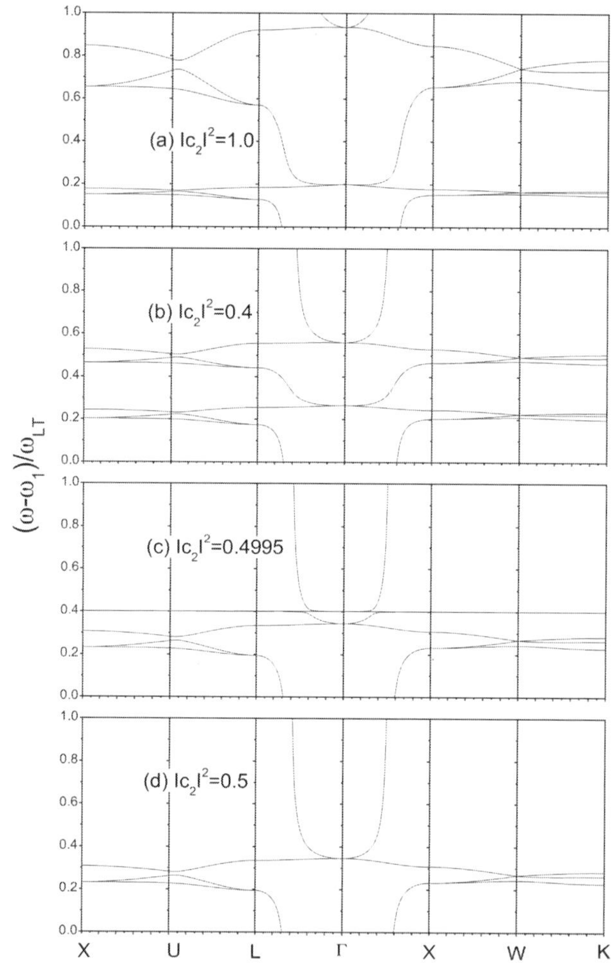

Fig. 4.12 Energy dispersion relationships of QD dimer systems in the fcc lattice. $R_1 = a/4$, $R_2 = a/5$, $\hbar\omega_1 = 1.5$ eV, $\hbar\omega_2 = 1.503$ eV, $\hbar\omega_{\text{LT}} = 5$ meV, $a = 0.95a_{\text{Bragg}}$, $\boldsymbol{\tau} = (a/2, a/2, a/2)$. $|c_1|^2 = 0$. (a) $|c_2|^2 = 1.0$ (complete excitation), (b) $|c_2|^2 = 0.4$, (c) $|c_2|^2 = 0.4995$, (d) $|c_2|^2 = 0.5$ (type-II QDs are total transparent). Y. Fu, Photonic energy band structure of excitonic quantum dot dimer system, J. Appl. Phys. vol. 106, p. 054302(5), 2009. Copyright (2009) by The American Institute of Physics.

QDs ($\hbar\omega_1 = 1.5\,\mathrm{eV}$) occupy the normal fcc lattice sites, while the excited type-II QDs ($\hbar\omega_2 = 1.503\,\mathrm{eV}$) are displaced by $\boldsymbol{\tau} = (a/2, a/2, a/2)$. $\hbar\omega_{\mathrm{LT}} = 5\,\mathrm{meV}$. Here we observe the modification of the photonic band structure of the QD dimer system by pumping one type of the QDs (type-II QDs), which evolves from the one of only type-I QDs ($|c_1|^2 = 0$, i.e., at their ground exciton state) when type-II QDs are transparent ($|c_2|^2 = 0.5$, Fig. 4.12d) to the compound system ($|c_2|^2 = 1.0$, Fig. 4.12a). We can observe modified but still characteristic features of the photonic dispersions of individual type-I and type-II QDs in their separate photonic crystal formats in the compound system. More specifically, the resonance features of type-I QDs around $(\omega - \omega_1)/\omega_{\mathrm{LT}} = 0.3$ in Fig. 4.12d becomes compressed by the radiatve interaction between type-I QDs and type-II QDs; they are also shifted down to around 0.18 in Fig. 4.12a.

It has been obtained that the numerical solutions of Eq. 4.119 are symmetric with respect to $|c_2|^2$ and $1 - |c_2|^2$ when $c_1 = 0$. Thus, Fig. 4.12a represents also the photonic dispersion of the QD dimer system when both type-I and type-II QDs are all initially at their ground exciton states, i.e., $|c_1|^2 = |c_2|^2 = 0$. It is concluded that the modification of the exciton state (from ground state to excited state) in one type of QDs in the dimer system does not affect the feature of the dispersion structure of the dimer system.

Note that the above conclusions are obtained under the periodic boundary condition. In other words, they are rigorous only when the loss is infinitely small. The QD dimer system was proposed in the previous sub-section to achieve lossless negative dielectric constant by using the Maxwell-Garnett formalism, which could amount to a mean-field approximation Numerical FDTD calculation[37] cannot resolve the issue of long-distance transport properties of the electromagnetic field in the lossless negative-dielectric QD dimer system because of the limited simulation space it can handle. This section thus complements the previous section by showing that the transport of an external electromagnetic field is normal in the QD dimer system when one type of the QDs becomes highly excited. In other words, for the QD dimer system at the wavelength of zero loss, the propagation of the electromagnetic field in the QD dimer system will be normal even if the dielectric constant of the dimer system becomes effectively negative. This section suggests that achieving a lossless isotropic material at a single frequency is possible by using the QD dimer system.

Note that Fig. 4.12 shows that the effect of the excited type-II QDs at $\boldsymbol{a} + \boldsymbol{\tau}$ persists even when the excitation rate is as close as to be 0.4995. This is due to the periodic boundary conditions. Furthermore, Fig. 4.12d is exactly the same when type-II QDs located at $\boldsymbol{a} + \boldsymbol{\tau}$ are physically removed; see Fig. 4.6 in sub-Section 4.5.2.

References

[1] A. Kavokin, "Exciton-polaritons in microcavities: Present and Future," Appl. Phys. A, **89**, 241–246, 2007.

[2] B Alén, A. Högele, M. Kroner, S. Seidl, K. Karrai, R. J. Warburton, A. Badolato, G. Medeiros-Ribeiro, and P. M. Petroff, "Absorptive and dispersive optical responses of excitons in a single quantum dot," Appl. Phys. Lett., **89**, 123124(3), 2006.

[3] F. Wang, J. Shan, M. A. IslamS, I. P. Herman, M. Bonn, and T. F. Heinz, "Exciton polarizability in semiconductor nanocrystals," Nat. Mate. **5**, 861–864, 2006.

[4] E. Merzbacher, *Quantum Mechanics*, 3rd edn, John Wiley & Sons, New York, 1998.

[5] D. G. Gevaux, P. Atkinson, D. Anderson, A. J. Bennett, S. P. Bremner, J. Griffiths, D. J. P. Ellis, R. M. Stevenson, G. A. C Jones, D. A. Ritchie, and A. J. Shields, "Controlling spontaneous emission from quantum dots using photonic crystal microcavities," Phys. Stat. Sol. (c), **3**, 3676–3679, 2006.

[6] J. P. Reithmaier, G. Sek, A. Löffler, C. Hofmann, S. Kuhn, S. Reitzenstein, L. V. Keldysh, V. D. Kulakovskii, T. L. Reinecke, and A. Forchel, "Strong coupling in a single quantum dot-semiconductor microcavity system," Nature, **432**, 197–200, 2004.

[7] G. Ramon, U. Mizrahi, N. Akopian, S. Braitbart, D. Gershoni, T. L. Reinecke, B. D. Gerardot, and P. M. Petroff, "Emission characteristics of quantum dots in planar microcavities," Phys. Rev. B, **73**, 205330(11), 2006.

[8] L. C. Andreani, D. Gerace, and M. Agio, "Exciton-polaritons and nanoscale cavities in photonic crystal slabs," Phys. Stat. Sol. (b), **242**, 2197–2209, 2005.

[9] E. M. Kessler, M. Grochol, and C. Piermarocchi, "Light-mass Bragg cavity polaritons in planar QD lattices," Phys. Rev. B, **77**, 085306(9), 2008.

[10] M. S. Ünlü, S. Strite, "Resonant cavity enhanced photonic devices," J. Appl. Phys., **78**, 607–639, 1995.

[11] C. Weisbuch, M. Nishioka, A. Ishikawa, and Y. Arakawa, "Observation of the coupled exciton-phonon mode splitting in a semiconductor quantum microcavity," Phys. Rev., Lett., **69**, 3314–3417, 1992.

[12] R. Houdre, R. P. Stanley, U. Oesterle, M. Ilegems, and C. Weisbuch, "Room temperature cavity polaritons in a semiconductor microcavity," Phys. Rev., **B49**, 16761–16764, 1994.

[13] T. B. Norris, J. K. Rhee, C. Y. Sung, Y. Arakawa, M. Nishioka, and C. Weisbuch, "Time resolved vacuum Rabi oscillations in a semiconductor quantum microcavity," Phys. Rev., **B50**, 14663–14666, 1994.

[14] L. C. Andreani, V. Savona, P. Schwendimann, and A. Quattropani, "Polaritons in high reflectivity microcavities: semiclassical and full quantum treatment of optical properties," Superlatt. Microstruct. **15**, 453–458, 1994; V. Savona, L. C. Andreani, P. Schwendimann, and A. Quattropani, "Quantum well excitons in semiconductor microcavities: unified treatment of weak and strong coupling regimes," Solid State Commun., **93**, 733–739, 1995.

[15] S. Pau, G. Björk, J. Jacobson, H. Cao, and Y. Yamamoto, "Stimulated emission of a microcavity dressed exciton and suppression of phonon scattering," Phys. Rev., **B51**, 7090–7100; "Microcavity exciton-polariton splitting in the linear regime," 14437–41447, 1995.

[16] E. L. Ivchenko, M. A. Kaliteevski, A. V. Kavokin, and A. I. Nesvizhskii, "Reflection and absorption spectra from microcavities with resonant Bragg quantum wells," J. Opt. Soc. Am. **B13**, 1061–1068, 1996.

[17] Y. Fu, M. Willander, E. L. Ivchenko, and A. A. Kiselev,"Four-wave mixing in microcavities with embedded quantum wells," Phys. Rev., **B55**, 9872–9879, 1997.

[18] R. Shimano, S. Inouye, M. Kuwata-Gonokami, T. Nakamura, M. Yamanishi, and I. Ogura,"Efficient phase conjugation wave generation from a GaAs single quantum well in a microcavity," Jpn. J. Appl. Phys., **34**, L817–L820, 1995.

[19] F. Quochi, G. Bongiovanni, A. Mura, S. Gürtler, C. Dill, R. Houdré, and J. L.Staenli,"Four-wave mixing response of a semiconductor microcavity: the influence of strong excitonic inhomogeneous broadening," 23rd International Conference on the Physics of Semiconductor, World Scientific, Singapore, 1996, 3151–3154.

[20] *Photonic Band Gaps and Localization*, ed. C. M. Soukoulis, NATO Advanced Study Institutes, Ser.B, vol. 308, Plenum Press, New York, 1993.

[21] Y. Fu, M. Willander, and E. L. Ivchenko,"Photonic dispersions of semiconductor-quantum-dot-array-based photonic crystals in primitive and face-centered cubic lattices," Superlatt. Microstruct. **27**, 255–264, 2000.

[22] E. L. Ivchenko, Y. Fu, and M. Willander,"Exciton polaritons in quantum-dot photonic crystals," Phys. Solid State, **42**, 1756–1765, 2000.

[23] Y. A. Vlasov, V. N. Astratov, O. Z. Karimov, A. A. Kaplyanskii, V. N. Bogomolov, and A. V. Prokofiev,"Existence of a photonic pseudogap for visible light in synthetic opals," Phys. Rev., **B55**, R13357–R13360, 1997.

[24] Y. A. Vlasov, A. K. Luterova, B. I. Pelant, B. Honerlage, and V. N. Astratov,"Enhancement of optical gain of semiconductors embedded in three-dimensional photonic crystals," Appl. Phys. Lett., **71**, 1616–1618, 1997.

[25] Y. A. Vlasov, Y. Nan, and D. J. Norris,"Synthesis of photonic crystals for optical wavelengths from semiconductor quantum dots," Adv. Mater., **11**, 165–169, 1999.

[26] M. M. Sigalas, C. M. Soukoulis, C. T. Chan, and K. M. Ho,"Electromagnetic-wave propagation through dispersive and absorptive photonic-band-gap materials," Phys. Rev., **B49**, 11080–11087, 1994.

[27] V. G. Veselago,"The electrodynamics of substances with simultaneously negative values of ϵ and μ," Sov. Phys. Uspekhi, **10**, 509–514, 1968.

[28] J. B. Pendry,"Negative refraction makes a perfect lens," Phys. Rev. Lett., **85**, 3966–3969, 2000.

[29] M. Yang and K. J. Webb,"Poynting vector analysis of a superlens," Opt. Lett., **30**, 2382–2384, 2005.

[30] P. B. Johnson and R. W. Christy,"Optical constants of noble metals," Phys. Rev. B, **6**, 4370–4379, 1972.

[31] K. J. Webb and J. Li,"Analysis of transmission through small apertures in conducting films," Phys. Rev. B, **73**, 33401, 2006.

[32] K. J. Webb and J. Li,"Waveguide cavity surface-enhanced Raman scattering," Phys. Rev. B, **73**, 73404, 2006.

[33] J. C. Maxwell-Garnett,"Colours in metal glasses, in metallic films, and in metallic solutions. II," Philos. Trans. R. Soc. Lond., **205**, 237–288, 1906.

[34] J. I. Gittleman and B. Abeles,"Comparison of the effective medium and the Maxwell-Garnett predictions for the dielectric constants of granular metals," Phys. Rev. B, **15**, 3273–3275, 1977.

[35] R. W. Cohen, G. D. Cody, M. D. Coutts, and B. Abeles,"Optical properties of granular silver and gold films," Phys. Rev. B, **8**, 3689–3701, 1973.

[36] Y. Fu, L. Thylén, and H. Ågren,"A lossless negative dielectric constant from quantum dot exciton polaritons," Nano Lett., **8**, 1551–1555, 2008.

[37] A. Bratkovsky, E. Ponizovskaya, S.-Y. Wang, Petter Holmström, L. Thylén, Y. Fu, and H. Ågren,"A metal-wire/quantum-dot composite metamaterial with negative ϵ and compensated optical loss," Appl. Phys. Lett. **93**, 193106(3), 2008.
[38] M. I. Stockman,"Criterion for negative refraction with low optical losses from a fundamental principle of causality," Phys. Rev. Lett., **98**, 177404(4), 2007.
[39] D. Smith, S. Schultz, P. Markos, and C. Soukoulis,"Determination of effective permittivity and permeability of metamaterials from reflection and transmission coefficients, Phys. Rev. B, **65**, 195104(5), 2002.

Chapter 5

Optoelectronic Devices

In semiconductor technologies the geometric miniaturization of individual components and the development of multi-function devices have been scaling according to "Moore's Law." On the other hand, the miniaturization of photonic components such as waveguides, emitters, and switches is a totally different matter as it is limited by the diffraction of light. For instance, diode lasers and optical guided wave structures must confine the light within their cavities and core layers, respectively, so that the minimum feature sizes of these optical structures are limited to around one wavelength. The waveguides mostly used in today's photonics are low index contrast optical fibers, which have a cross-section of tens of dielectric wavelengths squared, with extensions that are several thousands of wavelengths for a device such as a coupler. There are also integrated photonic circuits with high index contrast waveguides. However, the minimal lateral field confinement of these systems is still approximately one dielectric wavelength, and the length of a coupler is several hundreds of wavelengths. Waveguide couplers based on refractive-index-modulation photonic crystals, reducing this length to tens of wavelengths, have been fabricated.[1,2] However, the lateral feature of these becomes significantly increased to several dielectric wavelengths in order to establish the Bragg boundary. This imposes severe limitations on packing density and construction of high-density photonic integrated circuits needed for future photonics. Those will require increased integration of photonic devices of up to 1000×1000 input and output channels on a single substrate for increased data transfer rates and for improved photolithography with feature lengths well below one wavelength.[3]

By the unique physical properties thus discussed, exciton and exciton–polariton in quantum dots (QDs) have already become an enabling science in the form of optoelectronics and in many application fields such as photonics (including optical information technology), health care, life sciences, displays, security, sensing, lighting, materials science, and manufacturing. In this chapter, we present major optoelectronic devices based on nanostructures at the electronic and photonic engineering levels. Despite the extensive and successful research and technical development in this area both fundamental and technical difficulties remain to be solved. The major fundamental difficulty is that the unique optical properties of excitons and exciton–polaritons in nanostructures are based on quantum confinement, whereas for all practical applications we need to have direct interactions between excitons in nanostructures and the external circuit. For example, it is difficult to extract carriers from QDs into the circuit in solar cells. Confined excitons in nanostructures, by definition, are completely surrounded by insulating materials and a direct conducting path would break the confinement responsible for their special properties.

The realization of potential applications is intimately tied to the control of geometry and location of the nanoscale features. A precise positioning of the nanostructures not only allows for reproducible communication with the environment, but is also necessary for defining the interaction between the nanostructures. Solid-state-based nanostructures (e.g., self-assembled QDs) provide excellent communication to the environment. The geometry control of the QD size is, however, still a limiting factor. The size of colloidal II–VI QDs can be very well controlled during the chemical synthetic processes and many such QDs used in bioimaging applications are core-multishells formed so that possible effects of the surface and external environment on the exciton in the central core region are reduced. Again the interconnection to the environment becomes a major issue.

The use of exciton QD-based quantum bits (qubits) faces a number of difficult issues such as noise, decoherence, and problems referring to their fabrication. Noise in the environment seems unavoidable, and charge qubit states are easily affected by various sources of noise (thermal, electronic, acoustic, etc) present in the semiconductor material. Spin qubits are better isolated from the environment. The spin degree of freedom couples to higher-order fluctuations in the electric and magnetic fields. Although this is an advantage as such, it makes it more arduous to control the device. Spin transport is also a problem as decoherence may be a roadblock, like for most quantum systems. Fabrication is very demanding concerning purity and precision, and no two fabricated qubits seem to become alike and each of them would have to be individually characterized.

5.1 LIGHT EMITTING DIODE

Quantum information technology requires triggered emission of single photons and photon pairs that have been realized by InAs/GaAs QDs at low temperatures ($<50\,K$)[4] and by CdSe/ZnSe QDs at temperatures up to 200 K.[5] Optical quantum enhanced interferometry was proposed based on "triggered" entangled photon pairs by the radiative decay of the biexciton state in a semiconductor QD.[6] A planar photonic crystal microcavity design specially tailored for cavity quantum electrodynamics with a single QD emitter embedded in a semiconductor was designed for strong optical coupling and lasing with a single QD exciton.[7] Single-photon emission on demand was also demonstrated at around 510 nm with sub-nano-second accuracy from single CdSe QDs on ZnSe grown by molecular-beam epitaxy.[8]

Arakawa and Sakaki predicted in the early 1980s that QD lasers should exhibit a performance that is less temperature-dependent than existing semiconductor lasers, and that, in particular, will not degrade at elevated temperatures.[9] The first demonstration of a QD laser was reported in 1994.[10] In addition to utilizing their quantum size effects in edge-emitting lasers, self-assembled QDs have also been incorporated within vertical cavity surface-emitting lasers. Newly developed QD lasers achieve high-speed operation of 10 gigabits per second (Gbps) across a temperature range of 20–70°C without electrical current adjustments, and has minimal output fluctuations caused by changes in temperature.[11, 12]

Because of an almost exact balance between absorption and stimulated emission in QDs excited with single excitons, optical gain can only occur in QDs that contain at least two excitons. A complication associated with this multiexciton nature of light amplification refers to fast optical-gain decay induced by nonradiative Auger recombination. Core-shell type-II hetero-nanocrystals were engineered to spatially separate electrons and holes between the core and the shell; single-exciton optical gain and thus amplification in these QDs were demonstrated.[13]

A large amount of both theoretical and experimental works has been published concerning the resonant tunneling structure (RTD) leading to a broad range of electrical[14, 15] as well as optical[16, 17] applications. Photoluminescence (PL) characterization of the resonant-tunneling light-emitting diode (RTLED) consists of mainly the recombination of electrons and holes that each tunnel from the opposite contact layers into the central RTD active layer. If the two contact layers are both n^+-type, the electrons are majority carriers and holes are photogenerated minority carriers (if the contacts are p^+-type, the roles of electrons and holes are exchanged);[18] if one contact is n^+-type and the other p^+-type, both electrons and holes are majority carriers.[19]

There are two types of commonly investigated RTLEDs.

(1) Electrons or holes. The sample is prepared in the sequence of an n^+-type (p^+-type) GaAs substrate as the emitter, followed by an n-type (p-type) or not-intentionally doped (NID) GaAs spacer, AlAs barrier, GaAs well, AlAs barrier, n-type (p-type) or NID GaAs spacer, and finally n^+-type (p^+-type) GaAs cap as the collector.

(2) Electrons and holes. The emitter is an n^+-type GaAs substrate followed by an n-type or NID GaAs spacer, then AlAs barrier, GaAs well, AlAs barrier, p-type or NID GaAs spacer, and finally p^+-type GaAs cap as the collector.

The energy band structure of the RTLED is described by the combination of the Schrödinger and Poisson equations. The results are presented in Fig. 5.1.

Let the sample growth direction be the z-direction. The wave functions of carriers in both the emitter and the collector are described by propagating waves in the z-direction. The envelope function in the xy-plane is in the form of $e^{i\boldsymbol{k}_i\cdot\rho}$ because of the translational symmetry there, where $\boldsymbol{k}_i$ and ρ are the wave vector

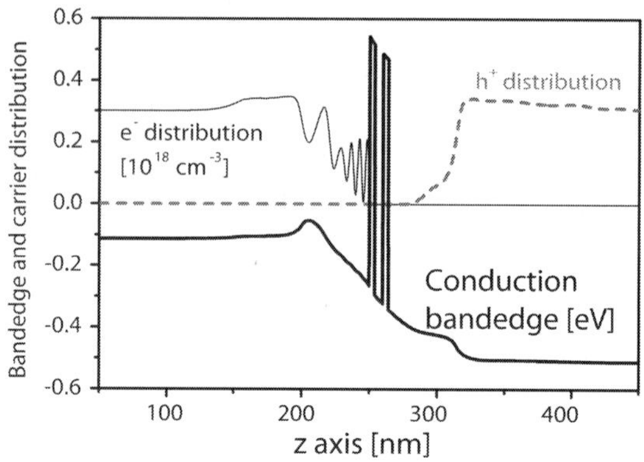

Fig. 5.1 Energy band diagram of the resonant tunneling light-emitting diode (RTLED) when the external bias applied across the RTLED is 0.4 V.

and coordinate in the xy-plane, respectively. In the central active region including the double barriers the wave function is denoted as $\psi_i(z)$. The total wave function is of course

$$\psi_i(z)u_i(\boldsymbol{r})\mathrm{e}^{i\boldsymbol{k}_i\cdot\rho} \tag{5.1}$$

where $u_i(\boldsymbol{r})$ is the Bloch function of band i.

The kinetic energy of the carriers in the effective-mass approximation is expressed as

$$\frac{1}{2}\boldsymbol{p}\bar{W}\boldsymbol{p} \tag{5.2}$$

where $\boldsymbol{p}$ is the momentum operator and $\bar{W}$ is the 3×3 inverse effective-mass tensor. The interaction between an electromagnetic field and the electron wave is

$$e\boldsymbol{A}\bar{W}\boldsymbol{p} \tag{5.3}$$

where $\boldsymbol{A}$ is the vector potential of the radiation field. (Refer to sub-section 5.3.1 about detailed discussion of the inverse effective-mass tensor.)

The optical transition matrix element between two electronic states $\psi_i(z)u_i(\boldsymbol{r})\mathrm{e}^{i\boldsymbol{k}_i\cdot\rho}$ and $\psi_j(z)u_j(\boldsymbol{r})\mathrm{e}^{i\boldsymbol{k}_j\cdot\rho}$ is

$$\begin{aligned} w_{ij} &= \langle\psi_i(z)u_i(\boldsymbol{r})\mathrm{e}^{i\boldsymbol{k}_i\cdot\rho}|\boldsymbol{A}\bar{W}\boldsymbol{p}|\psi_j(z)u_j(\boldsymbol{r})\mathrm{e}^{i\boldsymbol{k}_j\cdot\rho}\rangle \\ &= \delta_{\boldsymbol{k}_i,\boldsymbol{k}_j}\langle\psi_i u_i|\boldsymbol{A}\bar{W}|u_j\boldsymbol{p}\psi_j + \psi_j\boldsymbol{p}u_j + \hbar\boldsymbol{k}_j\psi_j u_j\rangle \end{aligned} \tag{5.4}$$

which indicates that optical transitions are vertical in the $\boldsymbol{k}$-space (the momentum is conserved in the xy-plane). The last term is zero because of the wave function orthogonal property so that

$$w_{ij} = \delta_{\boldsymbol{k}_i,\boldsymbol{k}_j}\Big(\langle u_i|u_j\rangle\langle\psi_i|\boldsymbol{A}\bar{W}\boldsymbol{p}|\psi_j\rangle + \langle u_i|\boldsymbol{A}\bar{W}\boldsymbol{p}|u_j\rangle\langle\psi_i|\psi_j\rangle\Big) \tag{5.5}$$

It is thus concluded that optical transitions occur in two ways: the band-to-band transition $\langle u_i|\boldsymbol{p}|u_j\rangle$ and between the envelope functions $\langle\psi_i|\boldsymbol{p}|\psi_j\rangle$ in the z-axis.

The above equation can be more explicitly expressed as

$$w_{ij} = \delta_{\boldsymbol{k}_i,\boldsymbol{k}_j}\left(\langle u_i|\boldsymbol{A}\bar{W}\boldsymbol{p}|u_j\rangle\langle\psi_i|\psi_j\rangle + \langle u_i|u_j\rangle\sum_l A_l W_{lz}\langle\psi_i|p_z|\psi_j\rangle\right) \tag{5.6}$$

because ψ is only z-dependent. Here $l = x, y, z$.

For those materials whose band structure is parabolic ($W_{il} = 0$ if $i \neq j$ and $W_{ii} = 1/m^*$), Eq. 5.6 becomes

$$w_{ij} = \frac{\delta_{\boldsymbol{k}_i,\boldsymbol{k}_j}}{m^*}\Big(\langle u_i|\boldsymbol{A}\cdot\boldsymbol{p}|u_j\rangle\langle\psi_i|\psi_j\rangle + A_z\langle u_i|u_j\rangle\langle\psi_i|p_z|\psi_j\rangle\Big) \tag{5.7}$$

Here the first term describes the light emission due to the band-to-band recombination, while the second one results from the recombination of the electronic sublevels along the sample growth direction.

As we shall see similarly when studying the absorption of incident radiation by one-dimensional quantum well (QW) detector, a quantum optical selection rule is observed here that the photon emitted from the recombination of the electronic sublevels along the sample growth direction (z-direction) in a one-dimensional system (xy-plane is translationally symmetric) is polarized and propagates in the xy-plane, i.e.,

$$A_z \neq 0$$

by Eq. 5.7, when the energy band structure of the semiconductor material is parabolic. This profoundly affects the device structures of QW light-emitting diode and photodetector.

The transition probability per unit time for the emission of a light quantum $\hbar\omega$ is

$$W \propto \sum_{ij}\frac{\Gamma\omega(N_\omega+1)}{(E_j - E_i - \hbar\omega)^2 + \Gamma^2}|w_{ij}|^2 f_i(1-f_j) \tag{5.8}$$

where N_ω is the number of photons, f_i is the occupation of electron state i, and Γ is the life time of electron. The term $f_i(1-f_j)$ means that the transition happens between occupied state i and empty state j.

When considering electron and hole recombination,

$$W \propto \sum_{ij}\frac{\Gamma\omega(N_\omega+1)}{(E_{ei} - E_{hj} - \hbar\omega)^2 + \Gamma^2}\Big|p_{1,ij}\langle u_c|u_v\rangle + p_{2,ij}\langle u_c|\nabla|u_v\rangle\Big|^2 f_{ei}f_{hj} \tag{5.9}$$

where

$$\begin{aligned} p_{1,ij} &= \langle\psi_{ei}|\frac{\partial}{\partial z}|\psi_{hj}\rangle \\ p_{2,ij} &= \langle\psi_{ei}|\psi_{hj}\rangle \end{aligned} \tag{5.10}$$

and u_c and u_v are CB and VB Bloch functions, respectively.

It is easy to understand that when wave function overlapping is zero, $p_1 = p_2 = 0$. By Fig. 5.1 we see that the wave function overlapping in the region outside the

QW between two barriers can then be neglected so that we only need to concentrate our attention on the optical transitions in the QW region.

Because of $\delta_{\boldsymbol{k}_i,\boldsymbol{k}_j}$, the sum over i and j is reduced to either i or j. At low temperature where thermal excitation is negligible,

$$\sum_{ij} f_{ei} f_{hj} = \min(n_e, n_h) \tag{5.11}$$

where n_e and n_h are two-dimensional electron and hole concentrations in the QW region. The use of "min" here is due to the demand that both the electron and hole states involved must be occupied. It is demonstrated experimentally that PL intensity sharply increases to reach its maximal value at resonance when the amplitudes of the carrier wave functions are maximal in the QW region. The maximal amplitudes of the wave functions indicate naturally a maximal carrier concentration.

The wave functions of electrons can be approximated as

$$\left(-\frac{\hbar^2}{2m_e^*}\frac{\partial^2}{\partial z^2} + \alpha z\right)\psi_{ei} = E_{ei}\psi_{ei} \tag{5.12}$$

in the QW region when the barrier is infinitely high, while for holes,

$$\left(\frac{\hbar^2}{2m_h^*}\frac{\partial^2}{\partial z^2} + E_g - \alpha z\right)\psi_{hj} = E_{hj}\psi_{hj} \tag{5.13}$$

where α is the electric field induced by the external bias applied across the QW. In Fig. 5.2, we show the numerical results of E_{e1}, E_{h1}, and the matrix elements p_1 and p_2 as functions of α.

When α is very small, the bottom of the QW is rather flat. In this case both the electron and hole subbands are of sinusoid form so that p_1 is very small, as

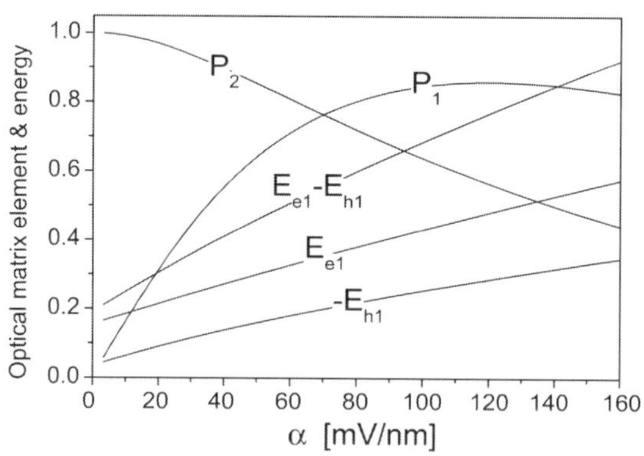

Fig. 5.2 The first electron and hole subbands and the optical transition matrix elements as function of the electric field strength in the quantum well. Y. Fu and M. Willander, Theoretical studies on the optical characteristics of the resonant tunneling light-emitting diode, J. Appl. Phys. vol. 76, p. 4225–30, 1994. Copyright (1994) by The American Institute of Physics.

demonstrated in Fig. 5.2. Increasing α pushes the electron to one side of the QW and the hole to the other side so that p_1 is increased. However, if α is too large, there will be no overlapping between the two wave functions and therefore p_1 reduces. The term p_2 decreases monotonously following the increase of α as a result of decreasing wave function overlapping.

For α less than 50 mV/nm, the decrease of p_2 is not significant as compared with the increase of p_1. It is thus concluded that for high optical transition, the electric field inside the QW should be large.

Since the effective mass of the hole is usually quite large compared with that of the electron, the PL spectrum of the RTLED is mainly determined by the electron resonances. For the n^+-NID-p^+ RTLED, the band diagram is almost symmetric. When the first electron resonance occurs, there is at least the first hole subband (in the QW region) already below the Fermi level of the p^+-contact. It is thus expected that the first hole subband ($h1$) related peak in the PL spectrum corresponding to the $h1$ resonance is missing because the first electron subband ($e1$) at the moment is not occupied. After the first $e1$ resonance there will be almost a one-to-one correspondence between peaks in PL spectrum and the resonances in I–V relationship. Maximal light output is expected when the electron and hole resonances occur at the same external bias.

5.2 LASER

The laser is a very special source of light. Modern optics effectively dates from the invention of the laser (Light Amplification by Stimulated Emission of Radiation) in 1960.[20] Lasers are extremely important in optoelectronics. The most commonly used visible-light laser is the helium-neon (He-Ne) laser. A tube with parallel mirrors at the ends is mounted providing a positive feedback to stimulate more photons. Part of the photons can emerge from the tube from the partial mirror in the form of laser:

(1) The energy of the emerged photons is accurately defined when the energy levels are sharp.
(2) The intensity is relatively large if the volume of the tube is large.
(3) It is well collimated since the light will get amplified within the tube only if it is able to bounce between the mirrors.
(4) The phase is well defined since the phase of the stimulated photon is locked to that of the stimulating photon.

Let us consider a two-energy-level atom (E_1 and E_2, $E_2 > E_1$).

- The atom can be raised from E_1 to E_2 by absorbing a photon

$$\hbar\omega_{12} = E_2 - E_1$$

(stimulated absorption $E_1 \rightarrow E_2$);

- The atom, after having been excited, will eventually, spontaneously, revert to its ground state E_1 by emitting a photon $\hbar\omega_{12}$ (spontaneous emission $E_2 \rightarrow E_1$);
- The atom at E_2 can be stimulated to decay to E_1 by another photon $\hbar\omega_{12}$ (stimulated emission $E_2 \rightarrow E_1$). (There can be no spontaneous "absorption," since this would violate the law of energy conservation.)

Consider an assembly of atoms being in equilibrium with each other and with the radiation in a closed system. The rate R_{12} (per unit volume) at which atoms are raised to E_2 can be expressed as

$$R_{12} = B_{12} N_1 n_{\text{ph}} \tag{5.14}$$

where N_1 is the density of atoms at state E_1, n_{ph} is the photon density. Similarly, the rate R_{21} at which atoms in state E_2 are stimulated to return to state E_1 is given by

$$R_{21} = B_{21} N_2 n_{\text{ph}} \tag{5.15}$$

where N_2 is the density of atoms in state E_2. The spontaneous emission from E_2 to E_1 is proportional to N_2

$$S_{21} = A_{21} N_2 \tag{5.16}$$

The coefficients A_{21}, B_{12}, and B_{21} are called the Einstein coefficients.

Clearly we must have

$$B_{12} N_1 n_{\text{ph}} = B_{21} N_2 n_{\text{ph}} + A_{21} N_2 \tag{5.17}$$

in equilibrium, since the total upward and downward transition probabilities must be equal.

It is easy to obtain

$$B_{12} = B_{21} = A_{21} \frac{\pi c^3}{2\hbar\omega_{12}^3} \tag{5.18}$$

which are known as the Einstein relations.

Here A_{21} is an atomic constant, representing the reciprocal of the spontaneous decay. By the Einstein relations, the ratio of stimulated to spontaneous emission from E_2 to E_1 is

$$\frac{R_{21}}{S_{21}} = \frac{1}{\exp\left(\frac{\hbar\omega_{12}}{k_B T}\right) - 1} \tag{5.19}$$

which is only about 2×10^{-27} for a He-Ne discharge at 370 K with $\lambda = 632.8$ nm. Stimulated emission is thus very unlikely for equilibrium systems.

One way to increase the level of stimulated emission is to inject radiation so that n_{ph} is increased above its equilibrium value, and the stimulated emission greatly exceeds the spontaneous emission

$$B_{21} N_2 n_{\text{ph}} \gg A_{21} N_2 \tag{5.20}$$

However, increasing $n_{\rm ph}$ does also increase the stimulated absorption in the way

$$B_{21}N_1n_{\rm ph} = B_{21}N_2n_{\rm ph} \tag{5.21}$$

which indicates $N_1 = N_2$ under these conditions. In other words, an incoming photon at $\hbar\omega_{12}$ is just as likely to cause a downward transition from E_2 to E_1 (stimulated emission) as it is an upward one from E_1 to E_2 (stimulated absorption). Hence we cannot increase the population N_2 above N_1 simply by pumping more radiation.

Consider a three-level rather than a two-level system, $E_1 > E_2 > E_3$. We inject light at $\hbar\omega_{13} = E_3 - E_1$ so that there is a large amount of stimulated absorption from E_1 to E_2. Spontaneous emissions occur in the ways of $E_3 \to E_2$ then $E_2 \to E_1$, with also $E_3 \to E_1$. When E_2 is a meta-stable state so that $E_3 \to E_2$ is fast and $E_2 \to E_1$ relatively much slower, atoms will accumulate at E_2. These atoms at E_2 are immune from stimulated emission by photons of $\hbar\omega_{13}$. We can thus ensure

$$N_2 > N_1 \tag{5.22}$$

so that we have an "inverted population" as a result of the "pump" at $\hbar\omega_{13}$. This inverted population can now be exploited to give optical amplification at $\hbar\omega_{12}$.

Suppose that under these conditions, the number of photon density at $\hbar\omega_{12}$ is n_{12}. The rate at which n_{12} increases depends on the difference between upward and downward transitions between E_1 and E_2

$$\frac{\mathrm{d}n_{12}}{\mathrm{d}t} = B_{21}N_2n_{12} - B_{12}N_1n_{12} \tag{5.23}$$

which gives us

$$n_{12} = n_{12}^0 \exp\left[B_{12}(N_2 - N_1)t\right] \tag{5.24}$$

If the injected wave is traveling at velocity c, we can transfer to a distance parameter via $x = ct$

$$n_{12} = n_{12}^0 \exp\left[\frac{B_{12}}{c}(N_2 - N_1)x\right] \tag{5.25}$$

As compared with the standard loss/gain relation for propagation in an interactive medium $I = I_0 \exp(gx)$, it is clear that the gain coefficient, defined as the ratio between the net power emitted per unit volume and the power per unit area, is identified as

$$\begin{aligned} \rho_{12} &= n_{12}\hbar\omega_{12} \\ \rho_{12} &= \rho_{12}^0 \exp\left[\frac{\hbar\omega_{12}}{c}B_{12}(N_2 - N_1)x\right] \\ g &= \frac{\hbar\omega_{12}}{c}B_{12}(N_2 - N_1) \end{aligned} \tag{5.26}$$

which is positive (gain rather than loss) if $N_2 > N_1$, as will be the case for an inverted population and the medium is an optical amplifier.

The laser, which is an acronym of "Light Amplification by Stimulated Emission of Radiation," is an optical resonator containing a medium that amplifies light generated by stimulated emission. The laser beams have several unique features compared with other light beams:

(1) they produce intense coherent monochromatic induced radiation;
(2) the incoherent spontaneous radiation component in the beams, called noise, is negligible;
(3) the beams can be highly collimated;
(4) the beams can be focused very sharply.

A laser consists principally of three key elements. One is the laser medium that generates and amplifies the light. The second is the power supply, which delivers energy to the laser medium to excite light emission. The third is an optical cavity, or resonator, which acts as a spectral filter and concentrates the light to the laser medium.

5.2.1 *The Generation of Laser Radiation*

Laser amplification can take place in a great variety of media including solids, gases, liquids, and plasmas. What material is used depends mainly on which wavelengths are to be produced and for which application. To obtain laser radiation, stimulated emission must occur in the laser medium. Stimulated emission together with spontaneous emission and absorption is the most common type of electronic transitions that atoms can undergo when interacting with photons.

Electronic transitions occur due to the fact that electrons occupy stationary states (eigenvalues and eigenfunctions of the system). When the system is exposed to radiation, interaction between the photons of the radiation field and electrons in the system occurs, resulting in radiative or nonradiative electronic transitions between two energy levels.

If an atom is initially in an excited state with electron energy E_2, it may fall to the lower energy state E_1 spontaneously, (Fig. 5.3a). When doing so, the electron releases its excess energy in the form of a photon of random direction, i.e., the light emitted is incoherent. The frequency of the emitted photon corresponds to the energy difference between the two energy states, that is

$$\hbar\omega = \frac{hc_0}{\lambda} = E_2 - E_1 \tag{5.27}$$

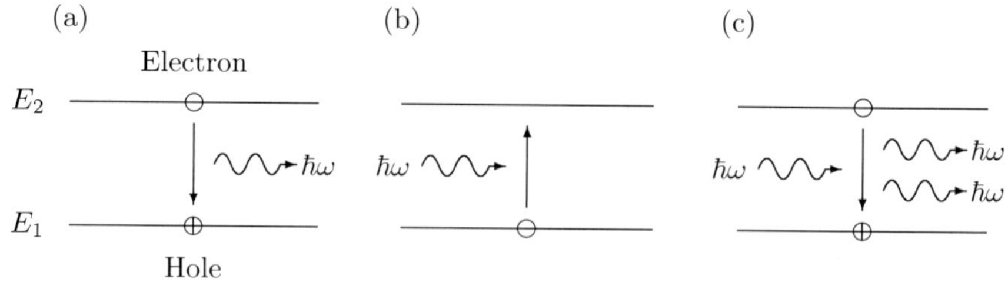

Fig. 5.3 Electronic transitions between the two energy levels E_1 and E_2. (a) spontaneous emission; (b) absorption; (c) stimulated emission. In all cases $\hbar\omega = E_2 - E_1$.

where h is Planck's constant, $\hbar = h/2\pi$, c_0 is the speed of light in vacuum, and λ is the wavelength of the emitted radiation. The process is called the spontaneous emission because the transition is not triggered by outside intervention.

Stimulated absorption is a transition induced by incident photons. In this process the electron is raised from a lower to a higher energy state, when it absorbs a photon having an energy that equals the energy difference between the two electronic states, (Fig. 5.3b).

The third transition that is induced by photons is the stimulated emission. Initially the electron is in an excited state. As it interacts with an incident photon, it is stimulated to emit a photon when it undergoes a downward transition, (Fig. 5.3c). The phenomenon that forms the basis of laser action is that the emitted photon has exactly the same characteristics as the stimulating photon; it has the same frequency, phase, polarization, and direction of propagation. Amplification of radiation has thus occurred in the medium.

All of the processes mentioned above take place when the material interacts with the radiation of appropriate frequency. Which process is dominating depends on the intensity of the radiation and on whether the material is at thermal equilibrium. At thermal equilibrium, the number of electrons, N_2, in an excited energy level is smaller than the number of electrons, N_1, at a lower energy level. The occupation of state E_i is in the form of Fermi–Dirac distribution

$$N_i = \frac{N_{\mathrm{DOS}}(E_i)}{1 + \exp\left(\dfrac{E_i - E_f}{k_B T}\right)} \tag{5.28}$$

where k_B is Boltzmann's constant and T is the temperature. $N_{\mathrm{DOS}}(E_i)$ is the density of states at E_i. E_f is the Fermi level at which the occupation probability is 0.5.

In this case, a photon is therefore more likely to be absorbed by electrons at the low energy level than to stimulate the optical transition of high-level electrons to low-energy state; spontaneous emission dominates over stimulated emission.

For stimulated emission to be the dominant process, a population inversion is needed. Population inversion is a state of nonthermal equilibrium in which N_2 is larger than N_1. In this situation, the total emission rate becomes larger than the absorption rate. This means that if a radiation with appropriate energy passes through the material, the radiation that leaves, as a result of stimulated emission, has more photons than the incident radiation. The result is thus a coherent amplification of radiation of this frequency, (Fig. 5.4), since the stimulated emission has the same characteristics as the stimulating radiation, i.e., the same frequency, phase, polarization, and direction of propagation.

The net amplification, defined as stimulated emission minus absorption, is called the optical gain, which is denoted as g. It is proportional to the population difference between N_2 and N_1, $g \propto (N_2 - N_1)$. To sustain a constant amplification, an external source must continuously supply the laser system with energy in order to maintain

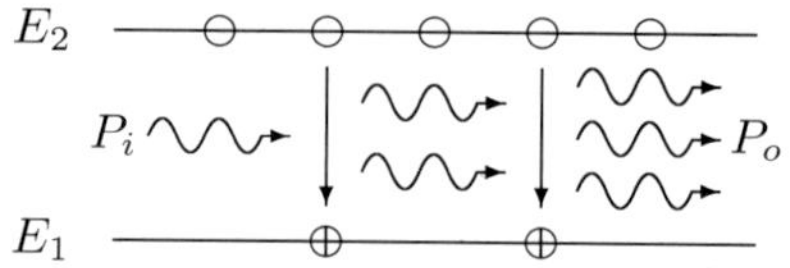

Fig. 5.4 If stimulated emission is the dominant process, the radiation is enhanced, since each photon of proper energy might lead to stimulated emission. This means that the output power, P_o, will exceed the input power, P_i, and optical gain, g, is achieved.

the population inversion. This process is called pumping and can be accomplished in many ways, e.g., by stimulated absorption, by electron collision excitation, or through chemical reactions.

The pump rate and the population inversion are linearly related until population inversion reaches its threshold value, N_{th}. When this level is reached at a threshold pump rate R_{th}, the stimulated emission will start to dominate over spontaneous emission and laser output will increase linearly with the pump rate. Below R_{th} spontaneous emission dominates, but it will stagnate to a constant level at pump rates above R_{th}.

In a steady-state situation, N_{th} remains constant regardless of the amount by which the threshold pump rate is exceeded. The gain required to support this steady state will also attain a threshold value, g_{th}, and a situation called the gain saturation is achieved.

5.2.2 *Optical Cavity*

When the medium, in which a population inversion has been created, is placed in an optical cavity (resonator), radiations at certain resonance frequencies will be confined and amplified in the cavity. There are several construction configurations for laser resonators. The simplest is the plane-parallel or Fabry–Perot resonator, in which two flat mirrors are placed at opposite ends of the cavity aligned parallel to each other and perpendicular to the cavity axis, (Fig. 5.5). One of the mirrors, or both, is slightly transparent so that part of the radiation can transmit through it.

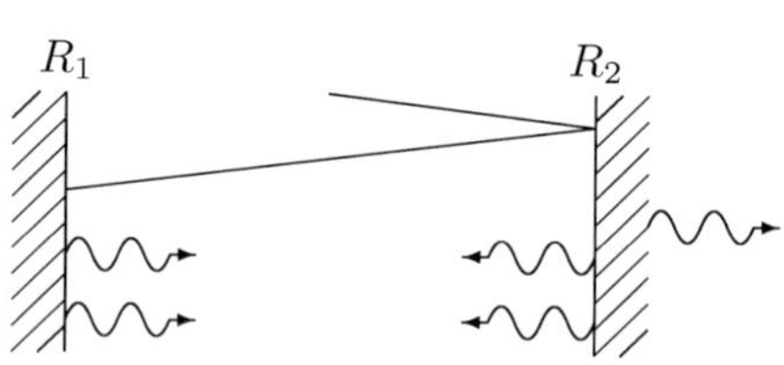

Fig. 5.5 The Fabry–Perot (plane-parallel) resonator. Light perpendicular to the mirrors with reflectance R_1 and R_2 will be repeatedly reflected within the cavity. Any inclined light will eventually escape the resonator.

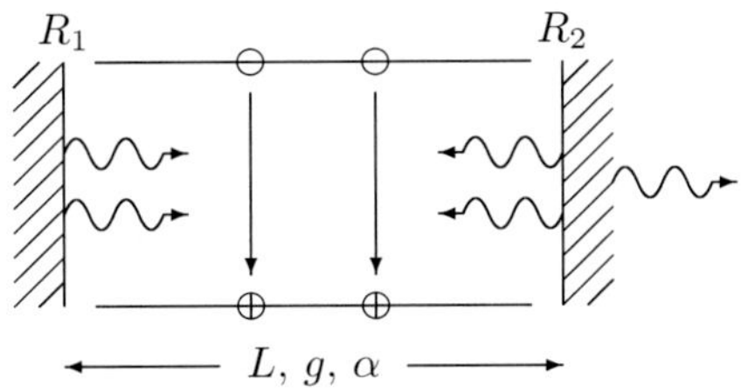

Fig. 5.6 An optical amplifier with an external power supply, placed in an optical resonator, constitutes the basic structure of a laser. Losses within the cavity, excluding stimulated absorption, are denoted as α.

Light rays perpendicular to the mirrors are reflected back and forth without escaping, but if they are inclined they will eventually escape. Most of the radiation that comes from spontaneous emission and thus travels in all directions will therefore escape from the open sides of the cavity. The exception is the radiation that travels along the direction normal to the cavity mirrors. This radiation will be amplified until saturation is reached, i.e., until the power emitted as laser radiation is balanced by the power input to sustain the population inversion. Since the amplification occurs only for one specific direction, the emitted radiation is confined to a very narrow, coherent, and uni-directional beam, so that the unique features of a laser beam are fulfilled.

To sum up, there are three key elements in a laser, the laser medium that produces optical gain, the power supply that creates population inversion, and the optical resonator or cavity that provides the system with optical feedback. Put together, (Fig. 5.6), these elements constitute the basis of a laser.

One way to achieve a wavelength-dependent loss profile is to use an external cavity. In an external cavity laser (ECL), the light from a diode laser is coupled to an external grating via a collimating lens. The distance from the far diode mirror to the grating constitutes the external cavity (Fig. 5.7).

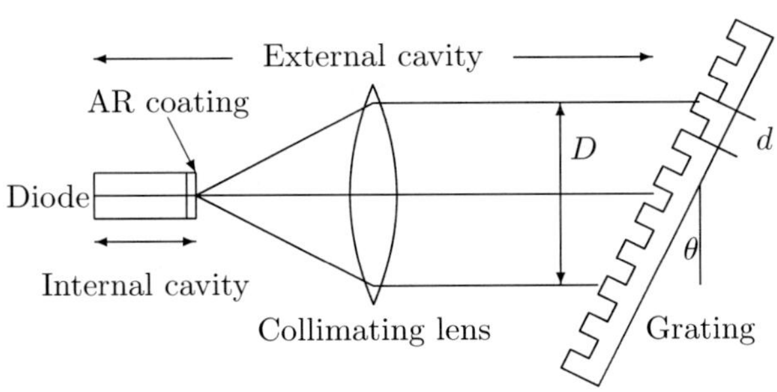

Fig. 5.7 In an external cavity laser, a collimating lens is used to collimate the diverging light from the diode laser. A grating is used as a spectral filter, which reflects one specific wavelength back to the diode.

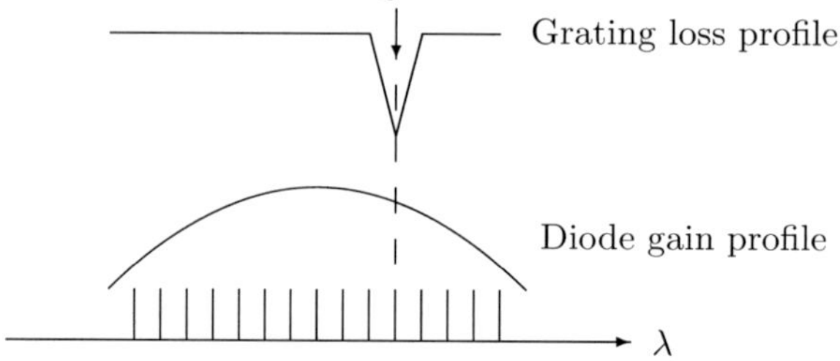

Fig. 5.8 In single longitudinal mode operation, the loss profile is wavelength dependent.

For a strong coupling between the diode and the grating, it is necessary to reduce the natural reflectivity of the cleaved facet facing the grating using an anti-reflection coating. These coatings consist of one or more dielectric layers that reduce reflection values over the entire emission bandwidth of the diode. This means that the diode laser emits light over its total emission bandwidth and only a small fraction of the power is reflected in the anti-reflection coated facet. When interacting with the grating, only light with a certain wavelength is reflected back to the diode where it will be amplified. All other wavelengths will be diffracted away from the optical axis and will hence not be amplified.

Which wavelength is reflected, i.e., the loss profile, is determined by the characteristics of the grating and its angle toward the diode. There will be a dip in the loss profile for the lasing mode, so that the loss profile becomes wavelength dependent (Fig. 5.8). The loss profile is sufficiently higher than the diode gain profile. This leads to single longitudinal mode operation, since there will only be one mode satisfying the gain condition.

The reflectivity of an anti-reflection-coated facet is typically 10^{-4}. Because the anti-reflection coating on a diode is not perfect, there will always be some reflection within the diode. This indicates the co-existence of the longitudinal modes from both the internal cavity and the external cavity. However, only a very small fraction of the radiation is reflected within the internal cavity; the intensity of the resulting longitudinal modes will be much lower than those from the external cavity.

Examination of the laser output reveals that it consists of discrete frequency components and that the intensity of the output beam has a spatial dependence. To appreciate how these frequency components arise and how the intensity is distributed, one has to examine the so-called longitudinal and transverse modes.

A consequence of the phase and gain conditions is that only certain frequencies are possible for laser oscillations. The radiation inside the cavity forms a standing wave at resonance, which occurs when the cavity length, L, equals an integer

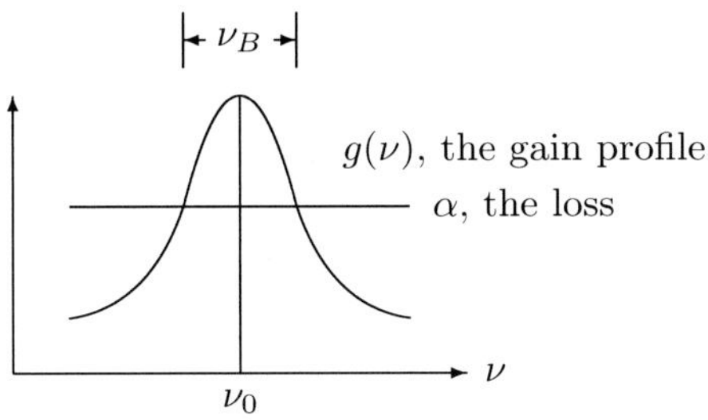

Fig. 5.9 The gain bandwidth, ν_B, is the frequency interval of possible oscillations where the gain exceeds the loss.

number, i, of the half wavelength ($\lambda/2$)

$$L = \frac{i\lambda}{2} = \frac{ic_0}{2n\nu} \tag{5.29}$$

where c_0 is the speed of light in vacuum, n is the refractive index of the cavity medium and $\nu = c_0/n\lambda$ is the frequency. Excluding semiconductor lasers, a typical cavity is tens or hundreds of thousands of wavelength long, $L \gg \lambda$, and might therefore seems to permit oscillations at a very large number of frequencies. Oscillations are, however, only possible at frequencies within the gain bandwidth, ν_B, of the laser medium. The gain bandwidth is the frequency interval where the gain exceeds the loss (Fig. 5.9).

The number of possible oscillation frequencies, or longitudinal modes, is therefore

$$I = \frac{\nu_B}{\Delta\nu} \tag{5.30}$$

where ν_B is the bandwidth and

$$\Delta\nu = \frac{\delta\nu}{\delta i} = \frac{c_0}{2nL} \tag{5.31}$$

is the spacing between two adjacent longitudinal modes. Oscillation will preferably occur at the longitudinal mode closest to the center of the gain profile peak, whereas theoretically, oscillations are possible at any frequencies within the bandwidth.

Longitudinal modes correspond to different resonances along the length of the laser cavity, which occur at different frequencies within the gain bandwidth of the laser. Transverse modes, on the other hand, exist in the cross-sectional profile of the laser beam, i.e., in its intensity pattern. The intensity pattern is associated with certain terms, according to TEM_{ij} (transverse electromagnetic), where i and j define the spatial distribution of the beam. Because of their different spatial distribution, different transverse modes undergo different gains and losses. The lowest order mode, TEM_{00}, is most widely used in the laser action. It is the most confined around the optical axis, the intensity is ideally Gaussian distributed across the beam's cross-section, and there are no phase shifts in the electric field as there are in other modes.

In many applications, the laser needs to be designed to operate in both single-spatial and single-longitudinal modes, i.e., that the TEM_{00} oscillates in a single longitudinal mode. A single-spatial mode of TEM_{00} is normally realized by controlling the width and thickness of the active layer. Single-longitudinal mode operation of a laser can be achieved in a number of ways. The basic idea is to filter out one of the allowed longitudinal modes by making the cavity loss wavelength-dependent. The longitudinal mode, which is closest to the gain peak and with the smallest cavity loss, reaches its threshold first and becomes the dominant mode. The surrounding modes are suppressed due to their high losses. Because the longitudinal modes decide the wavelength, they are often referred to as the spectral modes.

5.2.3 *Heterojunction Semiconductor Structure*

The semiconductor laser was invented in 1962. Today it is widely used in a broad spectrum of applications. It serves a central role in fiber-optic communication and data storage, primarily because of major advances in laser power, spectral purity, efficiency, wavelength range, ability to be modulated at extremely rapid rates, and long lifetimes.

Because they are used in such a large group of devices, semiconductor lasers are made out of a variety of materials with many different structures. Which material is used depends on which output wavelength is needed for its application, and the structure of the laser determines the characteristics of the laser beam.

When a p-type and an n-type semiconductor material are brought in contact, a p–n junction is formed. At thermal equilibrium, the Fermi level must be aligned across the p–n junction by the diffusions of electrons from the n side to the p side and of holes in the opposite direction. When a steady state is reached, further diffusion is prevented by the so-called built-in electric field.

When applying an external positive voltage to the p side of the junction, the built-in field is reduced and the junction is said to be forward biased. The field reduction in the depletion region induces further diffusion of electrons and holes across the junction and an electric current begins to flow. In the narrow depletion region of the junction, both electrons and holes are present simultaneously. They can recombine either radiatively or nonradiatively. The photons, which are emitted during radiative emission (spontaneous or stimulated emission), have an energy of $\hbar\omega \approx E_g$, where E_g is the energy band gap. These photons can be absorbed through a reverse process that generates electron–hole pairs. When the external voltage exceeds a critical value, a sufficient population inversion is achieved and the rate of photon emission exceeds that of absorption. Under these circumstances the p–n junction is capable of optically amplifying radiation at a certain wavelength and optical gain is achieved.

The p–n junction is a so-called homo-junction, which means that the junction is formed between p and n variants of the same semiconductor material. Because

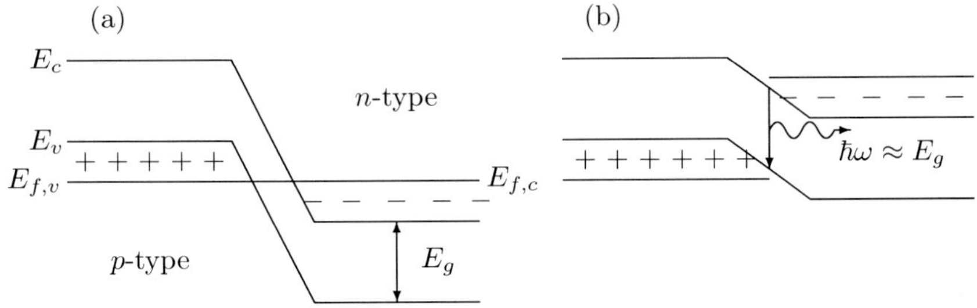

Fig. 5.10 A p–n homo-junction (a) at thermal equilibrium, and (b) is forward biased.

the excess electrons and holes distribute extensively in the depletion region, which is relatively quite wide, it is difficult to achieve the high carrier densities in order to obtain a sufficiently high optical gain. Moreover, the bandwidth can be rather wide (Fig. 5.10). The problems can be partially solved by using heterojunctions, which are junctions formed by different semiconductor materials.

In optical systems, the most common semiconductor laser structure is the index-guided structure. The index guiding is a light confinement mechanism due to the difference of Δn between the refractive indices of the active and cladding layers. Depending on the magnitude of Δn, the lasers are further classified into weakly and strongly index-guided. Cross-sections of the two types are presented in Fig. 5.11.

Fig. 5.11a shows a weakly index-guided laser, in which a ridge is etched from the p-layer, and a dielectric medium of SiO_2 is deposited on each side of the ridge. Outside the ridge, the dielectric medium will block the current flow from the contact to the active layer. Thus, the light will only be generated in the active layer just below the stripe. Because of the low refractive index of the dielectric medium, the light will be confined in the ridge region.

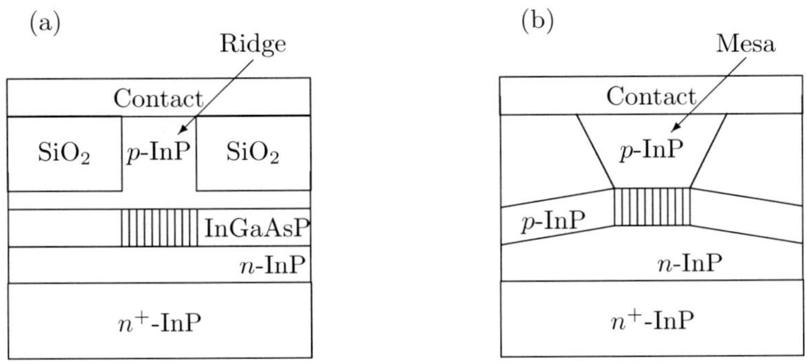

Fig. 5.11 Schematic cross-sections of index-guided semiconductor lasers. (a) Ridge-waveguide heterostructure for weak index guiding; (b) Etched-mesa buried heterostructure for strong index guiding. The shadowed area is the InGaAsP active region.

The strongly index-guided laser, presented in Fig. 5.11b, is a so-called buried-heterostructure laser. In such a laser, the active layer is buried on all sides by semiconductor materials having a lower refractive index. These layers are doped in a way to confine the current to the active region. The active layer will act as a waveguide. Because of the large built-in index step in both vertical and horizontal directions, the size of the waveguide determines the lasing characteristics of the buried-heterostructure laser.

The semiconductor laser waveguide is usually not symmetric. The horizontal and vertical directions are referred to as the lateral and transversal directions, respectively, and the TEM_{ij} modes are usually referred to as the spatial modes, where specifically the horizontal and vertical optical intensity distributions are called lateral and transversal modes, respectively. By altering the width of the active layer, the number of lateral modes, supported by the laser, can be controlled. A wider active layer can support a higher-order lateral mode than a narrower layer. However, single spatial mode lasers, lasers that emit lights in the lowest-order mode, TEM_{00}, are often the most desirable ones.

We have discussed the three key elements in a laser: the laser medium that produces the optical gain, the power supply that creates a population inversion, and the optical resonator that provides the system with optical feedback. In a semiconductor diode laser, the first two factors are established by the forward biased p–n junction (including the external voltage). For the optical resonator, the semiconductor wafers are cleaved along its crystal plane. Because of the differences in refractive index between the active medium and the surrounding air, the surfaces will act as mirrors and provide the optical feedback needed for laser operation. The reflectivity for a semiconductor/air interface is approximately 0.3. A semiconductor diode laser is typically $100 \times 500 \times 500$ μm^3.

In the so-called double heterojunction structure, a thin layer of a semiconductor material with a narrower band gap is grown between the p-type and n-type layers. Because of the band gap difference among the layers, the injected charge carriers are confined within the thin layer of the narrower band-gap material. This structure makes it easier to obtain sufficient inversion of the carrier population in order to produce optical gain by the stimulated radiative recombination. Because of the light generation within the thin narrow-band-gap layer, this layer is normally denoted as the active layer, whereas the surrounding layers of larger-band-gap materials are called cladding layers. The refractive index of the active layer is larger than that of the cladding layers because of its narrower band gap. Due to the refractive index difference, the active layer acts as a waveguide. The light emitted parallel to the active layer will be guided along the active layer, so that the light intensity will be the highest in the middle of the active layer, where the gain is the largest. This mechanism is often referred as the index guiding. Consequently, the use of a double heterojunction structure results in both the carrier and the optical confinement (Fig. 5.12).

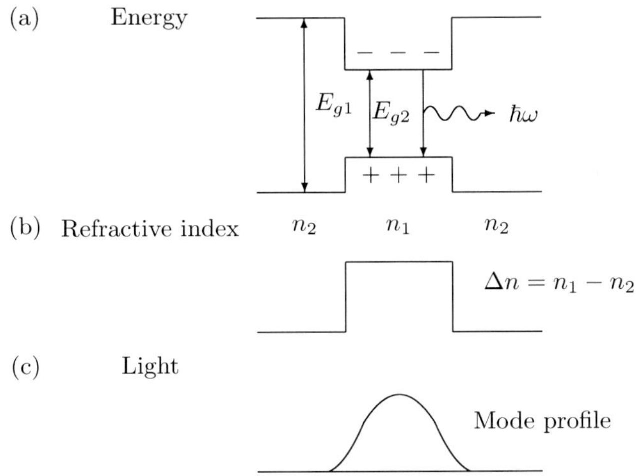

Fig. 5.12 Simultaneous confinements of charge carriers and generated light by the use of a double heterojunction structure. The active layer having a narrower band gap has a higher refractive index than the cladding layers.

As a result of the band-gap difference, electrons that are injected from the n-type region during the forward bias are prevented from diffusing beyond the energy barrier at the opposite heterojunction. Similarly, holes injected from the p side cannot diffuse beyond their opposite heterojunction. The charge carriers will thus be trapped in the active region where they recombine either radiatively or nonradiatively. The light emitted along the active layer will be guided along the layer because of the difference in refractive index, and a spatially confined beam of radiation is created. Because the semiconductor laser is based on a diode structure, it is often called a diode laser.

The heterojunction structure is crucially important for the development of semiconductor lasers. The first person who successfully produced a heterojunction structure with clear borders among layers was a Russian professor Zhores I. Alferov in 1969. In 1970, the research group submitted its report about a double heterostructure laser operating at room temperature. Alferov and a German professor Herbert Kroemer were awarded the Nobel Price in Physics in 2000 for the development of semiconductor heterostructures for high-speed electronics and optoelectronics.

5.2.4 *Quantum Wells*

A population inversion is needed for the stimulated emission to dominate over the spontaneous emission. In order to create a population inversion and thereby the necessary optical gain, the injection current must exceed a certain value, which is known as the transparency value. This value can be decreased by increasing the

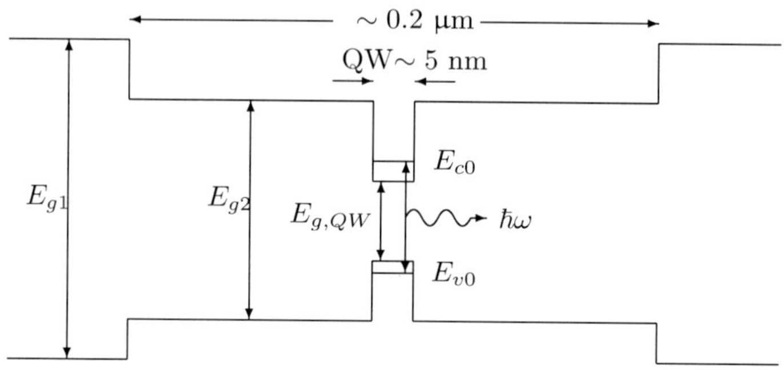

Fig. 5.13 A quantum well (QW) laser.

so-called differential gain coefficient (the gain cross section) of the medium. With a larger gain coefficient, the optical gain can be achieved with a lower injection current.

The differential gain coefficient of a QW semiconductor laser is in general larger than that of a conventional heterostructure laser. In a QW laser, the thickness of the active layer is approximately 5–10 nm, whereas it is about 0.1–0.3 μm in a conventional laser (Fig. 5.13). Because the layer is so thin, the electron and holes are effectively confined so that their kinetic energies are quantified. The density of states, which is the density of allowed energy states, in a QW differs from that of a normal bulk material (in bulk material the density of states forms a smooth parabolic curve). Because of the discrete distribution of energy levels in the QW, its density of states takes a staircase shape, where each step corresponds to one energy level. The energy of the radiation is $\hbar\omega = E_{g,\mathrm{QW}} + E_{v0} + E_{c0}$, where $E_{g,\mathrm{QW}}$ is the energy band gap of the QW material, E_{v0} and E_{c0} are ground sublevels of the CB electron and VB hole.

Because the carriers are spatially confined within such a small volume, the recombination rates are large and the spectral width of the light is substantially narrower than for the regular semiconductor laser. It is also easier to achieve population inversion in a QW laser because of the reduction in active layer thickness, which implies that a low injection current is needed. The reduction of the active layer thickness results in a reduced overlap with the optical mode. If the modal gain for a single QW laser is insufficient, a laser structure with multiple QWs is used.

5.2.5 *Buried Heterostructure Semiconductor Laser*

A buried heterostructure semiconductor laser consists of a core active layer with cladding material around it. Normally the two materials are lattice-matched to avoid strain, since strain causes defects and dislocations, which enhance the

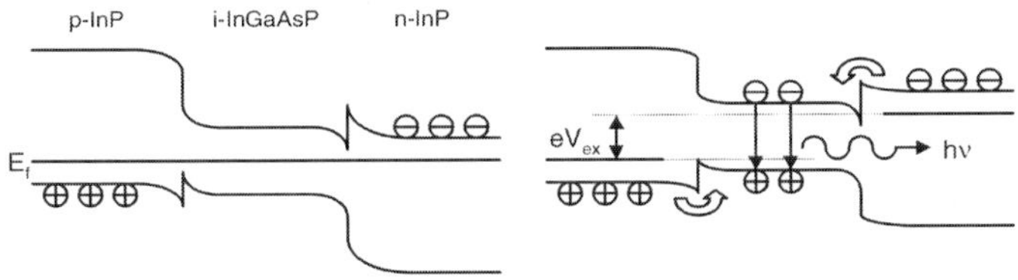

Fig. 5.14 Energy band structure of InP/InGaAsP/InP (a) without external bias and (b) with a bias V_{ex}.

nonradiative recombinations. The cladding material has a wide bandgap and a low refractive index than the active layer in order to confine carriers and light, respectively, to the active layer.

In Fig. 5.14, we show a combination of InGaAsP active layer, which is lattice-matched to the InP cladding. When we forward bias the *p-i-n* structure, it behaves a little like a forward biased *pn*-junction with a hole in the middle where the carriers are trapped for recombination. It is much due to the confinement of the carriers in the *i*-region that makes it possible to reach population inversion and net gain at a moderate forward bias.

When considering the optical transitions in a semiconductor from a VB state (E_v, k_v, m_v^*) to a CB state (E_c, k_c, m_c^*) in the form of

$$E_2 = E_c + \frac{\hbar^2 k_c^2}{2m_c^*} \quad , \quad E_1 = E_v - \frac{\hbar^2 k_v^2}{2m_v^*} \tag{5.32}$$

it is easy to obtain

$$g(\hbar\omega) = A\sqrt{\hbar\omega - E_g}[f_c(E_2) - f_v(E_1)] \tag{5.33}$$

where A is a constant and

$$E_g = E_c - E_v$$

is the energy bandgap. $f_c(E_2)$ and $f_v(E_1)$ are the occupations of CB state E_2 and VB state E_1,

$$\hbar\omega = E_2 - E_1$$

The resonant tunneling injection hot electron laser was proposed by Tolstikhin and Willander.[21] It is basically a resonant tunneling hot electron transistor, with an active layer and a waveguiding structure inserted between the base and collector. The basic idea is that we shall be able to control both the number of injected electrons via V_{be}, the base-emitter bias (which is proportional to the intensity of the optical field), and the kinetic energy that the electrons carry into the active layer via V_{cb}, the collector-base bias (the output optical energy is the sum of this kinetic energy and the energy bandgap).

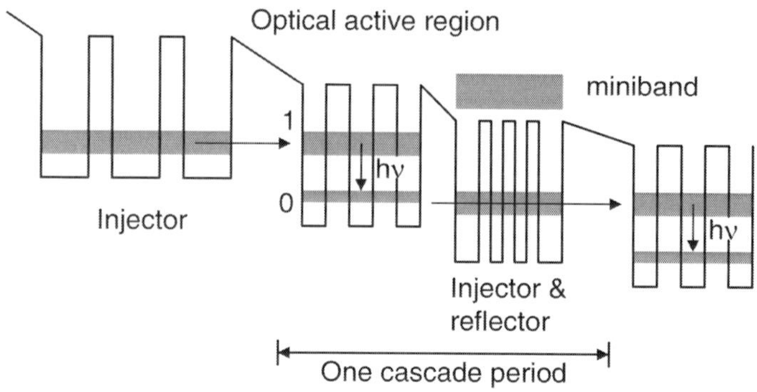

Fig. 5.15 Schematic conduction band profile of a portion of the GaInAs/AlInAs quantum cascade laser under positive bias.

5.2.6 *Quantum Cascade Laser*

In a quantum cascade laser, the population inversion between states of laser action is designed by tailoring the electron intersubband transition.[22] Since the initial report of the quantum cascade laser in 1994, the emission wavelengths in the 4–8.5 μm range have been demonstrated using AlInAs/GaInAs heterostructures.[23]

Figure 5.15 schematically represents the energy band profile of a quantum cascade laser. To prevent electron escape into the continuum, which will largely reduce the vertical transition, a superlattice is placed as an effective reflector for electrons in the excited state while simultaneously ensuring a swift escape from the lower state to the high state in the next cascade period via the miniband.

More recently, improvement of the quantum cascade laser has been reported with a thin barrier, inserted between the injector layer and the optical active region, to increase tunneling injection efficient. This improvement and substituting the AlInAs cladding layers with InP layer of higher thermal conductivity has led to room-temperature high peak power of 200 nW pulse operation at 5.2 μm.[24] Continuous-wave single-mode operation was also reported up to 140 K, as well as room-temperature pulse operation at 8.5 μm. This is the first semiconductor laser operating at room temperature in the mid-infrared regime. The current performance of the quantum cascade laser makes it a candidate for applications such as environmental sensing and pollution monitoring in the 3–5 and 8–13 μm atmospheric windows.

5.3 PHOTODETECTOR

Infrared photodetectors belong to the transducers that convert the infrared radiation into an electrical signal. They are usually classified as either thermal or photon devices.[25] The thermal detector raises its temperature by the light absorption; then the temperature-dependent parameters of the device are changed. However, for the photodetectors, some quantum events (such as electron interband or intraband transitions in semiconductors) will be induced by the infrared radiation absorption in the device. The main thermal detectors include thermopile, bolometer, Golay cell, and pyroelectric detectors, while the photon detector has two categories: photoconductors, such as CdS, PbS, PbSe, Si:As, HgCdTe; and photovoltaics, such as InSb and HgCdTe p–n junction.

A wide range of applications, including global and space surveillance and astronomy, deep space exploration and medical imaging, and remote monitoring of plants and the environment, has strongly stimulated the progress of the infrared detector. Moreover, both tactical and strategic use of infrared imaging systems requires the availability of high-performance focal plan array (FPA) detector operating at wavelengths around the 8–14 μm spectral band. The hybrid HgCdTe FPA has attracted a lot of work in the last two decades. A 256×256 element FPA with the noise equivalent temperature difference (NETD) of 5 mK has been achieved in recent years.

Thanks to the low-dimensional physics and the maturing of the III–V compound semiconductor technologies a major step has been made toward a new approach to design semiconductor structures with tailored electronic and optical properties for a new generation of long-wavelength QW infrared photodetector (QWIP).[26, 27] The QWIP FPA has been developed; 128×128-element with NETD= 30 mK in 1991[28] and 256×256-element with NETD= 25 mK in 1996[29] have been reported.

The QWIP has been compared with the HgCdTe[30–32] which shows that in the view of FPA device, the III–V semiconductor QWIP are expected to have the following advantages over the HgCdTe detectors: (1) a higher degree of uniformity; (2) technologically mature device fabrication; (3) radiation hardness; (4) easy of crystal growth, and (5) low cost. Moreover, it is important to notice the potential advantage in the FPA packaging for the III–V QWIP.

For QW detectors n-type material should be used instead of p-type material for better detector performance. The physical mechanism for normal absorption in n-type and p-type QWs is quite different, and from a technology point of view, normal incidence is obviously the simplest solution.

5.3.1 *Optimizing Material Growth Direction*

We consider the intraband optical transition among energy sublevels in a QW that is grown along the z-direction. In the effective-mass approximation, the Hamiltonian of the electrons in the system is written as

$$H = \frac{1}{2}\boldsymbol{p}\bar{W}\boldsymbol{p} + V(z) \tag{5.34}$$

where $\boldsymbol{p}$ is the momentum operator, $\bar{W}$ is the 3×3 inverse effective mass taking into account the nonspherical effect in the energy band structure of the QW material, and $W_{ij} = W_{ji}$. $V(z)$ is the potential energy. The wave function can be expressed as

$$\psi_m(\boldsymbol{k}) = \psi_m(z)u(\boldsymbol{r})e^{i\boldsymbol{k}\cdot\rho} \tag{5.35}$$

where $u(\boldsymbol{r})$ is the Bloch function and is the same for all values of the sublevel index m and the wave vector $\boldsymbol{k}$ when considering transitions among subbands. $\boldsymbol{k} = (k_x, k_y)$ is the wave vector in xy-plane, ρ is the corresponding spatial vector, and $\psi_m(z)$ is the envelope function in the z direction.

In the presence of a radiation $\boldsymbol{A}e^{i\omega t}$, the interaction between the incident radiation and the electron is

$$e\boldsymbol{A}\bar{W}\boldsymbol{p}$$

The matrix element of intersubband transition is

$$\langle\psi_m(\boldsymbol{k})|\boldsymbol{A}\bar{W}\boldsymbol{p}|\psi_n(\boldsymbol{q})\rangle = \delta_{\boldsymbol{k},\boldsymbol{q}}\sum_i A_i W_{iz}\langle\psi_m|p_z|\psi_n\rangle \tag{5.36}$$

In the principal coordinate system denoted as XYZ, for the two ellipsoids of X-valley electrons with major axis along the [001] direction,

$$\bar{W}_p = \begin{pmatrix} w_t & 0 & 0 \\ 0 & w_t & 0 \\ 0 & 0 & w_l \end{pmatrix} \tag{5.37}$$

where $w_t = 1/m_t$, $w_l = 1/m_l$, m_l and m_t are the longitudinal and transverse effective masses, respectively. The CB of Si consists of six ellipsoids described by a longitudinal effective mass $m_l = 0.9163$ and a transverse effective mass $m_t = 0.1905$ expressed in the unit of free electron mass m_0.[33] The equation of the ellipsoidal constant energy surface in the principal axis coordinate system (XYZ) is

$$W_{XX}k_X^2 + W_{YY}k_Y^2 + W_{ZZ}k_Z^2 - E = 0 \tag{5.38}$$

In the coordinate system of xyz whose z-axis is along the sample growth direction and the y-axis is perpendicular to the major axis of the energy valley under consideration, the equation of the ellipsoidal constant energy surface becomes

$$E = W_{xx}k_x^2 + W_{yy}k_y^2 + W_{zz}k_z^2 + 2W_{xy}k_xk_y + 2W_{yz}k_yk_z + 2W_{zx}k_zk_x \tag{5.39}$$

The first-order derivative with respect to z in the above equation can be eliminated and the envelope function in z-direction is determined by

$$\left[\frac{-W_{zz}\hbar^2}{2}\frac{\partial^2}{\partial z^2} + V(z)\right]\xi_m(z) = E_m\xi_m(z) \tag{5.40}$$

In the transformation of the coordinate systems, there are four mathematically invariable quantities

$$\begin{vmatrix} W_{xx} & W_{xy} & W_{xz} & 0 \\ W_{xy} & W_{yy} & W_{yz} & 0 \\ W_{xz} & W_{yz} & W_{zz} & 0 \\ 0 & 0 & 0 & -E \end{vmatrix} = \begin{vmatrix} w_t & 0 & 0 & 0 \\ 0 & w_t & 0 & 0 \\ 0 & 0 & w_l & 0 \\ 0 & 0 & 0 & -E \end{vmatrix} = -Ew_t^2 w_l$$

$$\begin{vmatrix} W_{xx} & W_{xy} & W_{xz} \\ W_{xy} & W_{yy} & W_{yz} \\ W_{xz} & W_{yz} & W_{zz} \end{vmatrix} = \begin{vmatrix} w_t & 0 & 0 \\ 0 & w_t & 0 \\ 0 & 0 & w_l \end{vmatrix} = w_t^2 w_l$$

$$W_{xx} + W_{yy} + W_{zz} = 2w_t + w_l$$

$$W_{xx}W_{yy} + W_{yy}W_{zz} + W_{zz}W_{xx} - W_{xy}^2 - W_{yz}^2 - W_{zx}^2 = w_t^2 + 2w_t w_l \tag{5.41}$$

By these four constants it is easy to see

$$\begin{aligned} W_{xx}W_{zz} - W_{xz}^2 &= w_t w_l \\ W_{xx} + W_{zz} &= w_t + w_l \end{aligned} \tag{5.42}$$

We can therefore write the following equations

$$\begin{aligned} W_{zz} &= (1-a)w_t + aw_l \\ W_{xz} &= \sqrt{a(1-a)}(w_t - w_l) \\ W_{yz} &= 0 \end{aligned} \tag{5.43}$$

by introducing a parameter a, which is to be determined later.

In the infinity high barrier approximation, the absorption coefficient is calculated as

$$\begin{aligned} \alpha(\hbar\omega) &= \frac{SN_s}{\mu\omega L^3 A^2} \left| \sum_i A_i W_{iz} \right|^2 \\ S &= \frac{256\hbar^2 e^2}{9\Gamma c\epsilon_0} \end{aligned} \tag{5.44}$$

where L is the well width and Γ is the lifetime of the excited state; μ and c are the refraction index and the speed of light, respectively. To make it simpler, we omit irrelevant physical constants so that

$$\alpha = \frac{1}{W_{zz}^{3/2}} \left| \sum_i A_i W_{iz} \right|^2 \tag{5.45}$$

When considering the normal incident and nonpolarized radiation, i.e., $A_x = A_y$ and $A_z = 0$,

$$\alpha_1 = \frac{W_{xz}^2}{W_{zz}^{3/2}} \tag{5.46}$$

Table 5.1 The parameter a for X- and L-valleys. $r^2 = l^2 + m^2 + n^2$.

X valley		L valley	
Major-axis	ar^2	Major-axis	$3ar^2$
$[lmn] = [100]$	l^2	$[111]$	$(l+m+n)^2$
$[010]$	m^2	$[\bar{1}11]$	$(-l+m+n)^2$
$[001]$	n^2	$[1\bar{1}1]$	$(l-m+n)^2$
		$[11\bar{1}]$	$(l+m-n)^2$

which becomes

$$\alpha_1 = \frac{a(1-a)(w_t - w_l)^2}{[(1-a)w_t + aw_l]^{3/2}} \tag{5.47}$$

by inserting Eq. 5.43.

It is easy to obtain the condition of a maximal α_1, which is given by

$$a = \frac{(3w_t + w_l) - \sqrt{(w_t + w_l)^2 + 12w_t w_l}}{2(w_t - w_l)} \tag{5.48}$$

The results of a for different major axes and energy valleys are listed in Table 5.1.

For parallel incident radiations ($A_x = 0$ and $A_y = A_z$), we have

$$\alpha_2 = \sqrt{W_{zz}} = \sqrt{(1-a)w_t + aw_l} \tag{5.49}$$

Generally speaking, α_1 is larger than α_2. However, the absorption of the intersubband transition for normal incidence radiation can be larger than that for parallel incident radiation in some special growth direction if

$$m_l > 5.828 m_t$$

In numerical calculations, the total absorption coefficient is the sum of the ones from all occupied valleys in the CB. Therefore, it is necessary to investigate the degeneracies and the occupancies of different energy valleys. In the infinitely high barrier approximation, the ground sublevel energy E_0 in the z-direction is written as

$$E_0 = \frac{\pi^2\hbar^2 W_{zz}}{2L^2} = \frac{\pi^2\hbar^2}{2L^2}\left[(1-a)w_t + aw_l\right] \tag{5.50}$$

where we have used Eq. 5.43. Here it is seen that the value of E_0 is different for different energy valleys, since a is different for different valleys (Table 5.1). The valleys with the lowest E_0 are first occupied by electrons. Table 5.2 lists the lowest energy valleys and the corresponding degeneracies, g, of different energy valleys.

Table 5.2 Degeneracy g of the lowest-energy valleys.

X valley			L valley		
$[lmn]$	g	occupied valleys	$[lmn]$	g	occupied valleys
$l=m=n$	6	all	$l=m=0$	4	all
$l=m>n$	4	[100][010]	$n=0$	2	[111][11$\bar{1}$]
$l>m>n$	2	[100]	otherwise	1	[111]

The next task is to determine the Fermi level for determining the energy sublevels to be occupied by electrons. The energy $E_{\boldsymbol{k}}$ in the xy plane is

$$E_{\boldsymbol{k}} = \frac{\hbar^2}{2}\left[\left(W_{xx} - \frac{W_{xz}^2}{W_{zz}}\right)k_x^2 + \left(W_{yy} - \frac{W_{yz}^2}{W_{zz}}\right)k_y^2 + \left(W_{xy} - \frac{W_{xz}W_{yz}}{W_{zz}}\right)k_x k_y\right] \tag{5.51}$$

It is easy to obtain the two-dimensional (2D) density of states

$$N_{2D} = \frac{\sqrt{W_{zz}}}{\pi\hbar^2\sqrt{w_t^2 w_l}} = \frac{m_d}{\pi\hbar^2} \tag{5.52}$$

where m_d is the 2D density-of-states effective mass

$$m_d = \sqrt{\frac{(1-a)w_t + a w_l}{w_t^2 w_l}} \tag{5.53}$$

The sheet density of electrons occupying state E_0 is

$$\begin{aligned} N_s &= gN_{2\mathrm{D}} \int_{E_0}^{\infty} \frac{\mathrm{d}E}{1+\exp[(E_0 - E_f)/k_B T]} \\ &= gN_{2\mathrm{D}} k_B T \ln\left[1 + \exp\left(\frac{E_f - E_0}{k_B T}\right)\right] \end{aligned} \tag{5.54}$$

which becomes

$$N_s = gN_{2D}(E_f - E_0) \tag{5.55}$$

in the low-temperature limit. The energy difference $(E_{02} - E_{01})$ between ground states E_{02} and E_{01} of the two lowest-energy valleys is (energy unit: meV),

$$E_{02} - E_{01} = \frac{41.33(a - a')(w_t - w_l)}{(1-a)w_t + a w_l} \tag{5.56}$$

for a QW with detecting wavelength at $\lambda = 10\,\mu\mathrm{m}$, where a and a' are parameters for the two energy valleys, which can be evaluated from Table 5.1.

For low-index growth directions of AlAs/AlGaAs, AlGaSb/AlSb, Si/SiGe, and Ge/SiGe QWs with doping concentration of $10^{18}\,\mathrm{cm}^{-3}$, one can conclude from the above analysis that electrons populate only the lowest-energy valleys.

Table 5.3 Optimal growth directions of QW photodetectors and the total absorption coefficients.

Material	Growth direction	Occupied valley	α_{total} (arb. unit)
AlAs	[110]	[100][010]	3.50
	[111]	all	3.40
AlGaSb	[110]	$[111][11\bar{1}]$	3.23
	[100]	all	2.90
	[102]	$[111][1\bar{1}1]$	2.86
Si	[110]	[100][010]	3.23
	[111]	all	3.18
Ge	[110]	$[111][11\bar{1}]$	6.31
	[203]	$[111][1\bar{1}1]$	5.93
	[102]	$[111][1\bar{1}1]$	5.26

The corresponding Fermi level is (meV)

$$E_f - E_{01} = \frac{2.39}{gm_d} = \frac{2.39\sqrt{w_t^2 w_l}}{g\sqrt{(1-a)w_t + aw_l}} \tag{5.57}$$

and the total absorption coefficient for the normal incident radiation is

$$\alpha_{\text{total}} = g\frac{a(1-a)(w_t - w_l)^2}{[(1-a)w_t + aw_l]^{3/2}} \tag{5.58}$$

Table 5.3 shows results of the optimal growth directions of different QWs. It is found that in the listed QW structures, the most common optimal growth direction is [110]. The Ge/SiGe QW, which has the largest anisotropy of the effective mass, has the largest absorption coefficient for normal incident radiation. For high doping concentration or other detecting wavelengths, the conclusions may be different, but the physical principle is the same.

5.3.2 *Optical Diffraction Grating*

For most commonly used AlGaAs/GaAs materials, normal incident radiation absorption is not possible because of their isotropic effective masses of the active Γ electrons. Optical grating is required by such QWIPs to diffract the normal incident radiation into non-normal, higher-order, transverse electromagnetic modes. It results from the intersubband transition selection rule for MQWs which only allows the optical absorption of infrared radiation modes having an electric field component perpendicular to the QW layer. One of the key requirements for the FPA fabrication is to optimize the diffraction grating. In order to improve the grating coupling efficiency for the QWIP, a lot of works have been done, such as the studies on symmetric grating;[34] long-period gratings;[35] a waveguide combined grating coupler;[36] and metallic diffraction grating.[37] Random gratings[26] and

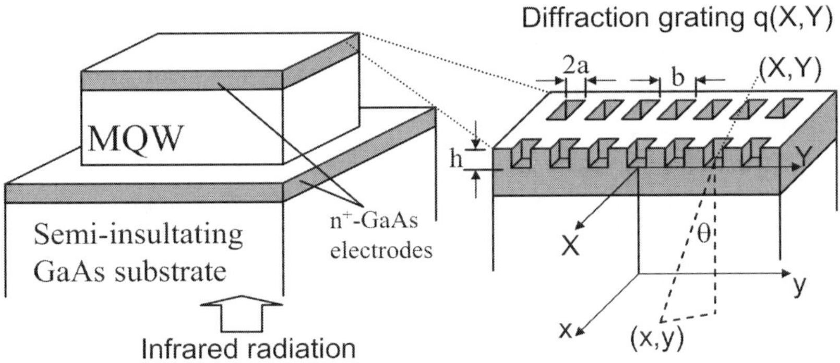

Fig. 5.16 Geometric structures of the quantum well infrared photodetector and the diffraction grating structure for optical coupling.

QWIP structure contained grating have also been proposed to improve the detector performance.

The investigations show that the grating optimization is accomplished by the proper design of the grating profile. The theoretical approach of transverse electric and magnetic polarization (TE and TM) by Andersson and Lundqvist is limited by the fact that when the medium is lossy (due to the optical absorption in the QWIP), there will be a longitudinal component of the electromagnetic field. Here we try to give an outline of the grating theory for QWIPs based on the well-known Huygen's principle and Kirchhoff formula to calculate the absorption coefficient for a general shaped reflection grating coupled QWIP.

The AlGaAs/GaAs QWIP structure is prepared normally as follows: semi-insulating (100) GaAs substrate, GaAs contact layer (1.0 μm, Si-doped $10^{18}\,\text{cm}^{-3}$), 50 periods of QWs, GaAs top contact layer (2.0 μm Si-doped 10^{18} cm^{-3}). The QW is made of $Al_{0.3}$ and $Ga_{0.7}As$ barrier (45–50 nm) and GaAs well (4–5 nm). The structure is schematically shown in Fig. 5.16.

The effective carriers to be photon-excited are Γ electrons. The energy sublevel structures of the QW, which are self-consistently calculated from the Schrödinger and Poisson equations, are plotted in Fig. 5.17. The barrier is 500 Å thick; the well is 45 Å thick. The central 25 Å thick layer in the well is doped at 10^{18} cm^{-3}. There is one localized sublevel ($E_0 = 92.48\,\text{meV}$) in the QW, and the Fermi level is $E_f = 62.89\,\text{meV}$ at room temperature and $E_f = 96.86\,\text{meV}$ at 77 K.

Since the electric potential (dotted line in Fig. 5.17) is very small as compared with the QW depth, the QW can be approximated as square. Let the well width be L and the depth V. We define the MQW direction as the z-axis, and the plane perpendicular to this direction as the xy-plane. Since the Hamiltonian H is translationally symmetric in the xy-plane, the wave function can be expressed as

$$\psi_m(\boldsymbol{k}) = \psi_m(z)u(\boldsymbol{r})e^{i\boldsymbol{k}\cdot\rho} \tag{5.59}$$

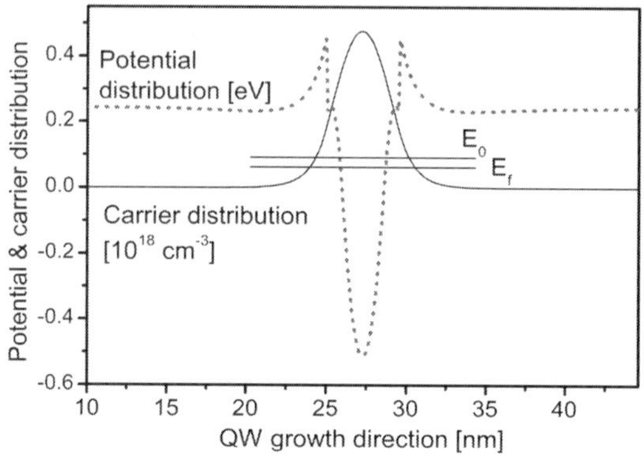

Fig. 5.17 Potential energy and carrier distribution in one period quantum well barrier at room temperature.

where the envelope function $\psi_m(z)$ is determined by

$$\left[\frac{-\hbar^2}{2m^*}\frac{\partial^2}{\partial z^2}+V(z)\right]\psi_m(z)=E_m\psi_m(z) \tag{5.60}$$

And the total energy of the electron is

$$E_m(\boldsymbol{k})=E_m+E_{\boldsymbol{k}} \quad , \quad E_{\boldsymbol{k}}=\frac{\hbar^2(k_x^2+k_y^2)}{2m^*} \tag{5.61}$$

where $\boldsymbol{k}=(k_x,k_y)$ and ρ are the wave vector and coordinate in the xy-plane.

The ground localized sublevel ψ_0 is described by

$$\psi_0=\begin{cases} A_0\exp\left(k_{0b}z\right) , & z<-\frac{L}{2} \\ B_0\cos\left(k_{0w}z\right) , & -\frac{L}{2}\le z\le\frac{L}{2} \\ A_0\exp\left(-k_{0b}z\right) , & z>\frac{L}{2} \end{cases} \tag{5.62}$$

The extended states over the barrier that can be optically coupled with ψ_0 are

$$\psi_1=\begin{cases} A_1\sin\left(k_{1b}z-\delta\right) , & z<-\frac{L}{2} \\ B_1\sin\left(k_{1w}z\right) , & -\frac{L}{2}\le z\le\frac{L}{2} \\ A_1\sin\left(k_{1b}z+\delta\right) , & z>\frac{L}{2} \end{cases} \tag{5.63}$$

Here

$$\frac{\hbar^2k_{0w}^2}{2m^*}=E_0$$

$$\frac{\hbar^2k_{0b}^2}{2m^*}=V-E_0$$

$$\frac{\hbar^2k_{1w}^2}{2m^*}=E_1$$

$$\frac{\hbar^2k_{1b}^2}{2m^*}=E_1-V \tag{5.64}$$

The normalization factors for wave functions of ground and extended states, $A_{0,1}$, $B_{0,1}$, are[38]

$$
\begin{aligned}
B_0 &= \sqrt{2}\left[L + \frac{1}{k_{0b}} + \frac{\sin(k_{0w}L)}{k_{0w}} + \frac{\cos(k_{0w}L)}{k_{0b}}\right]^{-1/2} \\
A_0 &= B_0 \cos\left(\frac{k_{0w}L}{2}\right)\exp\left(\frac{k_{0b}L}{2}\right) \\
A_1 &= \sqrt{\frac{1}{L+b}} \\
B_1 &= A_1\left[\sin^2\left(\frac{k_{1w}L}{2}\right) + \left(\frac{k_{1w}}{k_{1b}}\right)^2 \cos^2\left(\frac{k_{1w}L}{2}\right)\right]^{-1/2} \\
\delta &= \sin^{-1}\left[\frac{B_1}{A_1}\sin\left(\frac{k_{1w}L}{2}\right)\right] - \frac{k_{1b}L}{2}
\end{aligned} \tag{5.65}
$$

where b is the thickness of the barrier.

Since the effective mass of the Γ electron is isotropic, the interaction between the incident radiation and the Γ electrons in the QWIP layer is

$$\frac{e}{m^*}\boldsymbol{A}\cdot\boldsymbol{p} \tag{5.66}$$

for an incident light of $\boldsymbol{A}\mathrm{e}^{i\omega t}$.

It is easy to show that

$$
\begin{aligned}
\langle\psi_1(\boldsymbol{k})|\boldsymbol{A}\cdot\boldsymbol{p}|\psi_0(\boldsymbol{q})\rangle &= iA_z\hbar\delta_{\boldsymbol{k},\boldsymbol{q}}\langle\psi_1|\frac{\partial}{\partial z}|\psi_0\rangle \\
\langle\psi_1|\frac{\partial}{\partial z}|\psi_0\rangle &= \frac{-\hbar^2(A+B)}{m^*(E_1-E_0)}
\end{aligned} \tag{5.67}
$$

where

$$
\begin{aligned}
A = {} & \frac{B_0B_1}{2}k_{0w}\left\{(k_{1w}+k_{0w})\sin\left[\frac{(k_{1w}-k_{0w})L}{2}\right] - \right. \\
& \left. -(k_{1w}-k_{0w})\sin\left[\frac{(k_{1w}+k_{0w})L}{2}\right]\right\} \\
B = {} & A_0A_1k_{0b}\exp\left(\frac{-k_{0b}L}{2}\right) \\
& \left[k_{1b}\cos\left(\frac{k_{1b}L}{2}+\delta\right) - k_{0b}\sin\left(\frac{k_{1b}L}{2}+\delta\right)\right]
\end{aligned} \tag{5.68}
$$

Again we have observed that the momentum is conserved in the xy plane and the transition is between the envelope functions in the z-direction. The transition probability from state $(E_0,\boldsymbol{q})$ to $(E_1,\boldsymbol{q})$ is

$$w(\boldsymbol{q}) = \left|\frac{eA_z\hbar}{m^*}\langle\psi_1|\frac{\partial}{\partial z}|\psi_0\rangle\right|^2 \frac{2\Gamma}{\hbar[(E_1-E_0-\hbar\omega)^2+\Gamma^2]} \tag{5.69}$$

It is easy to see that $w(\boldsymbol{q})$ is actually $\boldsymbol{q}$-independent because of the momentum conservation and the isotropic effective mass. The probability of an electron scattered from sublevel E_0 to E_1 is

$$W = \int \frac{2d\boldsymbol{q}}{(2\pi)^2} w(\boldsymbol{q}) \{f[E_0(\boldsymbol{q})] - f[E_1(\boldsymbol{q})]\} \tag{5.70}$$

Since E_f is far below the barrier height, V for the QWIP structure under investigation so that

$$f[E_1(\boldsymbol{q})] = 0$$

$$n = \int \frac{2d\boldsymbol{q}}{(2\pi)^2} f[E_0(\boldsymbol{q})] \tag{5.71}$$

is the carrier density in the QW, the optical absorption coefficient of the QWIP is obtained as

$$\alpha = \frac{nw}{n_{\text{ph}}v} \quad , \quad n_{\text{ph}} = \frac{2\epsilon\omega|\boldsymbol{A}|^2}{\hbar} \tag{5.72}$$

Here we see that due to the isotropic effective mass of the active Γ electrons, normal incident photon absorption is not possible. A component of the optical electric field along the QWIP growth direction, A_z is required. For fabricating large arrays of optical QWIPs based on the AlGaAs/GaAs Γ electrons, diffraction gratings are needed for achieving large optical coupling.[26]

We consider a two-dimensional diffraction grating shown in Fig. 5.16. The size of the QWIP pixel is $L \times L$. The diffraction grating structure consists of $2a \times 2a$ rectangular apertures, or circular apertures with radius a periodically arranged on the optical pixel, at distance b from each other. The depth of the apertures is denoted as h.

For electromagnetic waves propagating in semiconductor materials with negligible electric conductivity, the wave equation for the electric field $\boldsymbol{E}$, Eq. 2.63, becomes

$$\nabla^2 \boldsymbol{E} = \epsilon\mu \frac{\partial^2 \boldsymbol{E}}{\partial t^2} \tag{5.73}$$

where ϵ is the permittivity or dielectric constant, and μ is the magnetic permeability of the medium. In our case we may assume $\mu = 1$. The velocity of the light waves in free space is $c = \sqrt{\epsilon_0\mu_0}$; it is $v = c/n$ in the medium, where $n = \epsilon/\epsilon_0$ in the refractive index.

For the diffraction radiation detection experiments with unpolarized incident radiation, the vector nature of the wave amplitude will not be important. The only consequence of the vector properties is multiplication of the diffracted intensity by a polarization factor, depending only on the angle of diffraction. Hence we may assume that we are dealing with a scalar function $\phi(\boldsymbol{r})$, which is a solution to the wave equation[39]

$$\nabla^2 \phi = \epsilon\mu \frac{\partial^2 \phi}{\partial t^2} \tag{5.74}$$

For purely elastic scattering, we need consider only a single frequency ω so that the wave and the wave equation are of the form

$$\nabla^2\phi + 4\pi^2k^2\phi = 0 \tag{5.75}$$

where

$$2\pi k = \sqrt{\epsilon\mu}\omega$$

is the wave number.

An important solution to Eq. 5.75 is

$$\phi_0 = \frac{\mathrm{e}^{-i2\pi k r_Q}}{r_Q} \tag{5.76}$$

which corresponds to a spherical wave radiating from a point source Q.

Huygen's principle indicates that the propagation of waves through space involve the generation of secondary waves at each point of a wave front so that the envelope of the secondary waves becomes a new wave front. By this the Kirchhoff formula is written in the form of: The wave field ϕ at point P due to any source wave ϕ_0 is given by integrating over any closed surface S containing P as[40]

$$\phi = \frac{1}{4\pi}\oint_S \left[\frac{\mathrm{e}^{-i2\pi kr}}{r}\cdot\nabla(\phi_0) - \phi_0\cdot\nabla\left(\frac{\mathrm{e}^{-i2\pi kr}}{r}\right)\right]\mathrm{d}S \tag{5.77}$$

where r is the distance from point P to the surface S.

The optical wave function for the refractive diffraction grating defined by its reflection coefficient $q(X,Y)$, which is perpendicular to the direction of radiation incidence z, becomes

$$\begin{aligned}\phi(x,y,R) = {} & \frac{1}{4\pi}\int\int \mathrm{d}X\mathrm{d}Y\ q(X,Y)\frac{\mathrm{e}^{-i2\pi kr}}{r}\cdot\frac{\mathrm{e}^{-i2\pi kr_Q}}{r_Q} \\ & \left[\left(\frac{1}{r}+i2\pi k\right)\cos\theta - \left(\frac{1}{r_Q}+i2\pi k\right)\cos\theta_Q\right]\end{aligned} \tag{5.78}$$

where

$$r^2 = (x-X)^2 + (y-Y)^2 + R^2$$

and R is the distance of the observation plane xy to the diffraction grating plane XY (in our QWIP structure, this is the thickness of the top n^+-GaAs contact). Here the surface S is the XY plane plus a closing surface at infinity and r_Q is the distance from the source Q to the point (X,Y). θ and θ_Q are the angles between the z axis and vectors $\boldsymbol{r}$ and $\boldsymbol{r}_Q$.

When the light source is far away from the diffraction plane XY (the illuminating light is in the form of plane wave propagating along the z-axis),

$$r_Q = \infty \gg k^{-1}\,,\ \theta_Q = 0$$

so that Eq. 5.78 becomes

$$\phi(x,y,R) = \beta(k)\int\int q(X,Y)\frac{\mathrm{e}^{-i2\pi kr}}{r}\left[\left(\frac{1}{r}+i2\pi k\right)\cos\theta - i2\pi k\right]\mathrm{d}X\mathrm{d}Y \tag{5.79}$$

where $\beta(k)$ is the normalization factor depending on the photon spectrum of the radiation source. If the intensity of the radiation is independent of its wave vector,

$$\int\int |\phi|^2 \mathrm{d}x\mathrm{d}y = \text{constant} \tag{5.80}$$

The intensity distribution of the optical field is

$$I(x, y, R) = |\phi(x, y, R)|^2 \tag{5.81}$$

Let us make simple analyses about Eq. 5.79. It is easy to observe that if $q(X, Y)$ is constant in the rather extended area, $\phi(x, y, R)$ is also constant, independent of (x, y),

$$\int_{-\infty}^{\infty}\int_{-\infty}^{\infty} \frac{e^{-i2\pi kr}}{r}\left[\left(\frac{1}{r} + i2\pi k\right)\cos\theta - i2\pi k\right]\mathrm{d}X\mathrm{d}Y = \text{constant} \tag{5.82}$$

In other words, when we simple cover the QWIP having a large surface area with a uniform refractive metal film, mirror reflection of the source radiation is obtained.

In general the extension of $q(X, Y)$ is limited. In an FPA system, $q(X, Y)$ is determined by the refractive grating structure on one optical pixel. In other words, $q(X, Y) = 0$ outside the optical pixel when considering only one single QWIP optical pixel. When the radiation wavelength is very small (large k, or high photon energy) as compared with the geometric dimension of the optical pixel and the distance between the observation plane and the diffraction grating plane (R),

$$|\phi(x, y, R)|^2 = |q(X, Y)|^2 \tag{5.83}$$

apart from a constant. The above result is due to the rapid oscillation factor $\mathrm{e}^{-i2\pi kr}$ so that only the radiation from the source at $(X, Y) = (x, y)$ is observed at (x, y). We thus obtain a mirror reflection from the reflective grating structure.

Now we apply Eq. 5.79 for the infrared radiation of $1/k = 10\,\mu\mathrm{m}$ to calculate the optical field diffracted from a reflective grating structure of $40 \times 40\,\mu\mathrm{m}^2$ ($X, Y \in -20, 20\,\mu\mathrm{m}$), corresponding to one optical pixel in the FPA system. The grating consists of 7×7 apertures arranged in the form of Fig. 5.16. $a = 2$, $b = 8$, $h = 1.5$, and $R = 2\,\mu\mathrm{m}$ (R is the thickness of the top n^+ GaAs contact layer, which separates the grating from the absorbing QWIP layer). The calculation results ($x, y \in 0, 50\,\mu\mathrm{m}$) of circular and square apertures are presented in Fig. 5.18a, b.

Periodicity is observed in the optical field ϕ in the xy-plane (Fig. 5.18). The period is determined by the wavelength of the incident radiation. To confirm such a numerical conclusion, we have calculated the optical field reflected from a simple uniform thin metal film ($h = 0$). The result is presented in Fig. 5.18c. Here the periodicity of $1/k$ is again observed. Another important result is the similar pattern of the optical fields diffracted from three different reflective metal grating structures. The above two conclusions are consequences of the long wavelength of the incident radiation, which is comparable with the geometric dimensions of the QWIP structure.

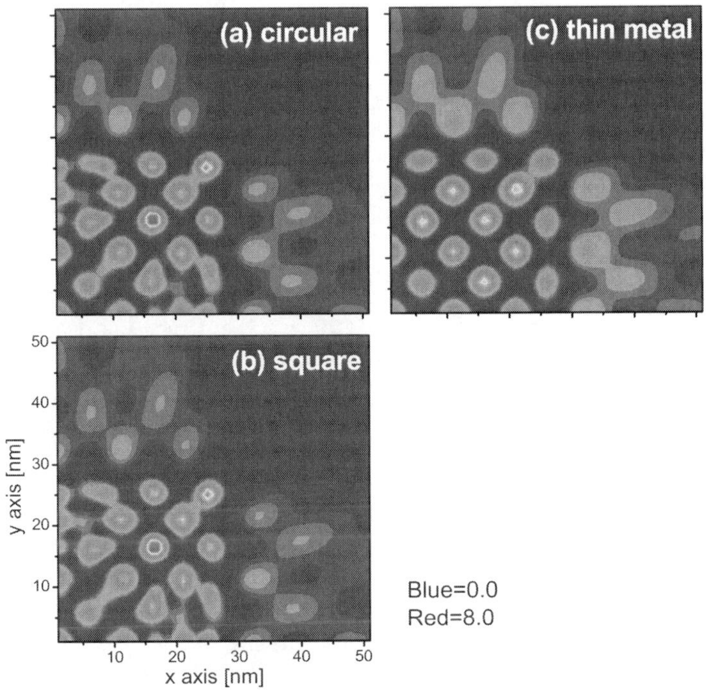

Fig. 5.18 Intensity distributions of the optical field for $1/k = 10$ μm with (a) circular diffraction apertures; (b) square diffraction apertures, and (c) thin metal reflection film ($h = 0$). See also Color Insert.

The optical field along the y-axis is presented for different radiation wavelengths in Fig. 5.19. We observe the periodicity of $1/k$. Such a periodicity is understood by considering one-dimensional grating structure so that the oscillation factor becomes

$$e^{-i2\pi kr} = e^{-i2\pi kx} e^{i2\pi kX}$$

It is thus concluded that the geometric shape of the grating structure can be made rather simple when the wavelength of the incident radiation is comparable with the geometric size of the photodetector. For the QWIP under investigation, a grating structure made of a uniformly thin reflective metal film is just as good for almost perfect optical coupling.

Fringing field (see Fig. 5.19 for $x > 20$ μm), on the other hand, is very strong due to the long wavelength of the incident radiation, which causes the cross talking between adjacent optical pixels on an FPA. The problem can be solved by reducing the size of the metal film on the QWIP pixel.

The solution of Eq. 5.75 can be expressed as

$$\phi(x, y, z) = e^{i2\pi kz} + \sum_{\boldsymbol{q}} t_{\boldsymbol{q}} e^{i2\pi(\boldsymbol{q}\cdot\rho - q_z z)} \tag{5.84}$$

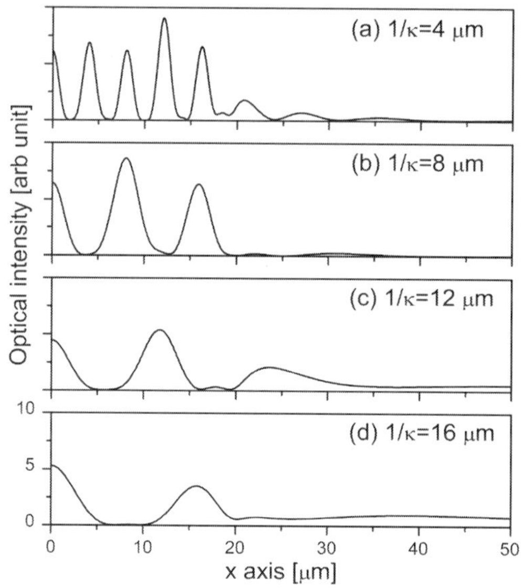

Fig. 5.19 Optical intensity distributions along the y-axis of the optical fields for different radiation wave vectors.

where $e^{i2\pi kz}$ is the incident light in the form of plane wave along the z direction,

$$t_{\boldsymbol{q}} = \int\int \phi(x, y, z)e^{-i2\pi \boldsymbol{q}\cdot\rho}dxdy \tag{5.85}$$

is the amplitude of the reflected wave diffracted from the grating surface, which propagates along $-z$ direction with wave vector $\boldsymbol{q} = (q_x, q_y)$ in the xy plane,

$$q_x^2 + q_y^2 + q_z^2 = k^2 \tag{5.86}$$

when considering elastic diffraction.

The design of the diffraction grating is to be guided by optimizing the optical intensities $t_{\boldsymbol{q}}$ with small q_z. The larger these optical intensities, the larger will be A_z, the component of the optical field along the z-direction, and thus a larger optical coupling between the incident photons and active electrons in the QWIP.

By considering Fig. 5.19 and the responding wave vector of the QWIP structure ($1/k = 10\,\mu$m), we can conclude that for $R \approx 1/k$,

$$t_{q_z} \approx \begin{cases} 1 \text{ if } q_z = 0 \\ 0 \text{ otherwise} \end{cases} \tag{5.87}$$

so that $A_z \approx A/2$, where A is the amplitude of the vector potential of the incident infrared radiation and A_z is the z-component of the vector potential of the radiation diffracted from the grating diaphragm.

After the determination of A_z, the optical absorption coefficient is calculated by Eq. 5.72 for our QWIP with a thin metal film grating and the result is presented in

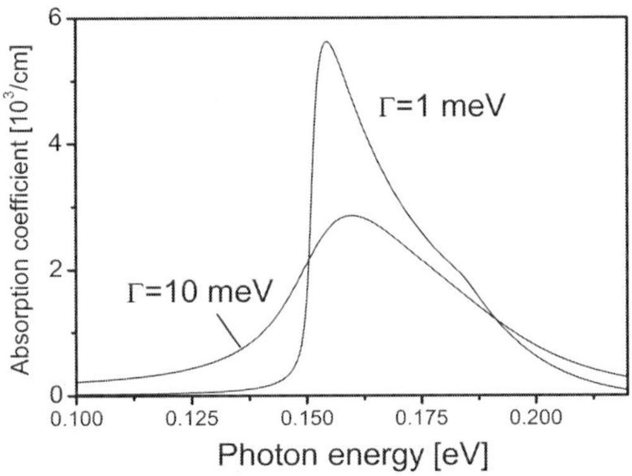

Fig. 5.20 Optical absorption coefficient at low temperature (solid line, $\Gamma = 1$ meV) and high temperature (dotted lines, $\Gamma = 10$ meV). Y. Fu, M. Willander, W. Lu, and Wenlan Xu, Optical coupling in quantum well infrared photodetector by diffraction grating, J. Appl. Phys. vol. 84, p. 5750–5, 1998. Copyright (1998) by The American Institute of Physics.

Fig. 5.20. The absorption coefficient depends naturally on the relaxation process in the system, which is defined by the relaxation energy Γ. Normally Γ increases with increasing temperature.

It is important to notice the magnitude of the absorption coefficient, about $1\,\mu\text{m}^{-1}$, which is comparable to the total thickness of the active GaAs/AlGaAs MQW. For the QWIP structure under investigation, the total thickness of the MQW is $50 \times (0.5+0.045) = 2.725\,\mu\text{m}$. The QWIP is thus very lossy for infrared radiations with energies around 0.15 eV. High quantum efficiency is thus expected.

Similar results have been derived by Andersson and Lundqvist based on the theoretical approach of TE and TM, which shows almost the same high quantum efficiency spectrum for different grating structures.[36]

5.3.3 *QD Solar Cells*

As discussed in Section 3.7, the experimental evidence of the QD effectiveness in solar energy applications has been mounting. Some researchers have made prototype solar cells from QDs. For instance, Qi *et al.* mixed a conductive, photosensitive polymer known as MEH-PPV with lead selenide QDs. Under visible light, a device incorporating QDs at only about 5% by weight generated 50% more current than expected if each photon yielded one exciton.[41] Lead sulfide nanocrystals in the conducting polymer poly(2-methoxy-5-(2-ethyl-hexyloxy)-*p*-phenylene vinylene) were reported.[42] Furthermore, QD-sensitized nanowire solar cells have been shown to exhibit short-circuit currents ranging from 1 to 2 mA/cm^2 and open-circuit voltages of 0.5–0.6 V when illuminated with 100 mW/cm^2 simulating the AM1.5 spectrum, and internal quantum efficiencies as high as 50–60% were also obtained.[43] Marti *et al.* discussed the possibility of engineering QDs with distinct energy levels inside

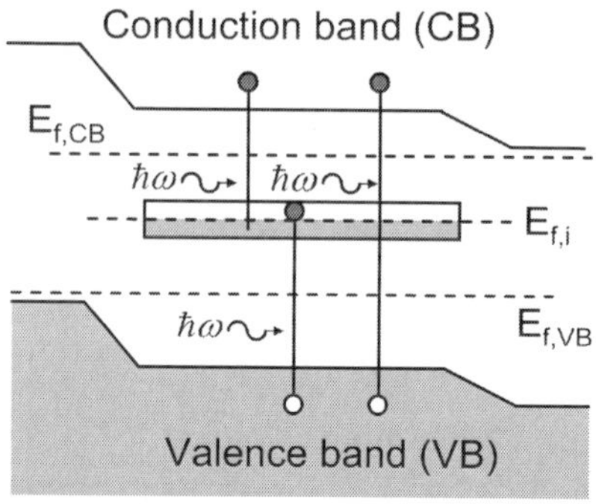

Fig. 5.21 Simplified bandgap of QD intermediate-state n–i–p cell.[44] $E_{f,\mathrm{CB}}$, $E_{f,\mathrm{VB}}$, and $E_{f,\mathrm{i}}$ denote the Fermi levels in the n, p, and i junctions, respectively.

the forbidden band gap of a bulk semiconductor p–n junction.[44] The generation of intermediate exciton states in the bandgap allows for the possibility of generation of carriers from the absorption of two photons. Because the density of states of QDs is narrow, thermalization from the intermediate states is avoided (Fig. 5.21). Among solid-state inorganic proposals, the work by Marti *et al.* appears to be the most promising. It essentially offers to generate a multiple band-gap equivalence through the use of QDs embedded in a p–i–n junction. This concept benefits from allowing a long time for radiative recombination in the QDs because it allows more time for a second incident photon to be absorbed.

Nonradiative resonant energy transfer channels between excitons in ligand-capped PbS QDs and confined states in embedding InGaAs QWs (QDs-in-QW buried in GaAs substrate) have been found experimentally and shown to be competitive with the QD exciton radiative decay, thus facilitating solar cell applications.[45] Long radiative lifetimes of colloidal QDs also help concerning photocarrier extraction.

Photocarrier extraction from QDs has been identified to be a major issue in solar cell applications. By assuming Coulomb interactions and wave function overlap as principal driving forces, several intriguing effects have been theoretically predicted, like inter-dot transport of excitons between two CdSe QDs,[46] resonant energy transfer for extended coherent exciton states in a linear chain of colloidal QDs[47] and QD face-centered cubic (fcc) lattice.[48] These studies are also important when applying QD arrays as qubits for quantum computation (see discussion below).

5.3.4 *Thermal Photovoltaics by Polymer Conjugated QDs*

Absorption of low-energy photons in the near infrared portion of the spectrum (800–2000 nm) forms the basis for the field of thermal photovoltaics. In this regime there is less power available from the solar spectrum. Generally thermal photovoltaics is envisioned as being paired with visual spectrum photovoltaics in a multiple band-gap structure. It is seen as a challenge to harvest the power available in the infrared in an economical fashion. One potentially low-cost approach for thermal voltaics has been examined in Ref.[49] The study was orientated around the use of PbS QDs embedded in conjugate polymers. An advantage of this technique lies in the low embedded energy involved in the QD synthesis. Together with the low cost of the basic materials one can infer that such QD structures may be fabricated in large quantities very inexpensively.

The minimum bandgap for PbS structures is 0.41 eV (3024 nm) for the bulk material. Through quantum confinement the bandgap for individual PbS QDs can be increased. A QD diameter of 6.5 nm corresponds to a structure with an absorption peak at 1300 nm. The efficiency reported by Sargent *et al.*[49] is three orders of magnitude larger than previous results using nanostructures for near infrared electro-optical applications. While this may represent a large improvement over previous results, the overall performance is still very poor. For an input of 207 mW the photocurrent was 91 μA at 5 V, corresponding to an external efficiency of 0.22%. Compared to solid-state systems, the efficiency is two orders of magnitude lower. The major route to achieving practical efficiency for PbS QDs lies in improving the conjugate polymer structure and its coupling to the embedded QDs along with better light trapping strategies to improve absorption.

5.3.5 *QD Bioimaging*

Highly fluorescent QDs have already been introduced successfully in many fluorescence-based optical imaging applications in the medical sciences. Polymer-coated QDs are water soluble and noncytotoxic and innocuous to normal cell physiology, exhibiting unique pharmacokinetics including minimal renal filtration. To achieve targeting abilities, polymer-coated QDs are conjugated to bio-affinity ligands such as monoclonal antibodies, peptides, or oligonucleotides using several approaches including passive adsorption, multivalent chelation, or covalent-bond formation.[50–54] Figure 3.4 shows fluorescence of chemically synthesized CdSe QDs under UV excitation. Confocal microscopic images of these QDs conjugated to VCAM-1 proteins are shown in Fig. 5.22. Exciton dynamics is an important factor in optimizing the visibility of such QDs.

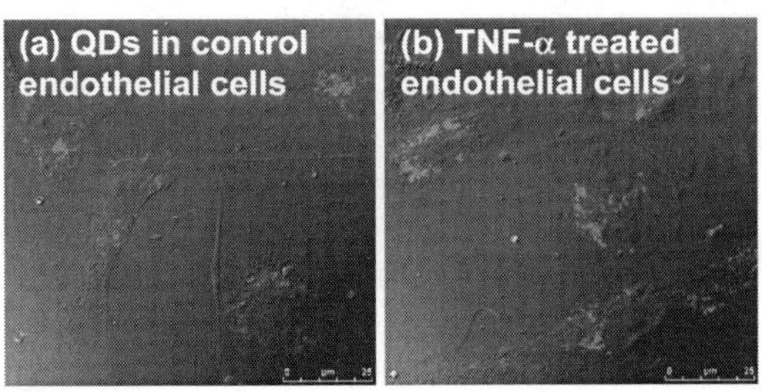

Fig. 5.22 Fluorescence of CdSe QDs (a) in control endothelial cells, and (b) in TNF-α-treated endothelial cells. See also Color Insert.

5.4 INTEGRATED CIRCUITS

Integrated circuits, that is assemblies of transistors that are the building blocks for all electronic devices, currently use electrons to ferry the signals. However, many communication devices use light, or photons, to send signals. The need to convert the signaling language from electrons to photons limits the speed of electronic devices. Along a similar strategy of electron-hole spatial separation, the latest development of so-called exciton integrated circuits can be controlled by the photoexcited excitons using voltage supplied by electrodes. These voltage gates create an energy barrier that can halt the movement of excitons or allow them to flow. Once that energy barrier is removed, the exciton can travel to the output of the device and transform to light, which could be fed directly into a communication circuit. Simple integrated circuits have been demonstrated by joining exciton transistors to form several types of switches that accurately direct signals along one or several pathways.[55]

CuCl QDs embedded in a NaCl matrix were constructed with many functionalities, such as an AND gate,[56, 57] surface plasmon–polariton condenser,[58] XOR logic gate,[59] or an optical condenser (optical nanofountain).[60] Optical NOT gates were demonstrated by utilizing resonant optical near-field energy transfer between CuCl QDs in NaCl and InAlAs QDs in AlGaAs, which further showed that QDs of III–V materials are suitable for nanophotonic devices using optical near fields.[61]

5.5 PHOTON STORAGE

Excitons confined in CdSe/ZnSe QDs on a solid-state substrate were probed through a nano-aperture using time-resolved PL. Two different decay processes of the PL peak were observed. The first has a short 450 ps lifetime, exhibiting a relatively broad emission line, and persists nearly to room temperature; the second exhibits a long (>4 ns) lifetime and is responsible for the sharp (about 100 μeV) lines seen at low temperatures (<60 K).[62] Two different exciton states were thus concluded and the extraordinarily long lifetime of the second state (4.0 ns) indicates that the overlap of the electron and hole wave functions is significantly weaker than for the short-lived state, suggesting that this exciton state may result from spatially indirect electron-hole transitions. The mechanism matches the exciton storage effect in which the optically generated excitons in one QD were dissociated and stored as separated electron-hole pairs in coupled QD pairs. The stored electron-hole pair is restored as an exciton by applying a bias to the device. This method allows for recording exciton lifetimes of several seconds i.e., more than a billion times the lifetimes of excitons in normal QDs (one QD in one InAs QD layer and the other QD in the adjacent QD layer, in a multiply stacked QD structure embedded in GaAs).[63] This mechanism has been utilized for photon storage.

Kraus *et al.* showed that CdSe/CdS QDs and nanorods can store excitons for up to 0.5 ms (10^5 times the natural lifetime which is in the order of ps of unperturbed nanocrystals) when using an electric field to quench the fluorescence de-excitation, and most importantly, at room temperature, opening up applications in optical memory elements.[64] Two QDs connected by a nanowire form a single nanostructure memory cell for storing excitons. Electrons and holes of the photogenerated excitons are separated by quantum confined Stark effect (QCSE), which inhibits their radiative recombination; photons can be retrieved by switching the electric field with a demonstrated storage time up to 30 ms at 7 K.[65]

5.6 WAVEGUIDES AND MODULATORS

As shown before, room-temperature PL measurement is a routine method to probe excitons in QDs because of the large exciton binding energies due to quantum confinements. The QCSE provides another way. An external electric field will pull the electron and hole, which form the exciton in the QD in opposite directions, resulting in modifications in the binding energy and of the optical matrix element of the exciton. A theoretical investigation was reported of QCSE in QDs by assuming an effective mass approximation for the CB electrons and 4×4 Luttinger-Kohn Hamiltonian for the VB holes and the effect on the change of refraction index by an applied field through the Kramers–Kronig relations.[66] Larger QCSEs were found in QDs as compared with QW, and a waveguide with embedded QDs leads to a high loss.[66] The high optical loss is a direct consequence of Eq. 4.32.

Equation (4.32) affects the reflection and transmission of the electromagnetic field through the QDs. To test this notion, a 10-plane InAs QD structure was grown and the sample edge polished for reflection measurement at room temperature. Room-temperature PL measurements were then also performed. The results are presented in Fig. 5.23 where a broad reflectance peak is observed at the wavelength of the PL peak, as expected.[67]

Excitons were shown to spread over a large spatial region of complex-shaped CdSe QD (multipods, trigonal and multipod pyramids) using ultrafast polarization transient grating measurements[68] which relate the relaxation rates to the QD size.

Excitons in asymmetrically coupled InAs QDs were tuned into resonance by an external electric field to form a coherent hole wave function that shows superposition of charge and spin configurations of the two QDs.[69] This demonstrates the possibility of optically coupling QDs for quantum information processing.

Coupled structures can be used for measuring propagating excitons. A layer of GaAs QDs was positioned at a distance of 500 nm away from a GaAs QW (acting as a waveguide), and the two layers were embedded in $Al_{0.3}Ga_{0.7}As$ (Fig. 5.24).

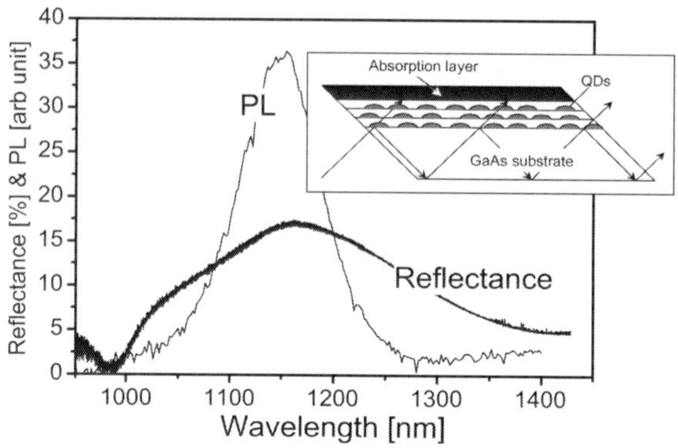

Fig. 5.23 Reflectance and PL spectra of the ten-layer InAs/InGaAs/GaAs QD sample. Inset shows the geometry of the edge polished sample. Y. Fu, H. Ågren, L. Höglund, J. Y. Andersson, C. Asplund, M. Qiu, and L. Thylén, Optical reflection from excitonic quantum-dot multilayer structure, Appl. Phys. Lett. vol. 93, p. 183117(3), 2008. Copyright (2008) by The American Institute of Physics.

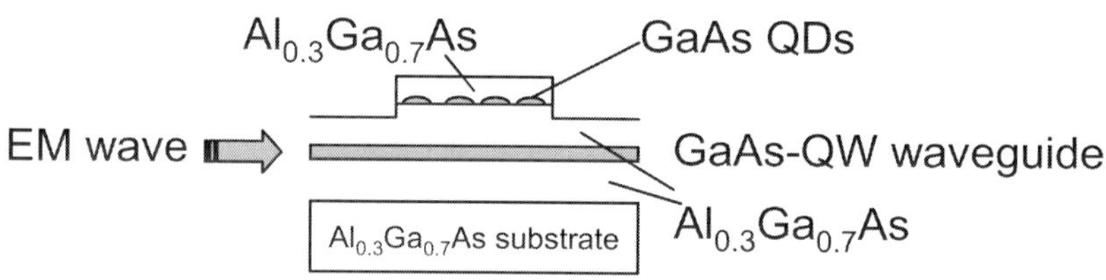

Fig. 5.24 Propagation of exciton-polariton with coupling to a QD structure generated by an incident electromagnetic (EM) wave. See also Color Insert.

Excitons in the GaAs QDs were shown theoretically and experimentally to be resonantly coupled with a propagating mode of an exciton–polariton in the GaAs QW waveguide.[70]

As discussed before, Eq. 4.32 affects the reflection and transmission of electromagnetic field through the QD lattice, as also demonstrated by Fig. 5.23 where a broad reflectance peak is observed at the wavelength of the PL peak. Figure 5.25 shows the reflection spectrum of an electromagnetic field propagating along the z-axis from a single-layer QD array extended in the xy-plane.[71]

A subwavelength QD waveguide of Fig. 5.26 was thus proposed, where excitons are first photogenerated in the QDs by the incident electromagnetic field; the exciton–polariton effect in the QD photonic crystal induces an extra optical dispersion in QDs. The high contrast ratio between the optical dispersion of the QDs and the background therefore creates clear photonic bandgaps. By carefully designing the QD size and the QD lattice structure, perfect electromagnetic field reflection can be obtained at a specific wavelength in the lossless case

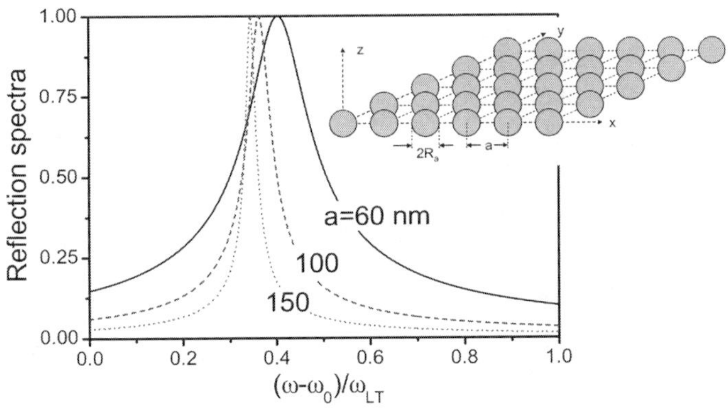

Fig. 5.25 Reflection spectrum of an electromagnetic field propagating along the z axis from a single-layer QD array extended in the xy-plane.

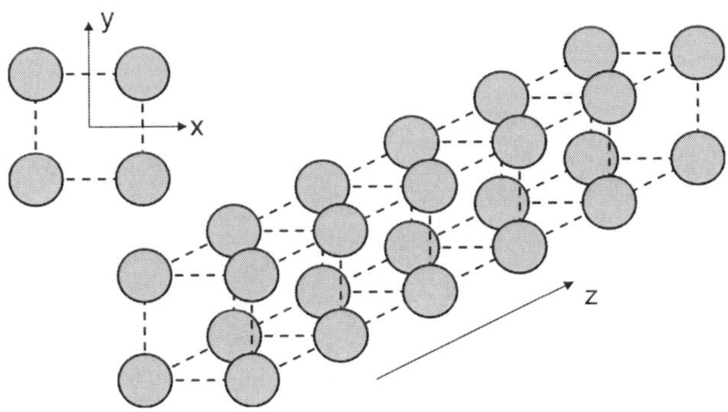

Fig. 5.26 Subwavelength optical waveguiding based on an exciton QD array. Y. Fu, E. Berglind, L. Thylén, H. Ågren, Optical transmission and waveguiding by excitonic quantum dot lattices, Journal of the Optical Society of America B, vol. 23, p. 2441–7, 2006. Copyright (2006) by The Optical Society of America.

(Fig. 5.27), thus providing a fundamental basis for ultra-dense optical waveguide applications.[72, 73]

Air-stable PbSe/PbS and PbSe/$PbSe_xS_{1-x}$ core-shell structures and the corresponding PbSe core QDs were used as passive Q-switches in an eye-safe laser of Er:glass, acting as saturable absorbers, with a relatively large ground-state cross-section of absorption (10^{-15} cm^2) and a fast effective lifetime (4.0 ps).[74]

Coupled resonator optical waveguides have been demonstrated by using polymer microspheres doped with CdSe QDs.[75] Electric field-induced modulation was reported in the absorption of PbS QDs (diameters of 5 and 7 nm) embedded in SiO_2.[76]

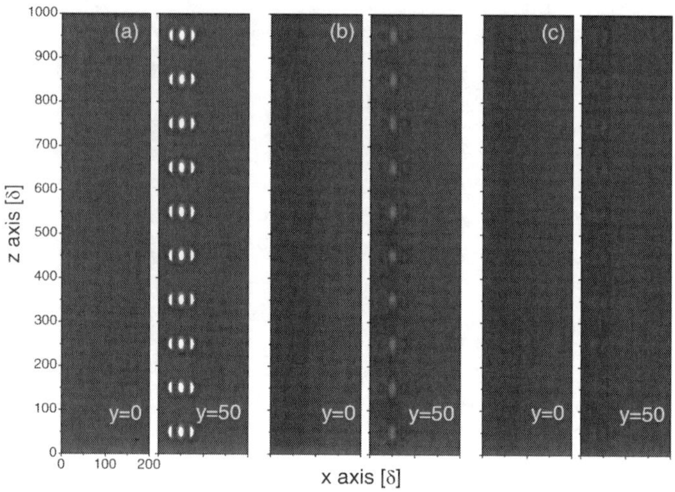

Fig. 5.27 Spatial distributions of the E_x-field amplitude. (a) $\omega = \omega_a + 0.1\omega_{\rm LT}$. (b) $\omega = \omega_a + 0.28\omega_{\rm LT}$. (c) $\omega = \omega_a + 0.5\omega_{\rm LT}$. $\hbar\omega = 1.0\,{\rm eV}$, $\delta = 2.2\,{\rm nm}$. GaAs material parameters are used for the numerical simulation, which corresponds to GaAs QDs embedded in AlGaAs barriers. See also Color Insert.

5.7 QUANTUM OPTICS

In the last two decades, several very valuable discoveries were made in this field, including nonlinear interaction of atoms with single electromagnetic mode via few photons (the one-atom maser),[77] quantum Zeno effect,[78] and suppression and redistribution of photon fluctuations.[79] On the other hand, "quantum opticians" have created powerful tools and high-sensitivity methods that permitted to discover new essences in other parts of physics; among them are electro-weak interactions in optical transitions (parity violation),[80] and the establishment of the upper limit for the electrical dipole momentum of electron and neutron.[81]

5.7.1 *Sensitivity and Resolution*

A number of measurement techniques in quantum optics, as well as in other parts of experimental physics, are based on a simple scheme: a degree of freedom is chosen and then its interaction with the object of interest is analyzed. It is evident that the higher the isolation of this degree of freedom, the higher has to be the resolution in the experiment. The level of isolation is usually described by the quality factor Q_e. For a Fabry–Perot resonator,

$$Q_e = \frac{2\pi L}{\lambda_e(1-R)} \tag{5.88}$$

where λ_e is the wavelength, L is the distance between the mirrors, and R is the reflectivity of each mirror. If a Fabry–Perot resonator is used to measure a small change ΔL in the separation between two mirrors, the minimal value $\Delta L_{\min}$ is[82]

$$\Delta L_{\min} = \frac{L}{Q_e}\sqrt{\frac{\hbar\omega_e}{W\tau}} = \frac{\lambda_e(1-R)}{2\pi\sqrt{N}} = \frac{\lambda_e}{2F\sqrt{N}} \tag{5.89}$$

where W is the pump power, τ is the averaging time, N is the number of "used" photons,

$$F \approx \frac{\pi}{(1-R)}$$

is the fitness of the mirrors,

$$\omega_e = \frac{2\pi c}{\lambda_e}$$

Equation 5.89 is valid if the pump radiation is in coherent state and $Q_e < \omega_e\tau$. At present, mechanical coordinate meters, based on Fabry–Perot resonators, allow to resolve $\Delta L_{\min} \simeq 10^{-15}$ cm/$\sqrt{\text{Hz}}$.[83] By present moment, $(1-R) = 10^{-6}$ has been demonstrated,[84] which is three orders of magnitude smaller than that available to experimentalists two decades ago. And it is important to emphasize that until now, there is no formulated fundamental limit for the minimal value of $(1-R)$. It is thus reasonable to expect further substantial improvement in the fitness and quality factor of optical resonators.

Better isolation of a probe degree of freedom is not the only way to improve sensitivity and resolution in quantum optics measurements. Another important factor is the stability of the available reference oscillator. At present, the achieved level of residual frequency instability approaches the value

$$\frac{\Delta\omega}{\omega} \simeq 10^{-18}$$

At the same time, the quantum analysis of intrinsic fluctuations of frequency in self-oscillators demonstrates that there is a substantial reserve for further improvement of frequency stability by several orders of magnitude.[85, 86] For example, if the resonator of the self-oscillator has the volume Ω and is made of a dielectric with Young's modulus Y, the limit for the frequency instability during averaging time τ must be

$$\frac{\Delta\omega}{\omega} \simeq \sqrt{\frac{\hbar}{Y\Omega\tau}} \simeq 10^{-22} \times \sqrt{\frac{10^{12}\ \text{dyn/cm}^2}{Y} \times \frac{10^3\ \text{cm}^3}{\Omega} \times \frac{10^2\ \text{s}}{\tau}} \tag{5.90}$$

which is usually called the standard quantum limit. It is important to note that this limit is not universally minimal; the frequency fluctuations can be further reduced if a nonlinear material is used.[86]

5.7.2 *Quantum Nondemolition Measurement*

The idea of quantum nondemolition (QND) measurement is simple: experimentalist has to choose the observable whose operator is commuting with itself in time. In this case it is possible to repeat the measurement and obtain the same result, if there were no external action on the object under study except the back-action of the QND meter itself.[87]

The simplest example of QND measurement is measuring the energy E stored inside an, e.g., resonator via measuring the ponderomotive force acting on one of the resonator walls that is made flexible. E is a QND observable, and the ponderomotive pressure is proportional to E. It is shown that the minimal error in measuring E under optimal conditions, with τ as the averaging time, can be

$$\Delta E = \frac{\hbar}{\tau} \tag{5.91}$$

It is evident that the motion of a resonator wall will produce perturbation of the phase ϕ of oscillations in the resonator. The uncertainty of the number of quanta $\Delta N = \Delta E/\hbar\omega$ satisfies the relation

$$\Delta N \Delta\phi = \frac{1}{2} \tag{5.92}$$

Combining the above two formulae we obtain that the error in measuring the photon number will be much smaller than unity $\Delta N \ll 1$, and the field in the resonator will be converted into N-state (Fock state). It is therefore possible to measure the number of quanta in the resonator without absorbing any of them (unlike the case in usual photodetectors, where all photons producing the readout are absorbed).

Quantum theory predicts not only the existence of N-states in a resonator, but also the so-called phase-squeezed state in which the uncertainty of the phase and photon number are "opposite" to Fock states[88]

$$\Delta\phi \simeq \frac{1}{2N}\ , \quad \Delta N \simeq N \tag{5.93}$$

These new states may permit us to "use" much smaller numbers of photons N than it is required when using coherent states with

$$\Delta\phi \simeq \frac{1}{2\sqrt{N}}$$

This advantage may be illustrated by the example of detecting gravitational radiation. The burst of gravitational radiation is a nonstationary perturbation of the metric and produces a very small change ΔL_{gr} in the separation L between two free test masses. In case of optimal orientation

$$\Delta L_{gr} \simeq \frac{hL}{2} \tag{5.94}$$

where h is the amplitude of the metric perturbation. By the Fabry–Perot optical resonator, it is necessary to register the small phase shift $\Delta\phi_{gr}$ of oscillations in the resonator

$$\Delta\phi_{gr} = \Delta\omega\tau_{gr} = \omega\frac{\Delta L_{gr}}{L}\tau_{gr} = \frac{h\omega\tau_{gr}}{2} \tag{5.95}$$

where τ_{gr} is the duration of the burst. If the Fabry–Perot resonator is directly illuminated by a laser to produce coherent state, it is then necessary to "use" the number of photons N that satisfies the inequality

$$\Delta\phi_{gr} \geq \Delta\phi \simeq \frac{1}{2\sqrt{N}}$$

In advanced version of laser interferometric gravitational wave antenna, $L \simeq 4 \times 10^5$ cm, $\omega \simeq 3 \times 10^{15}$ rad/s, $\tau_{gr} = 10^{-3}$ s, the required projected sensitivity has to be at the level $h \simeq 10^{-22}$. This would require to "spend" $N \simeq 10^{20}$ photons within the measurement time $\tau \simeq \tau_{gr}$ that corresponds to enormous optical power circulating in the interferometer. On the other hand, if the resonator is equipped with a quantum phase meter for a squeezed state, $\Delta\phi_{gr} \simeq \Delta = 1/2N$, and therefore for the same sensitivity $h \simeq 10^{22}$, the necessary number of probe photons may be strongly reduced: $N \simeq 10^{10}$.

5.7.3 *Quantum Information*

The elementary unit of quantum information is referred to as a quantum bit (qubit), and can be represented by any two-level quantum system. One possible approach is to use exciton excitations in the ground state of a QD as basis for a two-level system. The time evolution of an exciton qubit (QD) and a photon qubit (photonic cavity) coupled with each other was studied theoretically in Ref.[89] showing that finite entanglement remains in a steady state, which can be used for preparing the Bell state between photon and exciton qubits. An all-optical quantum gate was designed by controlling a biexciton in a single QD.[90] The system has four states: a ground state containing two unexcited electrons; two states containing one exciton; and a state containing two excitons. The two single-exciton states can be distinguished from each other because the excitons have different polarization. Rabi oscillations can be driven between the ground state and the one-exciton states, and also between the one-exciton states and the biexciton state, with lasers; thus, the QD system behaves like a controlled-NOT gate in which the value of one qubit is reversed (the NOT operation) if—and only if—the value of the other qubit is 1. By employing the power dependence of the two colors, a two-qubit logic gate operation was succeeded in controlling the population of an artificial exciton molecule of InGaAs coupled QDs in GaAs.[91] Biolatti *et al.* theoretically designed biexcitons via energy-selected interband optical transitions in a QD array with an inter-QD distance of 5 nm for quantum-information encoding and manipulation using ultrafast sequences of multicolour laser pulses.[92, 93]

A single exciton in a QD can serve as a charge qubit. Presence (absence) of an exciton corresponds to a qubit state $|1\rangle$ ($|0\rangle$). Excitons can be created optically by ultrafast laser pulses; controlling the pulse parameters allows creating exciton superposition states. By doping a QD with a single electron, a spin qubit can be realized based on the spin state of the single electron. Operations on the spin qubit are performed by creating a "trion" (a charged exciton made of the original electron and an exciton). With multilevel optical Bloch equations, leakage to the wetting layer and Auger capture were shown to be two main decoherence mechanisms of QD exciton states.[94]

Self-assembled $In_{0.5}Ga_{0.5}As$ QDs in a GaAs substrate was processed into single QD photodiodes.[95] The single QD photodiode was further developed to transfer optical excitations into deterministic electric currents in the context of quantum information processing.

References

[1] S.-Y. Lin, E. Chow, V. Hietala, P. R. Villeneuve, and J. D. Joannopoulos, "Experimental demonstration of guiding and bendings of electromagnetic waves in a photonic crystal," Science, **282**, 274–276, 1998.

[2] T. Zijlstra, E. van der Drift, M. J. A. de Dood, E. Snoeks, and A. Polman, "Fabrication of two-dimensional photonic crystal waveguides for 1.5 μm in silicon by deep anisotropic dry etching," J. Vac. Sci. Technol. B, **17**, 2734–2739, 1999.

[3] M. Ohtsu, K. Kobayashi, T. Kawazoe, S. Sangu, and T. Yatsui, "Nanophotonics: design, fabrication, and operation of nanometric devices using optical near fields," IEEE J. Sel. Top. Quantum Electron. **8**, 839–862, 2002.

[4] R. M. Stevenson, R. M. Thompson, A. J. Shields, I. Farrer, B. E. Kardynal, D. A. Ritchie, and M. Pepper, "Quantum dots as a photon source for passive quantum key encoding," Phys. Rev. B, **66**, 081302(4), 2002.

[5] S. Strauf, S. M. Ulrich, K. Sebald, P. Michler, T. Passow, D. Hommel, G. Bacher, and A. Forchel, "Single-photon and photon-pair emission from CdSe/Zn(S,Se) quantum dots," phys. stat. sol. (b), **238**, 321–324, 2003.

[6] R. M. Stevenson, A. J. Hudson, R. J. Young, P. Atkinson, K. Cooper, D. A. Ritchie, and A. J. Shields, "Biphoton interference with a quantum dot entangled light source," Optics Express, **15**, 6507–6512, 2007.

[7] J. Vuckovic and Y. Yamamoto, "Photonic crystal microcavities for cavity quantum electrodynamics with a single quantum dot," Appl. Phys. Lett., **82**, 2374(3), 2003.

[8] T. Aichele, V. Zwiller, O. Benson, I. Akimov, and F. Henneberger, "Single CdSe quantum dots for high-bandwidth single-photon generation," J. Opt. Soc. Am. B, **20**, 2189–2192, 2003.

[9] Y. Arakawa and H. Sakaki, "Multidimensional quantum well laser and temperature dependence of its threshold current," Appl. Phys. Lett., **40**, 939(3), 1982.

[10] N. N. Ledentsov, V. M. Ustinov, A. Yu. Egorov, A. E. Zhukov, M. V.; Maksimov, I. G. Tabatadze, and P. S. Kop'ev, "Optical properties of heterostructures with InGaAs-GaAs quantum clusters," Semiconductors, **28**, 832–834, 1994.

[11] N. Hatori, K. Otsubo, M. Ishida, T. Akiyama, Y. Nakata, H. Ebe, S. Okumura, T. Yamamoto, M. Sugawara, and Y. Arakawa, "20°C–70°C temperature independent 10 Gb/s operation of a directly modulated laser diode using p-doped quantum dots," European Conference and Exhibition on Optical Communication ECOC 2004, Stockholm, 5–9 September, 2004. T. Yamabana, S. We, K. Mori, T. Yamamoto, M. Kawai, M. Sugawara, M. Ishida, Y. Arakawa, "Temperature independent transmission for 10 Gbps 300 m - MMF using low driving-current quantum dot laser," Optical Fiber Communication Conference, 2006 and the 2006 National Fiber Optic Engineers Conference, OFC 2006, 5–10 March 2006.

[12] M. Sugawara, T. Yamamoto, and H. Ebe, "Quantum-dot-based photonic devices," Fujitsu Sci. Tech. J., **43**, 495–501, 2007.

[13] V. I. Klimov, S. A. Ivanov, J. Nanda, M. Achermann, I. Bezel, J. A. McGuire, and A. Piryatinski, "Single-exciton optical gain in semiconductor nanocrystals," Nature, **447**, 441–446, 2007.

[14] J. P. Sun, R. K. Mains, W. L. Chen, J. R. East, and G. I. Haddad, "C-V and I-V characteristics of quantum well varactors," J. Appl. Phys. **76**, 2340–2346, 1992.

[15] F. Capasso, S. Sen, F. Beltram, L. M. Lunardi, A. S. Vengurlekar, P. R. Smith, N. J. Shah, R. J. Malik, and A. Y. Cho, "Quantum functional devices: resonant-tunneling transistors, circuits with reduced complexity, and multiple valued logic," IEEE Trans. Electron. Devices, **36**, 2065–2082, 1989.

[16] P. E. England, J. E. Golub, L. T. Florez, and J. P. Harbison, "Optical switching in a resonant tunneling structure," Appl. Phys. Lett., **58**, 887–889, 1991.

[17] C. R. H. White, M. S. Skolnick, L. Eaves, and M. L. Leadbeater, "Electroluminescence and impact ionization phenomena in a double-barrier resonant tunneling structure," Appl. Phys. Lett., **58**, 1164–1166, 1991.

[18] C. Van. Hoof, and G. Borghs, and E. Goovaerts, "Optical detection of light- and heavy-hole resonant tunneling in p-type resonant tunneling structures," Appl. Phys. Lett., **59**, 2139–2141, 1991.

[19] C. Van Hoof, J. Genoe, R. Mertens, G. Borghs, and E. Goovaerts, "Electroluminescence from bipolar resonant tunneling diodes," Appl. Phys. Lett., **60**, 77–79, 1992.

[20] T. H. Maiman, "Stimulated optical radiation in ruby," Nature, **187**, 493–494, 1960.

[21] V. I. Tolstikhin and M. Willander, "Resonant tunneling injection hot electron laser: An approach to picosecond gain-switching and pulse generation," Appl. Phys. Lett., **67**, 2684–2686, 1995.

[22] J. Faist, F. Capasso, C. Sirtori, D. L. Sivco, A. L. Hutchinson, and A. Y. Cho, "Quantum cascade laser," Science, **264**, 553–556, 1994.

[23] C. Sirtori, J. Faist, F. Capasso, D. L. Sivco, A. L. Hutchinson, S. N. Chu, and A. Y. Cho, "Continuous wave operation of midinfrared (7.4-8.6 μm) quantum cascade lasers up to 110 K temperature," Appl. Phys. Lett., **68**, 1745–1747, 1996.

[24] J. Faist, F. Capasso, C. Sirtori, D. L. Sivco, J. N. Baillargeon, A. L. Hutchinson, S. N. Chu, and A. Y. Cho, "High power midinfrared ($\lambda \sim 5$ μm) quantum cascade lasers operating above room temperature," Appl. Phys. Lett., **68**, 3680–3682, 1996.

[25] J. Wilson and J. F. B. Hawkers, *Optoelectronics an Introduction.* 3rd ed., Prentice Hall Europe 1998.

[26] B. F. Levine, K. K. Choi, C. G. Bethea, J. Walker, and R. J. Malik, "New 10 μm infrared detector using intersubband absorption in resonant tunneling GaAlAs superlattices," Appl. Phys. Lett., **50**, 1092–1094, 1987; B. F. Levine, "Quantum-well infrared photodetectors," J. Appl. Phys., **74**, R1–R81, 1993.

[27] S. C. Shen, "Comparison and competition between MCT and QW structure material for use in IR detectors," Microelectronics J., **25**, 713–739, 1994.

[28] L. J. Kozlowski, G. M. Williams, G. J. Sullivan, C. W. Farley, R. J. Anderson, J. Chen, D. T. Cheung, W. E. Tennant, and R. E. DeWames, "LWIR 128 × 128 GaAs/AlGaAs multiple quantum well hybrid focal plane array," IEEE Trans. Electron Devices, **38**, 1124–1130, 1991.

[29] S. D. Gunapala, J. K. Liu, M. Sundaram, S. V. Bandara, C. A. Shott, T. Hoelter, P. D. Maker, and R. E. Muller, "Long wavelength 256*256 QWIP hand held camera," Proc. SPIE, **2746**, 124–133, 1996.

[30] *Semiconductor Quantum Wells and Superlattices for Long-Wavelength Infrared Detectors*, ed. M. O. Manasreh. Artech House, Boston, 1993.

[31] W. Lu, H. J. Ou, M. H. Chen, X. L. Huang, S. C. Shen, R. H. Gu, and L. B. Ye, "Application of GaAs/GaAlAs multiple-quantum well infrared detector array in the infrared imaging camera," Int. J. Infrared Millimeter Waves, **15**, 137–144, 1994.

[32] L. J. Kozlowski, J. M. Arias, and W. E. Tennant, "Experimental comparison of staring IR sensor technologies including PV HgCdTe, PV InGaAs, and quantum well GaAs/AlGaAs," Proc. of The SPIE, the International Society for Optical Engineering, **2373**, 354–360, 1995.

[33] J. C. Hensel, H. Hasegawa, and M. Nakayama, "Cyclotron resonance in uniaxially stressed silicon. II. Nature of the covalent bond", Phys. Rev, vol. 138, p. A225–38, 1965.

[34] J. Y. Andersson and L. Lundqvist, "Grating coupled quantum well infrared detectors: theory and performance," J. Appl. Phys., **71**, 3600–3610, 1992.

[35] D. Pan, J. M. Li, Y. P. Li, and M. Y. Kong, "Long period two-dimensional gratings for 8-12 μm quantum well infrared photodetectors," J. Appl. Phys., **80**, 7069–7071, 1996

[36] J. Y. Andersson and L. Lundqvist, "Near-unity quantum efficiency of AlGaAs/GaAs quantum well infrared detectors using a waveguide with a doubly periodic grating coupler," Appl. Phys. Lett., **59**, p. 857–859, 1991; J. Y. Andersson, L. Lundqvist, and Z. F. Paska, "Quantum efficiency enhancement of AlGaAs/GaAs quantum well infrared detectors using a waveguide with a grating coupler," Appl. Phys. Lett. vol. 58, p. 2264–2266, 1991; J. Y. Andersson, L. Lundqvist, Z. F. Paska, J. Borglind, and D. Haga, "Efficiency of grating coupled AlGaAs/GaAs quantum well infrared detectors," Appl. Phys. Lett., **63**, 3361–3363, 1993.

[37] K. W. Goossen, S. A. Lyon, and K. Alavi, "Grating enhancement of quantum well detector response," Appl. Phys. Lett., **53**, 1027–1029, 1988; K. W. Goossen, and S. A. Lyon, "Performance aspects of a quantum well detector," J. Appl. Phys. vol. 63, 5149–5153, 1988; L. S. Yu, S. S. Li, Y. H. Wang, and Y. C. Kao, "A study of the coupled efficiency versus grating periodicity in a normal incident GaAs/AlGaAs multi-quantum-well infrared detector," J. Appl. Phys., **72**, 2105–2109, 1992.

[38] W. Xu, Y. Fu, and M. Willander, "Oscillator strength of intersubband transition in n-type AlAs/GaAlAs quantum well for the normal incident absorption," J. Infrared Millimeter Waves, **16**, 86–92, 1997.

[39] K. Kishino and S. Arai, Chapter 11 Integrated lasers, *Handbook of Semiconductor Lasers and Photonic Integrated Circuits*, Chapman & Hall, London, 1994, p. 350.

[40] For example, see J. C. Stover, *Optical Scattering: Measurement and Analysis*, McGram-Hill, New York, 1990, p.51; J. Cowley, *Diffraction Physics*, Elsevier, Amsterdam 1995, p. 11.

[41] D. Qi, M. Fischbein, M. Drndic, and S Selmic, "Efficient polymer-nanocrystal quantum-dot photodetectors," Appl. Phys. Lett., **86**, 093103(3), 2005.
[42] A. A. R. Watt, D. Blake, J. H. Warner, E. A. Thomsen, E. L. Tavenner, H. Rubinsztein-Dunlop, and P. Meredith, "Lead sulfide nanocrystal: conducting polymer solar cells," J. Phys. D: Appl. Phys., **38**, 2006–2012, 2005.
[43] K. S. Leschkies, R. Divakar, J. Basu, E. Enache-Pommer, J. E. Boercker, C. B. Carter, U. R. Kortshagen, D. J. Norris, and E. S. Aydil, "Photosensitization of ZnO nanowires with CdSe quantum dots for photovoltaic devices," Nano Lett., **7**, 1793–1798, 2007.
[44] A. Martí, N. López, E. Antolín, E. Cánovas, C. Stanley, C. Farmer, L. Cuadra, and A. Luque, "Novel semiconductor solar cell structures: the quantum dot intermediate band solar cell," Thin Solid Films, 511–512, 638–644, 2006.
[45] S. Lu and A. Madhukar, "Nonradiative resonant excitation transfer from nanocrystal quantum dots to adjacent quantum channels," Nano Lett., **7**, 3443–3451, 2007.
[46] A. Franceschetti and A. Zunger, "Exciton dissociation and interdot transport in CdSe quantum-dot molecules," Phys. Rev. B, **63**, 153304(4), 2001.
[47] A. N. Al-Ahmadi and S. E. Ulloa, "Extended coherent exciton states in quantum dot arrays," Appl. Phys. Lett., **88**, 043110(3), 2006.
[48] G. W. Bryant, "Exciton states in quantum dot solids: excitation transfer and dynamic decorrelation," Physica B, **314**, 15–19, 2002.
[49] S. A. McDonald, G. Konstantatos, S. Zhang, P. W. Cyr, E. J. D. Klem, L. Levina, and E. H. Sargent, "Solution-processed PbS quantum dot infrared photodetectors and photovoltaics," Nat. Mater., **4**, 138–142, 2005; E. H. Sargent, "Infrared quantum dots," Adv. Mater., **17**, 515–522, 2005.
[50] X. Peng, L. Manna, W. Yang, J. Wickham, E. Scher, A. Kadavanich, and A. P. Alivisatos, "Shape control of CdSe nanocrystals," Nature, **404**, 59–61, 2000.
[51] D. Gerion, F. Pinaud, S. C. Williams, W. J. Parak, D. Zanchet, S. Weiss, and A. P. Alivisatos, "Synthesis and properties of biocompatible water-soluble silica-coated CdSe/ZnS semiconductor quantum dots," J. Phys. Chem. B, **105**, 8861–8871, 2001.
[52] X. Michalet, F. F. Pinaud, L. A. Bentolila, J. M. Tsay, S. Doose, J. J. Li, G. Sundaresan, A. M. Wu, S. S. Gambhir, and S. Weiss, "Quantum dots for live cells, in vivo imaging, and diagnostics," Science, **307**, 538–544, 2005.
[53] D. R. Larson, W. R. Zipfel, R. M. Williams, S. W. Clark, M. P. Bruchez, F. W. Wise, and W. W. Webb, "Water-soluble quantum dots for multiphoton fluorescence imaging in vivo," Science, **300**, 1434–1436, 2003.
[54] S. Courty, C. Luccardini, Y. Bellaiche, G. Cappello, and M. Dahan, "Tracking individual kinesin motors in living cells using single quantum-dot imaging," Nano Lett., **6**, 1491–1495, 2006.
[55] A. A. High, E. E. Novitskaya, L. V. Butov, M. Hanson, and A. C. Gossard, "Control of exciton fluxes in an excitonic integrated circuit," Science, **321**, 229–231, 2008.
[56] T. Kawazoe, K. Kobayashi, S. Sangu, M. Ohtsu, "Demonstration of a nanophotonic switching operation by optical near-field energy transfer," Appl. Phys. Lett., **82**, 2957(3), 2003.
[57] T. Kawazoe, K. Kobayashi, S. Sangu, M. Ohtsu, "Excitation dynamics in a three-quantum dot system driven by optical near-field interaction: towards a nanometric photonic device," J. Microsc., **210**, 247–251, 2003.
[58] W. Nomura, T. Yatsui, M. Ohtsu, "Nanodot coupler with a surface plasmon polariton condenser for optical far/near-field conversion," Appl. Phys. Lett., **86**, 181108, 2005.

[59] S. Sangu, K. Kobayashi, A. Shojiguchi, M. Ohtsu, "Logic and functional operations using a near-field optically coupled quantum-dot system," Phys. Rev. B, **69**, 115334(13), 2004.

[60] T. Kawazoe, K. Kobayashi, M. Ohtsu, "Optical nanofountain: a biomimetic device that concentrates optical energy in a nanometric region," Appl. Phys. Lett., **86**, 103102, 2005.

[61] T. Kawazoe, K. Kobayashi, K. Akahane, M. Naruse, N. Yamamoto, and M. Ohtsu, "Demonstration of nanophotonic NOT gate using near-field optically coupled quantum dots," Appl. Phys. B, **84**, 243–246, 2006.

[62] L. M. Robinson, H. Rho, J. C. Kim, H. E. Jackson, L. M. Smith, S. Lee, M. Dobrowolska, and J. K. Furdyna, "Quantum dot exciton dynamics through a nanoaperture: evidence for two confined states," Phys. Rev. Lett., **83**, 2797–2800, 1999.

[63] T. Lundstrom, W. Schoenfeld, H. Lee, and P. M. Petroff, "Exciton storage in semiconductor self-assembled quantum dots," Science, **286**, 2312–2314, 1999.

[64] R. M. Kraus, P. G. Lagoudakis, A. L. Rogach, D.V. Talapin, H. Weller, J. M. Lupton, and J. Feldmann, "Room temperature exciton storage in elongated semiconductor nanocrystals," Phys. Rev. Lett., **98**, 017401(4), 2007.

[65] H. J. Krenner, C. E. Pryor, J. He, and P. M. Petroff, "A semiconductor exciton memory cell based on a single quantum nanostructure," Nano Lett., **8**, 1750–1755, 2008.

[66] R. Prasanth, J. E. M. Haverkort, and J. H. Wolter, "Electro-refraction in QDs: dependence on lateral size and shape," IEEE Tran. Nanotechnol., **3**, 270–274, 2004.

[67] Y. Fu, H. Ågren, L. Höglund, J. Y. Andersson, C. Asplund, M. Qiu, and L. Thylén, "Optical reflection from excitonic quantum-dot multilayer structures," Appl. Phys. Lett., **93**, 183117(3), 2008.

[68] J. Kim, P. S. Nair, C. Y. Wong, and G. D. Scholes, "Sizing up the exciton in complex-shaped semiconductor nanocrystals," Nano Lett., **7**, 3884–3890, 2007.

[69] E. A. Stinaff, M. Scheibner, A. S. Bracker, I. V. Ponomarev, V. L. Korenev, M. E. Ware, M. F. Doty, T. L. Reinecke, and D. Gammon, "Optical signatures of coupled quantum dots," Science, **311**, 636–639, 2006.

[70] H. Takagi, H. Tanaka, M. Yamaguchi, and N. Sawaki, "Resonant coupling of an exciton in a quantum disc with a propagating mode of an exciton-polariton in a nearby waveguide," J. Phys. D, **38**, 1551–1555, 2005.

[71] Y. Zeng, Y. Fu, M. Bengtsson, X.-S. Chen, W. Lu, and H. Ågren, "Finite-difference time-domain simulations of exciton-polariton resonances in quantum-dot arrays," Opt. Exp., **16**, 4507–4519, 2008.

[72] Y. Fu, E. Berglind, L. Thylén, H. Ågren, "Optical transmission and waveguiding by excitonic quantum dot lattices," J. Opt. Soc. Am. B, **23**, 2441–2447, 2006.

[73] Y. Fu, Y. Zeng, E. Berglind, L. Thylén, H. Ågren, "Nanoscale excitonic-plasmonic optical waveguiding by metal-coated quantum dots (invited talk)," Proceedings of International Symposium on Biophotonics, Nanophotonics and Metamaterials, Hangzhou, China 16–18 October, 2006. 426–431.

[74] E. Lifshitz, M. Brumer, A. Kigel, A. Sashchiuk, M. Bashouti, M. Sirota, E. Galun, Z. Burshtein, A. Q. Le Quang, I. Ledoux-Rak, and J. Zyss, "Air-stable PbSe/PbS and PbSe/PbSe$_x$S$_{1-x}$ core-shell nanocrystal quantum dots and their applications," J. Phys. Chem. B, **110**, 25356–25365, 2006.

[75] U. Woggon, "Single semiconductor nanocrystals: Physics and applications," J. Appl. Phys., **101**, 081727, 2007.

[76] E. J. D. Klem, L. Levina, and E. H. Sargent, "PbS quantum dot electroabsorption modulation across the extended communications band 1200-1700 nm," Appl. Phys. Lett., **87**, 053101(3), 2005.

[77] D. Meschede, H. Walther, G. Muller, "One-atom maser," Phys. Rev. Lett., **54**, 551–554, 1985; G. Rempe, H. Walther, N. Klein, "Observation of quantum collapse and revival in a one-atom maser," Phys. Rev. Lett., **58**, 353–356, 1987.

[78] W. M. Itano, D. J. Heinzen, J. J. Bollinger, D. J. Wineland, "Quantum zero effect," Phys. Rev., **A41**, 2295–2300, 1990.

[79] Y. Yamamoto, N. Imoto, S. Mashida, "Amplitude squeezing in a semiconductor laser using quantum nondemolition measurement," Phys. Rev., **A33**, 3243–3261, 1986; G. Rempe, F. Schmidt-Kaller, H. Walther, "Observation of sub-Poissonian photon statistics in a micromaser," Phys. Rev. Lett., **64**, 2783–2786, 1990.

[80] M. A. Bouchiat, J. Guena, and L. Pottier, "Atomic parity violation measurements in the highly forbidden $6S_1^2$-$7S_1^2$ caesium transition. I. Theoretical analysis, procedure and apparatus," J. Phys., **46**, 1897–1924, 1985; M. A. Bouchiat, J. Guena, and L. Pottier, "Atomic parity violation measurements in the highly forbidden $6S_1^2$-$7S_1^2$ caesium transition. II. Analysis and control of systematic effects," J. Phys., **47**, 1175–1202, 1986; M. A. Bouchiat, J. Guena, L. Pottier, and L. hunter, "Atomic parity violation measurements in the highly forbidden $6S_1^2$-$7S_1^2$ caesium transition. III. Dta acquisition and processing. Results and implications," J. Phys., **47**, 1709–1730, 1986.

[81] M. Arndt, S. I. Kanorsky, A. Weiss, T. W. Hansh, "Can paramagnetic atoms in superfluid helium be used to search for permanent electric dipole moments," Phys. Lett., **A174**, 298–303, 1993; J. P. Jacobs, W. M. Klipstein, S. K. Lamoreaus, B. R. Heckel, E. N. Fortson, "Testing time-reversal symmetry using ^{199}Hg," Phys. Rev. Lett., **71**, 3782–3785, 1993.

[82] V. B. Braginsky, M. L. Gorodetsky, V. S. Ilchenko, S. P. Vyatchanin, "On the ultimate sensitivity in coordinate measurements," Phys. Lett., **A179**, 244–248, 1993.

[83] A. Abramovici, W. E. Althouse, R. W. P. Drever, Y. Gürsel, S. Kawamura, F. J. Raab, D. Shoemaker, L. Sievers, R. E. Spero, K. S. Thorne, R. E. Vogt, R. Weiss, S. E. Whitcomb, and M. E. Zucker, "LIGO: the laser interferometer gravitational-wave observatory," Science, **256**, 325–333, 1992.

[84] G. Rempe, R. J. Thompson, H. J. Kimble, "Measurement of ultralow losses in an optical interferometer," Opt. Lett., **17**, 363–365, 1992.

[85] V. B. Braginsky, S. P. Vyatchanin, V. I. Panov, "Limiting stability of the frequency of self-excited oscillators," Sov. Phys. Doklady, **24**, 562–563, 1979.

[86] V. B. Braginsky, S. P. Vyatchanin, "Nondestructive measurement of the energy of optical quanta," Sov. Phys. Doklady, **26**, 686–687, 1981.

[87] V. B. Braginsky, F. Ya. Khalili, *Quantum Measurement*, ed. K. S. Thorne, Cambridge: Cambridge University Press, 1992.

[88] H. J. Kimble, *Quantum Fluctuations in Quantum Optics, Squeezing and Related Phenomena*, in Les Houches Lecture Notes, Session LIII, J. Dalibard, J.-M. Raimond, and J. Zinn-Justin, eds. (North-Holland, Amsterdam, 1992).

[89] Y. Rikitake, H. Imamura, and H. Ebisawa, "Time evolution of an exciton and a photon qubit coupled with each other," J. Phys. Soc. Jpn., **72**, 3020–3023, 2003.

[90] X. Li, Y. Wu, D. Steel, D. Gammon, T. H. Stievater, D. S. Katzer, D. Park, C. Piermarocchi, and L. J. Sham, "An all-optical quantum gate in a semiconductor quantum dot," Science, **301**, 809–811, 2003.

[91] K. Goshima, K. Komori, and T. Sugaya, "Optical control of 2-qubit exciton states in a coupled quantum dot," Jpn. J. Appl. Phys., **47**, 3111–3114, 2008.

[92] El. Biolatti, R. C. Iotti, P. Zanardi, and F. Rossi, "Optical quantum gates with semiconductor nanostructures," Int. J. Circ. Theor. Appl., **29**, 137–150, 2001.

[93] F. Rossi, "The excitonic quantum computer," IEEE Trans. Nanotechnol., **3**, 165–172, 2004.

[94] H. J. Zhou, S. D. Liu, M. T. Cheng, Q. Q. Wang, Y. Y. Li, and Q. K. Xue, "Rabi oscillation damped by exciton leakage and Auger capture in quantum dots," Opt. Lett., **30**, 3213–3215, 2005.

[95] A. Zrenner, S. Stufler, P. Ester, S. M. de Vasconcellos, M. Hübner, and M. Bichler, "Recent developments in single dot coherent devices," Phys, Status Solidi B, (b), **243**, 3696–3708, 2006.

Chapter 6

Basics of Plasmonics

We now look at the necessary theory behind plasmonics. A plasma is an neutrally charged group of positively and negatively charged particles (approximately the same number of each). A plasma oscillation is when one group of charges moves relative to the other, such as in a metal with electrons moving with respect to the positive ions. The quantum of the plasma oscillation is a quasi particle referred to as a plasmon. It is a boson, so it follows Bose–Einstein statistics and has integer spin, although we will not look at it in that much detail in waveguide analysis.

At optical frequencies, the Drude model of the permittivity of the plasma (or metal) predicts a negative permittivity. This is not quite the same concept as negative refraction, which requires both negative permittivity and negative permeability to create an electively negative index of refraction. At certain frequencies light may freely propagate through the metal while at others it will be sharply attenuated and surface modes may exist at the interface with a dielectric.

The field of plasmons has matured to the point where an introduction to the optical properties of metals and sometimes even surface plasmons (SPs) is included in many standard textbooks in solid-state physics, optics, and electrodynamics. Some of the material in this chapter is thus found scattered in one place or another in varying notations and systems of units, and not specifically focused on waveguiding. A number of works are also devoted to surface plasmon polaritons (SPPs; note that SP is confined to the interface between two dielectric materials; SPP is a kind of quasi particle composed of photon and SP; see details below) themselves, but these works are now of limited availability. The goal here is to bring together the background necessary for proper introduction to SPP waveguides, and place it into an updated and consistent notation generalized for lossy, anisotropic media.

As discussed in the Introduction of Chapter 1, today's electronics industry has been following Moore's law over the past four decades. Miniaturization of photonic components such as waveguides, emitters, and switches is, however, limited fundamentally by diffraction of light. For instance, diode lasers and optical guided wave structures must confine the light within their cavities and core layers, respectively, and so that the minimum feature sizes of these optical structures are approximately one wavelength. Working in optics in the subwavelength regime is not trivial. This restriction of optics in the subwavelength regime has, in the past, been a constraint for researchers. It has become more and more crucial for the industry. Microchips and contemporary optical data storage media are being produced at the limit of the conventional optical possibilities. New ways of modulating photons have to be found. SPP has been believed to be one of the best ways.[1,2]

In physics, the plasmon is the quasi particle resulting from the quantization of plasma oscillations, which are density waves of the charged carriers, often referred to as "Langmuir waves" or "plasma waves," in a conducting medium such as a metal, semiconductor, or plasma. Bulk plasmons are longitudinal waves. SPs, i.e., plasmons confined to surfaces, interact strongly with light, resulting in a polariton, called SPP. SPPs have both longitudinal and transverse components. Plasmons play a large role in the optical properties of metals. Light of frequency below the plasma frequency is reflected because the electrons in the metal screen the electric field of the light. Light of frequency above the plasma frequency is transmitted because the electrons cannot respond fast enough to screen it. In most metals, the plasma frequency is in the ultraviolet regime, making them shiny in the visible range, though some, such as gold and copper, have plasma frequencies in the visual range, yielding their distinct color. In doped semiconductors, the plasma frequency is usually in the infrared range.

SPPs occur and propagate in a wave-like fashion at the interface between a material with a negative permittivity, usually a metal or doped dielectric, and a dielectric material with a positive permittivity, where the free electrons of the negative-permittivity medium are collectively moving back and forth.[3] SPP has been studied since the turn of the century in the context of radio waves traveling along the surface of the earth.[4] In the 1940s, Fano studied these SPPs at optical frequencies to explain the Wood's anomalies in metal diffraction gratings.[5] Following the pioneering works of Ritchie[6] in 1957 and Stern and Ferrell[7] in 1960, SPP has been widely recognized in the surface science field. SPPs were first observed in electron energy loss experiments[8] and were later observed optically in attenuated total reflection experiments devised by Kretschmann[9] and Otto.[10]

SPP makes it possible to transform photons into plasmons, use their special properties to manipulate the properties of the original light wave, and reconvert the plasmons into photons. Because of its pure imaginary wave vector in the normal-to-the-interface direction, the amplitude of the SPP decays exponentially with increasing distance into each medium from the interface, which means that the SPP is bound to the interface. This feature provides us with the opportunity to confine light in subwavelength structures. It is this field confinement that leads to the enhancement of the electromagnetic (EM) field at the interface, which in its turn results in the extraordinary sensitivity of SPPs to the surface conditions. This induces many surface-enhanced optical phenomena such as Raman scattering, which has been used by biochemists to detect the presence of a molecule on a surface,[11,12] the second harmonic generation,[13,14] and SP-enhanced fluorescence.[15,16] In addition, because SSP has both the capacity of photonics and the miniaturization of electronics, it creates the ability to combine the superior technical advantages of photonics and electronics on the same chip.[2] Using SPP, different research groups have made the achievements of enhancing the transmission through nano apertures[17,18] for near-field optics and produced some basic nano optic devices such as mirrors, beam splitters, SPP interferometers,[8] and nano lenses.[7] These works are the first steps into a new branch of photonics using SPP, sometimes called plasmonics. SPP has thus attracted a significant attention due to the potential applications in scaling down optical, electro-optical, and photonic devices, to transmit light-like waves along nanoscale wires in very small chips.

Plasmonics has also been considered as a potential means of transmitting information on computer chips, since the plasmons propagate considerably faster than electrons but can be channeled with conductive paths similar to those on existing chips. The eventual result can be a new generation of powerful and ultra compact telecom and computing devices.[1]

6.1 CLASSICAL ELECTRODYNAMICS IN MATTER

Electrodynamics correctly describes the time and spatial evolution of EM fields in the presence of charges, currents, and any type of matter. Due to its inherent covariance it may even be used to describe relativistic effects. Only when the quantum nature of the EM field plays an important role, the classical concepts of electrodynamics have to be modified. However, in the context of the experiments described later such a quantum treatment is unnecessary.

The theory of electrodynamics developed by Maxwell in the 19th century very successfully describes the interaction of magnetic and electric fields and their propagation in free space. In its original form it fails, however, in the description of the interaction with matter. The internal structure of solids was largely unknown until the beginning of the 20th century, when the discovery of the electron, the crystal lattice structure, and the advent of quantum mechanics provided the background for the description of the interaction of light with matter. Among the most important researchers who developed this description were H. A. Lorentz and W. K. Heisenberg.[19]

6.1.1 *Limits*

The extended electrodynamical theory includes the interaction with matter. Although it describes correctly the reaction of matter to electric and magnetic fields, some material properties (for example, the complex dielectric function ϵ) enter as input parameters. Classical electrodynamics is therefore somewhat "inherently correct" as all unknown parameters are included in the material-dependent parameters. These material parameters are not explained in the context of electrodynamics —

they must be either taken from tabulated experimental data or explained by solid state theory. Only a microscopic perspective on the origin of the values of material properties gives insight into the "real" physical processes of the light interaction with the material, for example, electron-hole-pair excitation (see Chapter 3). It is further important to realize that any changes in the material properties induced by the interaction with EM fields cannot be explained or predicted by classical electrodynamics. Thus important optical effects such as light-induced transmission or luminescence are implicitly excluded if time-independent and field-independent material properties are used.

6.1.2 *Maxwell's Equations in Matter*

With these limitations in mind, we now briefly describe the fundamental equations governing the evolution of EM fields in the presence of charges, current, and polarizable media: the Maxwell's microscopic equations:

$$\begin{aligned}
\nabla \cdot \boldsymbol{E} &= \frac{n_{\text{tot}}}{\epsilon_0} \\
\nabla \cdot \boldsymbol{B} &= 0 \\
\nabla \times \boldsymbol{B} &= \mu_0 \boldsymbol{j}_{\text{tot}} + \epsilon_0 \mu_0 \frac{\partial \boldsymbol{E}}{\partial t} \\
\nabla \times \boldsymbol{E} &= -\frac{\partial \boldsymbol{B}}{\partial t}
\end{aligned} \tag{6.1}$$

In these equations n_{tot} represents the total, possibly both time- and space-dependent, electric charge, i.e., free (n_{free}) as well as induced (n_{ind}) (polarization) charges, and $\boldsymbol{j}_{\text{tot}}$ represents the total, possibly both time- and space-dependent, electric current, i.e., conduction currents (motion of free charges) as well as all atomistic (polarization, magnetization) currents.

To include situations involving matter, it is useful to distinguish external fields and fields produced by the response of the material. By denoting $\boldsymbol{P}$ as the local electric polarization of matter and $\boldsymbol{M}$ the local magnetization, the electric displacement field $\boldsymbol{D}$ and magnetic field strength $\boldsymbol{H}$ are given by

$$\boldsymbol{D} = \epsilon_0 \boldsymbol{E} + \boldsymbol{P} \quad , \quad \boldsymbol{H} = \frac{1}{\mu_0}\boldsymbol{B} - \boldsymbol{M} \tag{6.2}$$

where $\boldsymbol{E}$ and $\boldsymbol{B}$ are the electric field and magnetic flux density, respectively. μ_0 and ϵ_0 are the permeability and permittivity of free space, respectively. The sources of $\boldsymbol{D}$ and $\boldsymbol{H}$ are the charge density n_{free} and current density $\boldsymbol{j}_{\text{free}}$, respectively. The Maxwell's equations now read

$$\begin{aligned}
\nabla \cdot \boldsymbol{D} &= n_{\text{free}} \\
\nabla \cdot \boldsymbol{B} &= 0 \\
\nabla \times \boldsymbol{H} &= \boldsymbol{j}_{\text{free}} + \frac{\partial \boldsymbol{D}}{\partial t} \\
\nabla \times \boldsymbol{E} &= -\frac{\partial \boldsymbol{B}}{\partial t}
\end{aligned} \tag{6.3}$$

In order to solve real problems, relations describing the responses of the materials, the so-called constitutive relations, are needed

$$\boldsymbol{j} = \sigma \boldsymbol{E}, \quad \boldsymbol{B} = \mu\mu_0 \boldsymbol{H}, \quad \boldsymbol{P} = \epsilon_0 \chi \boldsymbol{E} \tag{6.4}$$

with the phenomenological coefficients σ, μ, and χ (conductivity, permeability, and electric susceptibility, respectively). They depend on the medium under consideration, but are assumed to be independent of the field strength (the medium is linear). They can also be frequency-dependent. The time dependence of the involved fields can be assumed to be harmonic, e.g., $\boldsymbol{E}(\boldsymbol{r}, t) = \boldsymbol{E}(\boldsymbol{r})e^{-i\omega t}$. More complicated time-dependent fields can be written as superposition of such fields (Fourier synthesis). Introducing all these equations into the Maxwell's equations,

$$\begin{aligned} \nabla \cdot (\epsilon_0 \epsilon \boldsymbol{E}) &= n_{\text{free}} \\ \nabla \cdot \boldsymbol{B} &= 0 \\ \nabla \times \boldsymbol{H} &= -i\omega \left(\epsilon + \frac{i\sigma}{\epsilon_0 \omega} \right) \epsilon_0 \boldsymbol{E} \\ \nabla \times \boldsymbol{E} &= i\omega\mu\mu_0 \boldsymbol{H} \end{aligned} \tag{6.5}$$

with the dielectric function $\epsilon = 1 + \chi$.

Note that the conductivity of a metal is frequency-dependent. If n electrons per unit volume all move with velocity $\boldsymbol{v}$, the current density is

$$\boldsymbol{j} = -ne\boldsymbol{v} \tag{6.6}$$

which is valid for both the free and induced charges. Taking τ as the relaxation time (or the average time between successive collisions), the average velocity acquired in the presence of a DC field $\boldsymbol{E}$ is $-e\boldsymbol{E}\tau/m^*$ so that

$$\boldsymbol{j} = \frac{ne^2\tau}{m^*} \boldsymbol{E} \tag{6.7}$$

where m^* is the effective mass of the charge under discussion.

Treating the electron as semiclassical particle and letting $\boldsymbol{p} = m^*\boldsymbol{v}$ be the total momentum of the electron, the equation of motion is

$$\frac{\mathrm{d}\boldsymbol{p}}{\mathrm{d}t} = -\frac{\boldsymbol{p}}{\tau} + \boldsymbol{f} \tag{6.8}$$

where $\boldsymbol{f} = -e(\boldsymbol{E} + \boldsymbol{v} \times \boldsymbol{B})$ is the Lorentz force due to the electric and magnetic fields.

To calculate the current in a metal driven by a time-dependent electric field, we write the field in the form $\boldsymbol{E}(t) = \Re\{\boldsymbol{E}(\omega)e^{-i\omega t}\}$, where ω is the angular frequency of oscillation, then seek a steady-state solution of the form $\boldsymbol{p}(t) = \Re\{\boldsymbol{p}(\omega)e^{-i\omega t}\}$ for the equation of motion, Eq. 6.8. We find that $\boldsymbol{p}(\omega)$ must satisfy

$$-i\omega\boldsymbol{p}(\omega) = -\frac{\boldsymbol{p}(\omega)}{\tau} - e\boldsymbol{E}(\omega) \tag{6.9}$$

i.e.,

$$\boldsymbol{p}(\omega) = -\frac{e\tau}{1 - i\omega\tau} \boldsymbol{E}(\omega)$$

so that the current density is $\boldsymbol{j}(t) = \Re\{\boldsymbol{j}(\omega)e^{-i\omega t}\}$,

$$\boldsymbol{j}(\omega) = -\frac{ne\boldsymbol{p}(\omega)}{m^*} = \frac{ne^2\tau/m^*}{1-i\omega\tau}\boldsymbol{E}(\omega) \tag{6.10}$$

One customarily writes this result as

$$\boldsymbol{j}(\omega) = \sigma(\omega)\boldsymbol{E}(\omega) \tag{6.11}$$

where $\sigma(\omega)$, known as the frequency-dependent conductivity, is given by

$$\sigma(\omega) = \frac{\sigma_0}{1-i\omega\tau}, \quad \sigma_0 = \frac{ne^2\tau}{m^*} \tag{6.12}$$

Together with the appropriate boundary conditions and the conservation of charge and energy we are ready to solve the time and spatial evolution of EM fields in the presence of matter.

6.2 SOLID STATE THEORY FOR OPTICAL CONSTANT OF METAL

The material properties needed for the electrodynamical calculations have to be calculated using solid state theory. A simple model for metals was developed by Drude (1900) based on the kinetic gas theory. It assumes independent and free electrons with a common relaxation time. Sommerfeld incorporated corrections originating from the Pauli exclusion principle. This so-called free-electron model was later modified to include minor corrections from the band structure of matter (effective mass) and termed quasi-free-electron model.

This Drude–Sommerfeld model describes very successfully many (but not all!) properties of metals despite its drastic assumptions. At optical frequencies it often fails due to the presence of interband transitions. Nevertheless, it is important to have the basic concepts of the Drude–Sommerfeld model in mind to analyze the experiments.

In 1958, Ferrell predicted the excitation of plasma oscillations in a thin metal film by fast electrons, which was incident normal to the thin metal film and inelastically scattered by plasma oscillations inside the metal film.[20] The prediction was verified shortly.[21–24] Stern attributed the surface roughness of the metal film as the reason for the observation of a peak of scattered radiation at the plasma frequency from the metal films irradiated with light.[25] The theory was modified by Kröger and Kretschmann.[26, 27]

More recently, different theoretical approaches about SPP have been reported, e.g., the reduced Rayleigh equations[28] and the Chandezon method.[29] A rigorous formulation for the scattering of SP from a one-dimensional surface defect was reported.[30] Periodic structures were studied by Salomon *et al.*[31, 32] Furthermore, quantitative structures concerning the SPP have been elucidated in terms of the contributions to the dielectric response from surface states and truncated bulk band states.[33]

Extended and advanced theoretical works about SPP in more and more complicated systems have been reported based on large-scale numerical analysis. However, the fundamental basics about the SPP remains largely unchanged. Consider a neutral plasma that consists of a gas of positively charged ions and negatively charged electrons. If one displaces by a tiny amount all of the electrons with respect to the ions, the Coulomb force pulls back, acting as a restoring force. It is easy to show that the charge density oscillates at the plasma frequency

$$\omega_p = \sqrt{\frac{ne^2}{\epsilon_0 m_0}} \tag{6.13}$$

where n is the density of electrons, e is the electric charge, and m_0 is the mass of the electron. Note that the above formula is derived under the approximation that the ion mass is infinite. This is generally a good approximation, since the electrons are so much lighter than ions. One must modify this expression in the case of electron–positron plasmas, often encountered in astrophysics.

If the characteristic wave vector k of a plasma oscillation is comparable to the product of the electron thermal speed v and the plasma frequency ω_p, then the plasma wave may propagate in the medium. These waves will obey a dispersion relation

$$\omega^2 - \omega_p^2 = 3k^2v^2 \tag{6.14}$$

called the Bohm–Gross dispersion relation.[34] If the electron thermal speed is comparable to the phase velocity v_{ph}, i.e.,

$$v \approx v_{\mathrm{ph}} = \frac{\omega}{k} \tag{6.15}$$

the plasma waves can accelerate electrons that are moving with speed nearly equal to the phase velocity of the wave. This process often leads to a form of collisionless damping, called Landau damping. In a metal or semiconductor, the effect of the ions' periodic potential must be taken into account. This is usually done by using the electron's effective mass m^* in place of m_0, see Section 1.5. This large speed of wave propagation thus lays down the basis for transmitting information on computer chips.[1, 35]

The dispersion relationship of a SPP at a dielectric–metal interface is described by the Maxwell's equations. Consider a simple system consisting of a dielectric material with a positive permittivity ϵ_{d} ($\epsilon_{\mathrm{d}} > 0$) in one half-space ($z < 0$) and a metal with a complex dielectric function ϵ_{m} in the other half-space ($z \geq 0$). The dispersion relationship of a transverse magnetic (TM) EM wave (also referred to as a p-polarized EM field) that propagates along the interface ($z = 0$) of this structure is given by

$$k = \frac{\omega}{c}\sqrt{\frac{\epsilon_{\mathrm{d}}\epsilon_{\mathrm{m}}}{\epsilon_{\mathrm{d}} + \epsilon_{\mathrm{d}}}} \tag{6.16}$$

where ω is the frequency of the EM wave and c is the speed of light in vacuum. This equation shows us that the frequency-dependent permittivity of the metal ϵ_{m} has to have the opposite sign than the $\epsilon_{\rm d}$ of the dielectric material to make the square root bigger than one. If we choose metals, we satisfy this condition because $\epsilon_{\rm m}$ of metals is negative and complex, which is normally expressed as

$$k_{\rm m} = \frac{\omega}{c}\sqrt{1 - \frac{\omega_p^2}{\omega^2}} \tag{6.17}$$

for a lossless metal. Figure 6.2 shows the dispersion relations of SPPs at the metal–vacuum and metal–dielectric interfaces. Note that the magnetic properties of dielectric and metal are both positive and equal to 1; SPP cannot exist under the transverse electric mode.[3]

Converting a photon into a SPP will confront us with the two most fundamental properties of the SPP. First, the EM field will decay exponentially with the distance to the surface. And second, there is a momentum mismatch between SPP and the exciting photon. In order to excite an SPP by TM polarized light incident on a planar metal surface from the adjacent dielectric medium, momentum conservation must be satisfied so that the component of the wave vector $k_{||}$ of the incident light parallel to the surface equals the SPP wave vector $k_{\rm SPP}$. However,

$$k_{\rm SPP} > k_{||} \tag{6.18}$$

so that light illuminating a negative-permittivity material surface through a positive-permittivity dielectric medium cannot directly couple to SPPs, and special experimental arrangements must be designed to provide the necessary momentum conservation.[3] To match the momentums, we can choose one of three principal techniques. The first technique uses a prism and the total internal reflection (Kretschmann[9] and Otto[10] configurations, see Fig. 6.1), which are still used today in surface characterization and measurement of the optical constants of metals;[36] The second one involves field scattering from a topological defect such as small holes in a thin film. The third technique makes use of periodic corrugations in the metal surface, where the periodicity of the structure leads to a quasi-phase matching, i.e., the wave vector of an excited SPP wave can be expressed as

$$\boldsymbol{k}_{\rm SPP} = \boldsymbol{k}_{||} + \frac{2m\pi}{L_x}\,\boldsymbol{u}_x + \frac{2n\pi}{L_y}\,\boldsymbol{u}_y \tag{6.19}$$

where $\boldsymbol{u}_x$ and $\boldsymbol{u}_y$ are unit vectors of the reciprocal lattice of the periodic structure in the xy plane, L_x and L_y are the primitive lattice constants in the x- and y-direction, respectively. m and n are integers determining the SPP propagation direction.

6.3 Drude–Sommerfeld model

The basic picture of the properties of metals in the framework of this theory is a gas of independent, point-like electrons. These electrons move freely in between independent collisions with unspecified collision centers (lattice ions, other electrons, defects, phonons, etc.), which occur with an average rate of $1/\tau$. Each collision leads

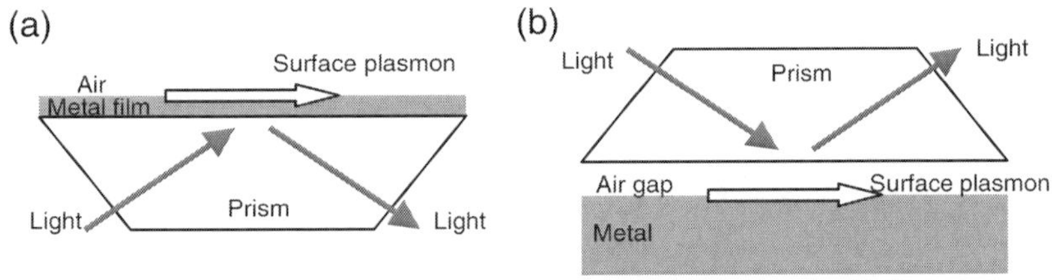

Fig. 6.1 (a) Kretschmann and (b) Otto configurations.

to a complete loss of directional information and results in a random orientation of the electron velocity afterwards.

In an external field, the electrons are accelerated in between collisions, resulting in a drift motion. In reality, only electrons near the Fermi level contribute because the Pauli exclusion principle does not allow deeper lying electrons to change their electronic state. Band-structure corrections lead to a modification of this motion. These corrections are customarily incorporated into an effective mass m^*, which is in general different from the free-electron mass m_0. Many (but not all) properties of real metals, including their optical properties as described by the frequency-dependent dielectric function $\epsilon(\omega)$, are surprisingly well predicted from this simple model.

Assume now that these electrons are subject to a time-harmonic electric field (optical wave) of the form $\boldsymbol{E}(t) = E_0 e^{-i\omega t}\boldsymbol{y}_0$, where $\boldsymbol{y}_0$ is the unit vector along the y axis. If treated classically as a free particle, the equation of motion of the electron is

$$\frac{\mathrm{d}^2 y}{\mathrm{d}t^2} + \frac{1}{\tau}\frac{\mathrm{d}y}{\mathrm{d}t} = -\frac{e}{m^*}E_0 \mathrm{e}^{-i\omega t} \tag{6.20}$$

by Eq. 6.8, where $p_y = m^* \mathrm{d}y/\mathrm{d}t$, m^* is the mass of the electron, and e is the elementary charge. The above equation has the solution $y(t) = y_0 \mathrm{e}^{-i\omega t}$, and upon substitution we find the position of the electron at time t to be

$$\boldsymbol{y}(t) = \frac{e\boldsymbol{E}(t)}{m^*\omega(\omega + i/\tau)} \tag{6.21}$$

The polarization $\boldsymbol{P}$ is defined as the induced dipole moment ($-e\boldsymbol{y}$ for one electron at position $\boldsymbol{y}$) per unit volume

$$\boldsymbol{P} = -ne\boldsymbol{y} = -\frac{ne^2}{m^*\omega(\omega + i/\tau)}\boldsymbol{E} \tag{6.22}$$

As Eq. 6.4, we also define the polarization for linear dielectrics in terms of the permittivity of free space ϵ_0, and the electrical susceptibility χ as

$$\boldsymbol{P} = \epsilon_0 \chi \boldsymbol{E} = \epsilon_0 \Big[\epsilon(\omega) - 1\Big]\boldsymbol{E} \tag{6.23}$$

$\epsilon(\omega)$, the relative permittivity, or dielectric constant of the material under investigation, is found to be

$$\epsilon(\omega) = 1 - \frac{\omega_p^2}{\omega(\omega + i/\tau)} = 1 - \frac{\omega_p^2}{\omega^2 + 1/\tau^2} + i\frac{\omega_p^2 \tau}{\omega(1 + \omega^2\tau^2)} \tag{6.24}$$

Table 6.1 Drude parameters for the noble metals Au, Ag, and Cu. DC-conductivity σ and electron density n from Kopitzki (1993), effective mass m^* from Johnson and Christy (1972).

	Au	Ag	Cu
n [m^{-3}]	5.90×10^{28}	5.76×10^{28}	8.45×10^{28}
σ [1/Ω·m]	4.9×10^{7}	6.6×10^{7}	6.5×10^{7}
m^* [m_0]	0.99	0.96	1.49
$\hbar\omega_p$ [eV]	9.1	9.1	8.8
τ [fs]	29	40	40

where $\omega_p = \sqrt{ne^2/m^*\epsilon_0}$ is known to be the plasma frequency. This is commonly called the Drude model for the dielectric constant of a metal, since it is based on the Drude theory of electrical conductivity.

If the positive ion core background has a dielectric constant labeled ϵ_∞ which is essentially constant up to frequencies well above ω_p for many common metal materials, then Eq. 6.24 becomes

$$\epsilon(\omega) = \epsilon_\infty \left[1 - \frac{\tilde{\omega}_p^2}{\omega(\omega + i/\tau)}\right] \tag{6.25}$$

where $\tilde{\omega}_p = \omega_p/\epsilon_\infty$.

Often it is useful to work backwards from Eq. 6.24 and find the plasma frequency and relaxation time in terms of the dielectric constant $\epsilon = \epsilon' + i\epsilon''$, resulting in

$$\begin{aligned} \tau &= \frac{1-\epsilon'}{\omega\epsilon''} \\ \omega_p &= \sqrt{(1-\epsilon')\left(\omega^2 + \frac{1}{\tau^2}\right)} \end{aligned} \tag{6.26}$$

Note that a heavily doped semiconductor will optically have the same behavior as a metal and a similar derivation can be done for its dielectric constant. It turns out to be similar to that of a metal, with the only alteration being the replacement of the "1" in the first real term in Eq. 6.24 by the relative permittivity of the undoped semiconductor. The reader can refer to Section 4.2 about the derivation of the effective permittivity of QD exciton polariton.

In Table 6.1, these so-called Drude parameters are provided for the noble metals Au, Ag, and Cu.

In the following sections we will discuss the consequences of a relative permittivity in the form of Eq. 6.24.

6.3.1 *Plasmon*

We now discuss the charge density and its oscillation. To avoid undue complications we consider only the high-frequency behaviors of such oscillations, ignoring the dynamical effects of the background ions. Denoting n as the electron density, the equilibrium charge density of background ions and electrons is $\pm n_0$; the total charge density is therefore $n_{\text{tot}} = n - n_0$. When this charge density has an oscillatory time dependence $e^{-i\omega t}$,

$$\nabla \cdot \boldsymbol{j}_{\text{tot}}(\omega) = i\omega n_{\text{tot}}(\omega) \tag{6.27}$$

by the equation of continuity $\nabla \cdot \boldsymbol{j} = -\partial n/\partial t$. The first equation of Eq. 6.5

$$\nabla \cdot \boldsymbol{E}(\omega) = \frac{n_{\text{tot}}(\omega)}{\epsilon_0} \tag{6.28}$$

In view of Eq. 6.11 we find that

$$i\omega n_{\text{tot}}(\omega) = \frac{\sigma(\omega) n_{\text{tot}}(\omega)}{\epsilon_0} \tag{6.29}$$

which has a solution provided that

$$1 = \frac{\omega_p^2}{\omega(\omega + i/\tau)} \tag{6.30}$$

which results in $\epsilon(\omega) = 0$ by Eq. 6.24, i.e., the onset of propagation of radiation in the metal. In the present context it emerges as the condition that the frequency must meet if the induced charged density wave is to propagate.

The nature of this charge density wave, known as a plasma oscillation or plasmon, can be understood in terms of a simple model. Let us image that we displace the entire electron gas, as a whole, through a distance y with respect to the fixed positive ion background. The resulting surface charge gives rise to an electric field of magnitude nye/ϵ_0, where ny is the number of charges at either end of the slab. Consequently the electron gas as a whole will obey the equation of motion

$$nm^* \frac{\mathrm{d}^2 y}{\mathrm{d}t^2} = -neE = -\frac{n^2 e^2}{\epsilon_0} y \tag{6.31}$$

which is the equation of motion of a simple harmonic oscillator of frequency $\omega_p = \sqrt{ne^2/m^*\epsilon_0}$, the plasma frequency.

We may excite a plasmon by passing an electron through a thin metallic film or by reflecting an electron from the thin metallic film. The charge of the passing or the reflecting electron will interact with the electric field fluctuations of the plasmon so that the reflected or transmitted electron will show an energy loss equal to integral multiples of the plasmon energy.

6.3.2 *Plane Waves in a Metal*

6.3.2.1 *Dispersion relation*

We first look at plane waves propagating freely in a metal, away from any boundary or surface. This can be more simply understood as an optical wave propagating through a material with a relative permittivity $\epsilon(\omega)$.

For the free-standing metal, free of net electric charge and current, we assume a plane wave traveling in the z-direction with wave vector k and frequency ω in the form

$$\boldsymbol{E} = E_0 \mathrm{e}^{i(kz-\omega t)} \boldsymbol{y}_0 \tag{6.32}$$

Eq. 6.5 then result in

$$-\nabla^2 \boldsymbol{E} = \epsilon(\omega)\epsilon_0 \mu \boldsymbol{E} \tag{6.33}$$

so that $k = \omega\sqrt{\epsilon(\omega)\epsilon_0\mu}$, μ is the permeability, which from now on is assumed to be the permeability of free space $\mu = \mu_0$ (nonmagnetic media) unless otherwise noted. If the relaxation time $\tau \to \infty$ (lossless approximation) then the relative permittivity of the metal simplifies to

$$\epsilon(\omega) = 1 - \frac{\epsilon_p^2}{\omega^2}$$

and

$$k = \frac{\omega}{c}\sqrt{1 - \frac{\omega_p^2}{\omega^2}} \tag{6.34}$$

where $c = 1/\sqrt{\epsilon_0\mu_0}$ is the speed of light in vacuum.

For $\omega > \omega_p$, $\epsilon(\omega)$ is real and positive; k is real so that the wave propagates along z. However, if $\omega < \omega_p$, $\epsilon(\omega)$ is real but negative so that k is imaginary and the fields attenuate as $\mathrm{e}^{-|k|z}$ along z. The wave is evanescent along z and therefore does not propagate.

Rearranging the above equation we find the dispersion relation

$$\omega = \sqrt{c^2k^2 + \omega_p^2} \tag{6.35}$$

from which we see that for all real k, $\omega > \omega_p$.

6.3.2.2 *Phase and group velocity*

From the dispersion relation 6.35, we can easily find the phase and group velocity. For the plane wave described in 6.32 the phase velocity is

$$v_p = \frac{\omega}{k} = c\left(1 - \frac{\omega_p^2}{\omega^2}\right)^{-1/2} \tag{6.36}$$

which is always greater than the speed of light in vacuum. The group velocity

$$v_g = \frac{\partial\omega}{\partial k} = c\sqrt{1 - \frac{\omega_p^2}{\omega^2}} \tag{6.37}$$

and is always less than c.

In the limit that $k \to \infty$ the frequency approaches the light line $\omega = ck$ and the wave becomes photon-like as it propagates identically to a photon in free space, i.e., the phase and group velocity converge to c. As $k \to 0$, the group velocity falls to zero and the wave reaches a resonance at the plasma frequency ω_p, the natural frequency of oscillation of the free electrons in the material. Between these extremes there is coupling between the two excitations (the plasma oscillation and the photon) and we refer to the excitation as a plasmon, although commonly it is referred to as a plasmon for all k.

6.4 SURFACE PLASMON POLARITON AT ONE INTERFACE

The situation of SPP is much different from that of plane waves in metals. A surface polariton is an EM wave at the interface between a material with a negative dielectric constant (surface active medium), and a material with a positive dielectric constant (passive medium), which induces a polarization in the active medium and couples to it. When the polarized medium is a metal, the surface polariton is termed a SPP because of the plasma contribution to the negative dielectric constant. Other types of surface polaritons exist such as exciton-polaritons (see Chapter 4) and phonon–polaritons.

There are two distinct types of solutions propagating along the z direction. One type has $E_z = 0$, i.e., the mode electric field only has components perpendicular (or transverse) to the propagation direction. These are the TE modes. The second type has $B_z = 0$, i.e., the mode magnetic field only has components perpendicular (or transverse) to the propagation direction. These are the TM modes. If we work with $\boldsymbol{E}$ and $\boldsymbol{B}$, this reduces the number of vector components we need to find from 6 to 5, but if we work with $\boldsymbol{A}$ and ϕ, as we do in Section 2.2, it reduces the number from 4 to 2. Refer to Eqs.2.53, which are repeated below

$$\begin{aligned}\boldsymbol{E} &= -\nabla\phi - \frac{\partial \boldsymbol{A}}{\partial t}\\ \boldsymbol{B} &= \nabla \times \boldsymbol{A}\end{aligned} \tag{6.38}$$

TE Modes: The condition that $E_z = 0$ turns out to require that $A_z = 0$ and $\phi = 0$. Hence, we only have to find A_x and A_y.

TM Modes: The condition that $B_z = 0$ turns out to require that $A_x = 0$ and $A_y = 0$. Hence, we only have to find A_z and ϕ.

TEM Modes: In a completely hollow guide along the z direction, waves with both $E_z = 0$ and $B_z = 0$ parallel to the axis of the guide are impossible. However, with a conductor along the axis these waves are possible. Their dispersion relation is simply $\omega = kc$ with k parallel to the axis of the guide. The electric field points outward from the central conductor and terminates on the outer surface of the guide while the magnetic field circulates around the central conductor and runs parallel to the outer conductor. Hence, they are just about like free space waves with $\boldsymbol{E}$, $\boldsymbol{B}$, and $\boldsymbol{k}$ mutually perpendicular. In a cylindrical guide (a coaxial cable) of radius a these waves are described, in cylindrical coordinates $(\boldsymbol{\rho}, z)$, by $\boldsymbol{k} = k\boldsymbol{z}_0$,

$$\begin{aligned}E_\rho &= E_0 \frac{a}{\rho}\, \mathrm{e}^{i(kz-\omega t)}\\ B_\theta &= \frac{E_0}{c}\frac{a}{\rho}\, \mathrm{e}^{i(kz-\omega t)}\end{aligned} \tag{6.39}$$

where E_0 is the electric field amplitude at $\rho = a$.

6.4.1 *TM Surface Plasmon–Polaritons*

We begin by solving for the dispersion relation of a TM (sometimes called p-polarized) surface polariton at one interface. We assume the coordinates with two infinite half spaces joined at $y = 0$ with ϵ_1 for $y > 0$ and ϵ_2 for $y < 0$, and solve for bound modes, i.e., the field goes to zero at $y = \pm\infty$. Those modes typically have discrete k values for a given geometry, in contrast to plane waves that take on a continuum of values.

For a TM mode propagating along z direction, only the fields H_x, E_y, and E_z are nonzero. Modes are solutions to Maxwell's equations, which maintain their shape when propagating through a geometry constant in the direction of propagation. The TM mode is solved in terms of the propagating H field in the form of

$$\begin{aligned} \boldsymbol{H}_1 &= H_0 \mathrm{e}^{i(kz-\omega t)-k_1 y} \boldsymbol{x}_0 \\ \boldsymbol{H}_2 &= H_0 \mathrm{e}^{i(kz-\omega t)+k_2 y} \boldsymbol{x}_0 \end{aligned} \tag{6.40}$$

where k is the wavenumber along z. k_1 and k_2 are the transverse wavenumbers in region 1 and 2, respectively, and the real parts of k_1 and k_2 are greater than zero to maintain evanescent fields in region 1 and and 2. H_0 is a normalization constant.

The fields must satisfy Maxwell's equations for time-harmonic fields in anisotropic media

$$\begin{aligned} \nabla \times \boldsymbol{E} &= i\omega\mu_0\mu \boldsymbol{H} \\ \nabla \times \boldsymbol{H} &= -i\omega\epsilon_0\epsilon_j \boldsymbol{E} \end{aligned} \tag{6.41}$$

where ϵ_j is the relative permittivity tensor.

We assume the materials are oriented so that ϵ_j is diagonal

$$\epsilon_j = \begin{pmatrix} \epsilon_{xj} & 0 & 0 \\ 0 & \epsilon_{yj} & 0 \\ 0 & 0 & \epsilon_{zj} \end{pmatrix} \tag{6.42}$$

where the elements are in general complex to account for material losses and the subscript j denotes the layer. Substituting Eq. 6.40 into the second equation of Eq. 6.41 gives the electric field components in terms of the H field. For $y > 0$,

$$\begin{aligned} \boldsymbol{E}_{y1} &= -\frac{H_0 k}{\omega\epsilon_0\epsilon_{y1}} \mathrm{e}^{i(kz-\omega t)-k_1 y} \boldsymbol{y}_0 \\ \boldsymbol{E}_{z1} &= \frac{iH_0 k_1}{\omega\epsilon_0\epsilon_{z1}} \mathrm{e}^{i(kz-\omega t)-k_1 y} \boldsymbol{z}_0 \end{aligned} \tag{6.43}$$

and for $y < 0$,

$$\begin{aligned} \boldsymbol{E}_{y2} &= -\frac{H_0 k}{\omega\epsilon_0\epsilon_{y2}} \mathrm{e}^{i(kz-\omega t)+k_2 y} \boldsymbol{y}_0 \\ \boldsymbol{E}_{z2} &= -\frac{iH_0 k_2}{\omega\epsilon_0\epsilon_{z2}} \mathrm{e}^{i(kz-\omega t)+k_2 y} \boldsymbol{z}_0 \end{aligned} \tag{6.44}$$

The transverse wavenumbers are found by substituting the fields $\boldsymbol{E}$ and $\boldsymbol{H}$ into the first equation of Eq. 6.41

$$k_j = \sqrt{\epsilon_{zj}\left(\frac{k^2}{\epsilon_{yj}} - \frac{\omega^2}{c^2}\right)} \tag{6.45}$$

The boundary conditions at $y = 0$ require continuity of the tangential fields. For the case $\boldsymbol{E}_{z1}(y = 0^+) = \boldsymbol{E}_{z2}(y = 0^-)$,

$$\frac{k_1}{\epsilon_{z1}} + \frac{k_2}{\epsilon_{z2}} = 0 \tag{6.46}$$

Since $k_1 > 0$ and $k_2 > 0$, this shows that the surface polariton can only exist at the interface between media with permittivities of opposite sign.

Now substituting k_1 and k_2 of Eq. 6.45 into Eq. 6.46 and solving for k leads to a new dispersion relation for single-interface surface polaritons ($\mu = 1$)

$$k = \frac{\omega}{c}\sqrt{\frac{(\epsilon_{z2} - \epsilon_{z1})\epsilon_{y1}\epsilon_{y2}}{\epsilon_{z2}\epsilon_{y2} - \epsilon_{z1}\epsilon_{y1}}} \tag{6.47}$$

Allowing both media to be isotropic leads to the more common result

$$k = \frac{\omega}{c}\sqrt{\frac{\epsilon_2\epsilon_1}{\epsilon_2 + \epsilon_1}} \tag{6.48}$$

Inserting the permittivity function for a metal into the above equation results in the dispersion relation for SPP. These dispersion relations place more conditions on ϵ_1 and ϵ_2 for lossless media, namely that in addition to the requirement of $\epsilon_2\epsilon_1 < 0$, it is also required that $\epsilon_2 + \epsilon_1 < 0$ so that the real part of k is not zero and the mode can propagate along z.

The range of possible k values can be found from the above equation by assuming $\epsilon_1 > 0$ and setting $\epsilon_2 = -(\epsilon_1 + \delta)$ with $\delta > 0$, since $\epsilon_1 + \epsilon_2 < 0$. For this, $k = k_0\sqrt{\epsilon_1^2/\delta + \epsilon_1}$. In the limit of $\delta \to 0^+$, $k \to \infty$, and in the limit of $\delta \to \infty$, $k \to k_0\sqrt{\epsilon_1}$. Therefore, for bound surface polaritons, $k > k_0\sqrt{\epsilon_1}$.

6.4.2 *Phase and Group Velocity*

As mentioned, the dispersion equation derived above holds for all surface polaritons. To apply it to SPP, we insert the relative permittivity value for the metal.

Substituting the lossless version of $\epsilon_2 = 1 - \omega_p^2/\omega^2$ into Eq. 6.48 we solve for the dispersion equation in a form with ω a function of k

$$2\omega^2 = \omega_p^2 + k^2c^2\left(1 + \frac{1}{\epsilon_1}\right) - \sqrt{\omega_p^4 + 2\omega_p^2k^2c^2\left(1 - \frac{1}{\epsilon_1}\right) + k^4c^4\left(1 + \frac{2}{\epsilon_1} + \frac{1}{\epsilon_1^2}\right)} \tag{6.49}$$

Note that ϵ_1 is also generally a function of ω for dispersive dielectrics. For a metal–vacuum interface $\epsilon_1 = 1$ and Eq. 6.49 reduces to

$$\omega = \left[\frac{1}{2}\left(2k^2c^2 + \omega_p^2 - \sqrt{4k^4c^4 + \omega_p^4}\right)\right]^2 \tag{6.50}$$

Using Eqs. 6.34, 6.48, and the constraint $\epsilon_1 + \epsilon_2 < 0$, we find an upper bound on the range of possible frequencies for lossless single-interface SPP

$$\omega < \frac{\omega_p}{\sqrt{\epsilon_1 + 1}} \tag{6.51}$$

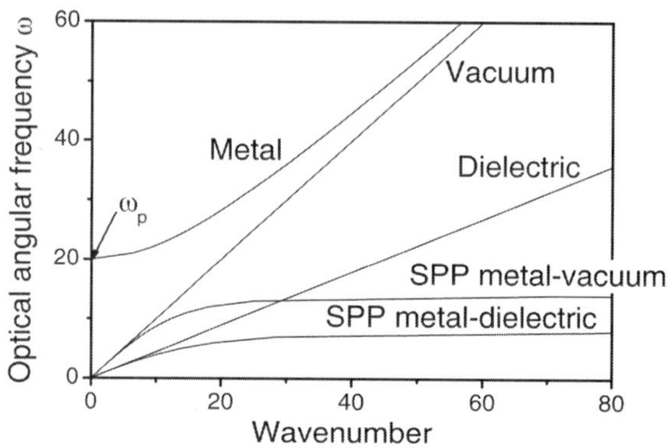

Fig. 6.2 Generic dispersion curves for plane waves and surface plasmons. "Metal": plane wave in a metal; "Vacuum": plane wave in vacuum; "Dielectric": plane wave in $\epsilon_1 > 1$.

This upper limit is the so-called SP frequency, ω_{sp}.

Note that for doped semiconductors the upper limit is

$$\omega < \frac{\omega_p}{\sqrt{\epsilon_1 + \epsilon_{\mathrm{s}}}} \tag{6.52}$$

where ϵ_{s} is the relative permittivity of the undoped semiconductor.

Figure 6.2 shows generic dispersion curves for plane waves and SPPs. Curve "Metal" shows the dispersion curve for the plane wave in a metal, and we see that as k increases the curve converges to that of the dispersion of light in free space ("Vacuum" $\omega = ck$), while decreasing k converges the dispersion to the plasma frequency ω_p. "SPP metal-vacuum" is Eq. 6.50 about SPP, which is now denoted as SPP at a metal–vacuum interface. At low k it converges to the light line in free space, while for high k it converges to $\omega_p/\sqrt{2}$. "Metal-dielectric" is the dispersion for the SPP at a metal–dielectric interface with $\epsilon_1 > 1$. Note that this curve always lies below that of the metal-vacuum SPP case. We have left ϵ_1 independent of ω here for this example.

We see from Fig. 6.2 that the phase velocity, $v_p = \omega/k$, and the group velocity $v_g = \partial\omega/\partial k$ for SPPs are always less than c. At low k they both converge to the speed of light in the material ϵ_1, and at high k they both converge to zero with the group velocity converging much more quickly.

Excitation: During any excitation, energy and momentum must be conserved. Referring to Fig. 6.2, we see that at a given frequency, the wavenumber of the SPP is always below the optical plane wave. This means that SPPs cannot be directly excited by light sources. SPPs can, however, be launched by end-fire excitation, attenuated (ATR) or frustrated (FTR) total reflection, and grating or roughness coupling.

Plasmon band gap: Figure 6.2 also shows that for SPPs with small k, $\omega \to ck/\sqrt{\epsilon_1}$, while for large k, $\omega \to \omega_p/\sqrt{\epsilon_1 + 1}$. SPPs are then limited to the frequencies $\omega < \omega_p/\sqrt{\epsilon_1 + 1}$, while plane waves (or radiating modes) are limited to the frequencies $\omega > \omega_p$. There exists then a so-called "plasmon band gap" in the region $\omega_p/\sqrt{\epsilon_1 + 1} < \omega < \omega_p$ in which no plasmon related fields can propagate. As ϵ_1 increases, the band gap increases. This is an idealization for lossless metals. If loss is included, we find that damped excitations are possible within the gap.

6.4.3 *TE Surface Modes*

A similar derivation can be made for the case of TE (s-polarized) surface modes. TE modes consist of only the fields E_x, H_y, and H_z. We begin by assuming electric fields of the form

$$\boldsymbol{E}_1 = E_0 \mathrm{e}^{i(kz-\omega t)-k_1 y} \boldsymbol{x}_0 \tag{6.53}$$

for $y > 0$ and

$$\boldsymbol{E}_2 = E_0 \mathrm{e}^{i(kz-\omega t)+k_2 y} \boldsymbol{x}_0 \tag{6.54}$$

for $y < 0$. Eventually the continuity of the fields leads to

$$\frac{k_1}{\mu_1} + \frac{k_2}{\mu_2} = 0 \tag{6.55}$$

However, for nonmagnetic media we have $\mu_1 = \mu_2 = \mu_0$ so the above equation reduces to $k_1 + k_2 = 0$, which cannot be satisfied for bound modes, since $k_1 > 0$ and $k_2 > 0$. As well naturally occurring materials have $\mu > 0$, so bound TE surface modes cannot be supported whether the material is magnetic or not. Recent progress in negative refraction and left-handed materials has led to composite materials that effectively have $\mu < 0$ and this may open up the possibility of TE surface modes.

6.5 SURFACE PLASMON–POLARITON COUPLED BETWEEN TWO INTERFACES

Imagine now an optically infinite (thick) metal having interfaces with a dielectric at $y = \pm\infty$. Two single-interface SPPs exist at $y \pm \infty$. If the thickness of the metal is decreased, eventually the fields of these two modes begin to couple through the metal, creating supermodes. We can guess that there will be two bound surface modes for this sort of structure, a symmetric supermode corresponding to both modes having fields with the same sign of amplitude (polarity) and an antisymmetric supermode corresponding to the modes having opposite polarity at the interfaces. These modes are denoted as s_b and a_b, respectively, with the subscript referring to the bound nature of the modes (leaky modes are also known to exist).

We now discuss a three-layer structure with a metal slab between two semi-infinite dielectrics. Quite often SPPs can be investigated using only three layers and we outline the modal solutions for that case later.

6.6 FINITE-WIDTH SPP WAVEGUIDE

Finite-width SPP waveguide consists of a metal strip, with a width w (along the x direction) and a thickness t (y direction) extended along the z direction. These waveguides confine the modes laterally. The main long-range (low-loss) mode is the symmetric and bound mode. It is symmetric along the horizontal and vertical directions, allowing it to couple well to optical fiber. This suggests the use of this mode in integrated optics.

In finite-width waveguides we must now include all six electric and magnetic field components. These modes are termed hybrid modes, since they are no longer TE or TM but a combination of the two. The fields in the cross-section of such a waveguide take the general form

$$\begin{aligned} \boldsymbol{E}(x,y,z) &= \boldsymbol{E}_0(x,y)\mathrm{e}^{i(kz-\omega t)} \\ \boldsymbol{H}(x,y,z) &= \boldsymbol{H}_0(x,y)\mathrm{e}^{i(kz-\omega t)} \end{aligned} \tag{6.56}$$

where $\boldsymbol{E}_0 = E_x\boldsymbol{x}_0 + E_y\boldsymbol{y}_0 + E_z\boldsymbol{z}_0$ and $\boldsymbol{H}_0 = H_x\boldsymbol{x}_0 + H_y\boldsymbol{y}_0 + H_z\boldsymbol{z}_0$.

An analytical derivation of the guided modes is no longer possible. One should resort to numerical methods that discretize both the permittivity function and field solution with a finite spatial resolution, to approximate the true guided modes. Two representative numerical methods are the finite-element method and the finite-difference method. Detailed discussions on these methods are, however, beyond the scope of the current book.

6.7 NUMERICAL APPROACHES

Research, development, and application of advanced SPP theories include the SPP formation in small nanoscale system and thin film, the coupling between the SPP and the external EM field, as well as the heavy numerical calculations at the most fundamental levels of starting from Maxwell's and Schrödinger equations and at the established microwave engineering approach. Although Maxwell's equations apply equally to macroscopic as well as microscopic worlds, new features distinguish nano optics from its traditional format. These features are as follows: (1) The size of a nano system is always much smaller than the wavelength of the incident light, which might suggest the long-wavelength approximation. In other words, the nano system is effectively homogeneous for those EM waves whose wavelengths are much longer than the feature size of the nano system. However, this approximation is not always valid because of the spatial confinement of the mobile electrons and the strong spatial confinement of the EM field in the vicinity of the interface. (2) A nanostructure has normally a low degree of symmetry so that the corresponding Maxwell's equations cannot be simplified by using spherical or cylindrical polar coordinates. (3) The evanescent near-field EM components play a very important role in most of the optical properties of a nano system, such as the perfect lens,[7,37] which leads to severe computational difficulties in space-differential approaches including the finite element method and the finite-difference time-domain (FDTD) method.

Several powerful and reliable methods have been developed to solve the Maxwell's equations in ultimately reduced and confined geometries. A far-from-complete list of these approaches includes (1) discrete dipole approximation (DDA) method, which is particularly useful to describe isolated nanoparticles with arbitrary geometric shapes and complex surrounding environments;[38] (2) classical Mie scattering theory, often used in the scattering of plane EM waves from an isotropic sphere or two-concentric spheres;[39, 40] (3) time-dependent local density approximation, which is a first-principles method capable of providing the dielectric function of metal nanoparticles;[41] (4) coupled dipole approximation (also called dipole–dipole approximation), which applies to the theoretical investigation of the image formation in the near-field optical microscopy;[42] (5) FDTD method, which is a direct time-domain solution of Maxwell's differential equations on spatial grids or lattices;[43] (6) Fourier modal method (also called couple modal method or modal method by Fourier expansion), frequently used to calculate the transmission spectrum of periodic nanostructures, such as metal gratings and metal-nanoparticle photonic crystals (PhCs).[44–46]

In the following section, we present a brief description of the Fourier modal method and the FDTD method, which are very general with respect to their application fields including the geometric features of the nanostructures, the spatial distributions of the EM fields, and the physical problems of interest.

6.7.1 *The Fourier Modal Method*

An important class of nanostructures is the periodic nanostructure, such as metal gratings[17] and metallic PhCs.[47–52] As stated by Eq. 6.19, a quasiphase matching exists in these periodic structures for the SPP. In addition, metallic PhCs, also named as metallic photonic bandgap materials, have many interesting optical properties for important potential applications in the optical, microwave, and infrared fields.[53–55] One of such important properties is to mold and control the flow and distribution of the SPP waves at the microscopic level. The synergetic interplay in these PhCs between the microcavity resonance of composite nanoparticles and the Bragg scattering resonance of the nanoparticle arrays leads to the formation of a photonic band gap (PBG), i.e., a range of frequencies in which no propagating SPP modes are allowed. Due to the presence of the PBG in the dispersion relationship of the EM field, the photonic density of states (DOS) in PhCs is suppressed over this PBG frequency window.[55]

The original Fourier modal method was first developed some 20 years ago to study one- and two-dimensional gratings. In order to treat nano-scale periodic structures, especially metal grating and metallic PhCs, a rigorous method has been developed to take into account the more complex and dispersive periodic structures.[45, 46] We summarize the essential numerical procedure of the modified method.

Consider a three-dimensional structure that is periodic in the xy plane. The whole structure is discretized into thin slices along the z direction. Within each slice the permittivity ϵ and the permeability μ are approximated as z-independent; they are however xy-dependent. Furthermore, it is assumed that each slice is surrounded by two thin air films of zero thickness. The EM field in the air films is expressed by Floquet harmonics

$$\begin{aligned} \boldsymbol{E}(\boldsymbol{r}) &= \sum_{mn} \boldsymbol{E}_{mn}(z)\mathrm{e}^{i(k_{mn,x}x+k_{mn,y}y)} \\ \boldsymbol{H}(\boldsymbol{r}) &= \sum_{mn} \boldsymbol{H}_{mn}(z)\mathrm{e}^{i(k_{mn,x}x+k_{mn,y}y)} \end{aligned} \tag{6.57}$$

where $\boldsymbol{k}_{mn} = (k_{mn,x}, k_{mn,y}) = (k_x, k_y) + m\boldsymbol{b}_x + n\boldsymbol{b}_y$ is the Bragg wave vector, and k_x and k_y are the x- and y-components of the wave vector of the incident plane wave. Note that $(k_x, k_y) = \boldsymbol{k}_{||}$ by Eq. 6.19. $\boldsymbol{b}_x$ and $\boldsymbol{b}_y$ are reciprocal lattice vectors of the corresponding grating, which are $\frac{2\pi}{L}\,\boldsymbol{u}_x$ and $\frac{2\pi}{L}\,\boldsymbol{u}_y$ by Eq. 6.19, and m and n are integers. $\boldsymbol{E}_{mn}(z)$ and $\boldsymbol{H}_{mn}(z)$ are expansion coefficients of the EM field. We can further write down the plane-wave expansion expressions for permittivity and permeability

$$\begin{aligned} \epsilon(\boldsymbol{r}) &= \sum_{mn} \epsilon_{mn}(z)\mathrm{e}^{i(mb_x x+nb_y y)} \\ \mu(\boldsymbol{r}) &= \sum_{mn} \mu_{mn}(z)\mathrm{e}^{i(mb_x x+nb_y y)} \end{aligned} \tag{6.58}$$

In addition, we expand the EM field in each slice by the eigenmodes in that slice, which are the eigenvectors of a matrix T_1T_2, where

$$\begin{aligned} T_1^{ij;mn} &= \frac{i}{k_0}\begin{pmatrix} k_{ij,x}\epsilon^{-1}_{ij;mn}k_{mn,y} & k_0^2\mu_{ij;mn} - k_{ij,x}\epsilon^{-1}_{ij;mn}k_{mn,x} \\ k_{ij,y}\epsilon^{-1}_{ij;mn}k_{mn,y} - k_0^2\mu_{ij;mn} & -k_{ij,y}\epsilon^{-1}_{ij;mn}k_{mn,x} \end{pmatrix} \\ T_2^{ij;mn} &= \frac{i}{k_0}\begin{pmatrix} -k_{ij,x}\mu^{-1}_{ij;mn}k_{mn,y} & k_{ij,x}\mu^{-1}_{ij;mn}k_{mn,x} - k_0^2\epsilon_{ij;mn} \\ k_0^2\epsilon_{ij;mn} - k_{ij,y}\mu^{-1}_{ij;mn}k_{mn,y} & k_{ij,y}\mu^{-1}_{ij;mn}k_{mn,x} \end{pmatrix} \end{aligned}$$

where k_0 is the amplitude of the wave vector of the incident EM field.

Using the boundary conditions among EM fields in the slice and two neighboring air films, a transfer matrix (referred also as the scattering matrix) can be obtained, from which the EM field across the whole system is determined. Since the independent variable in the numerical calculation is the frequency rather than the wave vector of the incident EM field, the method can effectively simulate dispersive periodic nanostructures; even their permeabilities and permittivities are all frequency-dependent and negative.[46] In order to obtain accurate and fast converging results, one must pay special attentions to the way of formulating the products of Fourier series involved in the eigenmode problem.[44]

6.7.2 *The Finite-Difference Time-Domain Method*

Started from the pioneering work of Yee in 1966,[56] the numerical technique of the FDTD has been used successfully and extensively to solve Maxwell's curl equations directly in the time domain on a space grid of many kinds of structures including metal waveguides, cavities, high-speed computer circuit boards, and PhCs.[43] FDTD is an accurate and robust approach because the sources of numerically errors are well understood and can be bounded. In addition, FDTD is a systematic approach. The problem of modeling a new structure using this method is reduced to a problem of mesh generation rather than the potentially complex reformulation of an integral equation. Based on these features, FDTD is often used to study nanostructures to obtain all kinds of information of interest. In the following, we describe this method in some extension.

The main part: Permitting also the possibility of magnetic losses that can dissipate EM fields in materials, we define an equivalent magnetic current to account for the magnetic loss mechanisms:

$$\boldsymbol{J}_m = \rho' \boldsymbol{H} \tag{6.59}$$

with which we obtain from Eq. 2.49 that

$$\begin{aligned} \frac{\partial \boldsymbol{H}}{\partial t} &= -\frac{1}{\mu}\nabla \times \boldsymbol{E} - \frac{\rho'}{\mu}\boldsymbol{H} \\ \frac{\partial \boldsymbol{E}}{\partial t} &= \frac{1}{\epsilon}\nabla \times \boldsymbol{H} - \frac{\sigma}{\epsilon}\boldsymbol{E} \end{aligned} \tag{6.60}$$

We now write out the vector components of the curl operator in the above two equations to yield the following system of six coupled scalar equations in the three-dimensional Cartesian coordinate system (x, y, z):

$$\begin{aligned} \frac{\partial H_x}{\partial t} &= \frac{1}{\mu}\left(\frac{\partial E_y}{\partial z} - \frac{\partial E_z}{\partial y} - \rho' H_x\right) \\ \frac{\partial H_y}{\partial t} &= \frac{1}{\mu}\left(\frac{\partial E_z}{\partial x} - \frac{\partial E_x}{\partial z} - \rho' H_y\right) \\ \frac{\partial H_z}{\partial t} &= \frac{1}{\mu}\left(\frac{\partial E_x}{\partial y} - \frac{\partial E_y}{\partial x} - \rho' H_z\right) \\ \frac{\partial E_x}{\partial t} &= \frac{1}{\epsilon}\left(\frac{\partial H_z}{\partial y} - \frac{\partial H_y}{\partial z} - \sigma E_x\right) \\ \frac{\partial E_y}{\partial t} &= \frac{1}{\epsilon}\left(\frac{\partial H_x}{\partial z} - \frac{\partial H_z}{\partial x} - \sigma E_y\right) \\ \frac{\partial E_z}{\partial t} &= \frac{1}{\epsilon}\left(\frac{\partial H_y}{\partial x} - \frac{\partial H_x}{\partial y} - \sigma E_z\right) \end{aligned} \tag{6.61}$$

The system of six coupled partial differential equations forms the basis of the FDTD numerical algorithm for the EM wave interactions with general three-dimensional objects. The FDTD algorithm need not explicitly enforce the Gauss's law relations indicating zero free electric and magnetic charge. This is because these relations are theoretically a direct consequence of the curl equations. However, the

FDTD space grid must be structured so that the Gauss's law relations are implicit in the positions of the electric and magnetic field vector components in the grid and the numerical space derivative operations upon these vector components that model the action of the curl operator.

We denote a space point in a uniform, rectangular lattice as

$$(x, y, z) = (i\Delta x, j\Delta y, k\Delta z) = (i, j, k) \tag{6.62}$$

Here, Δx, Δy, and Δz are, respectively, the lattice space increments in the x, y, and z directions. i, j, and k are integers. Further, we denote any function u of space and time evaluated at a discrete point in the grid and at a discrete point in time as

$$u(x, y, z, t) = u(i\Delta x, j\Delta y, k\Delta z, n\Delta t) = u^n_{i,j,k} \tag{6.63}$$

where Δt is the time increment, uniform over the observation time interval. n is an integer.

The first-order partial derivative of u in the x direction, evaluated at time $n\Delta t$, is given by

$$\frac{\partial u(x, y, z, t)}{\partial x} = \frac{u^n_{i+\frac{1}{2},j,k} - u^n_{i-\frac{1}{2},j,k}}{\Delta x} + O[(\Delta x)^2] \tag{6.64}$$

We note the $\pm 1/2$ increment in the i subscript (x-coordinate) of u, denoting a space finite-difference over $\pm\Delta x/2$. The first-order partial derivative of u of time, evaluated at (x, y, z):

$$\frac{\partial u(x, y, z, t)}{\partial t} = \frac{u^{n+\frac{1}{2}}_{i,j,k} - u^{n-\frac{1}{2}}_{i,j,k}}{\Delta t} + O[(\Delta t)^2] \tag{6.65}$$

Noted the $\pm 1/2$ increment in the n superscript (time coordinate) of u, denoting a time finite-difference over $\pm\Delta t/2$.

Apply the above ideas and notation to achieve a numerical approximation of Maxwell's curl equations of Eq. 6.61. The first equation becomes

$$\frac{H_x|^{n+\frac{1}{2}}_{i,j,k} - H_x|^{n-\frac{1}{2}}_{i,j,k}}{\Delta t} = \frac{1}{\mu_{i,j,k}} \left(\begin{array}{c} \dfrac{E_y|^n_{i,j,k+\frac{1}{2}} - E_y|^n_{i,j,k-\frac{1}{2}}}{\Delta z} \\ -\dfrac{E_z|^n_{i,j+\frac{1}{2},k} - E_z|^n_{i,j-\frac{1}{2},k}}{\Delta y} \end{array} \right) - \frac{\rho'_{i,j,k}}{\mu_{i,j,k}} H_x|^n_{i,j,k} \tag{6.66}$$

Note that all field quantities on the right-hand side are evaluated at time step n, including the magnetic field term H_x appearing due to the magnetic loss ρ'. Since H_x at time step n is not calculated in the FDTD scheme (Hs are defined at time step $n - 1/2$, whereas Es are defined at time step n), we use the semi-implicit approximation to estimate this term

$$H_x|^n_{i,j,k} = \frac{H_x|^{n+\frac{1}{2}}_{i,j,k} + H_x|^{n-\frac{1}{2}}_{i,j,k}}{2} \tag{6.67}$$

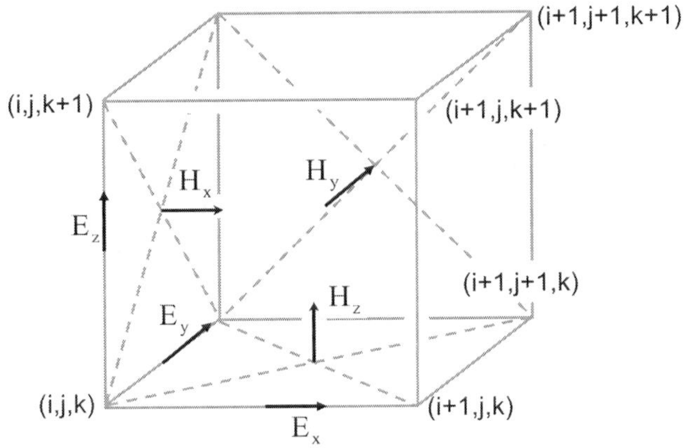

Fig. 6.3 Yee space lattice. ***E***-components are in the middle of the edges and the ***H***-components are in the center of the faces, in a rectangular coordinate system. See also Color Insert.

Finally, the explicit time-stepping relation for $H_x|_{i,j,k}^{n+\frac{1}{2}}$ is written as:

$$H_x|_{i,j,k}^{n+\frac{1}{2}} = \frac{1-\dfrac{\rho'_{i,j,k}\Delta t}{2\mu_{i,j,k}}}{1+\dfrac{\rho'_{i,j,k}\Delta t}{2\mu_{i,j,k}}} H_x|_{i,j,k}^{n-\frac{1}{2}} + \frac{\dfrac{\Delta t}{\mu_{i,j,k}}}{1+\dfrac{\rho'_{i,j,k}\Delta t}{2\mu_{i,j,k}}} \left(\begin{array}{c} \dfrac{E_y|_{i,j,k+\frac{1}{2}}^n - E_y|_{i,j,k-\frac{1}{2}}^n}{\Delta z} \\ -\dfrac{E_z|_{i,j+\frac{1}{2},k}^n - E_z|_{i,j-\frac{1}{2},k}^n}{\Delta y} \end{array} \right)$$

In a similar manner, we can derive finite-difference expressions based on Yee's algorithm for H_y, H_z, E_x, E_y, and E_z. In applying the Yee space lattice of Fig. 6.3, i.e., the ***E***-components are in the middle of the edges and the ***H***-components are in the center of the faces, in a rectangular coordinate system, $E_x|_{i,j,k} \equiv E_x|_{i+\frac{1}{2},j,k}$, $E_y|_{i,j,k} \equiv E_y|_{i,j+\frac{1}{2},k}$, $E_z|_{i,j,k} \equiv E_z|_{i,j,k+\frac{1}{2}}$, $H_x|_{i,j,k} \equiv H_x|_{i,j+\frac{1}{2},k+\frac{1}{2}}$, $H_y|_{i,j,k} \equiv H_y|_{i+\frac{1}{2},j,k+\frac{1}{2}}$, and $H_z|_{i,j,k} \equiv H_z|_{i+\frac{1}{2},j+\frac{1}{2},k}$, the above equation is re-formatted

$$H_x|_{i,j,k}^{n+\frac{1}{2}} = \frac{1-\dfrac{\rho'_{i,j,k}\Delta t}{2\mu_{i,j,k}}}{1+\dfrac{\rho'_{i,j,k}\Delta t}{2\mu_{i,j,k}}} H_x|_{i,j,k}^{n-\frac{1}{2}} + \frac{\dfrac{\Delta t}{\mu_{i,j,k}}}{1+\dfrac{\rho'_{i,j,k}\Delta t}{2\mu_{i,j,k}}} \left(\begin{array}{c} \dfrac{E_y|_{i,j,k+1}^n - E_y|_{i,j,k}^n}{\Delta z} \\ -\dfrac{E_z|_{i,j+1,k}^n - E_z|_{i,j,k}^n}{\Delta y} \end{array} \right)$$

Clearly, the new value of a field vector component at any space lattice point depends only on its previous value and the previous values of the components of the other field vector at adjacent points. Therefore, at any given time step, the computation of a field vector can proceed one point at a time.

Boundary conditions: The basic consideration of the FDTD approach to solve the EM wave equations is that many systems of interest are defined in "open" regions where the spatial domain of the field is unbounded in one or more spatial directions. A suitable boundary is therefore needed to minimize the spatial region involved in the numerical calculation. Many kinds of boundary conditions have been proposed including the perfect electrically (magnetically) conducting boundary conditions, the radiation boundary conditions, the Bloch boundary conditions, and absorbing boundary conditions.

(1) Perfectly matched layer (PML): One of the most important and powerful absorbing boundary conditions uses the PML, which is based on a splitting of electric or magnetic field components in the absorbing boundary region with the possibility of assigning losses to the individual split field components.[57] The net effect of this is to create a nonphysical absorbing medium adjacent to the outermost FDTD mesh boundary. The absorbing medium has a wave impedance that is independent of the field propagation direction and the frequency of outgoing scattered waves.

(2) Bloch boundary conditions: When the structure under the investigation is periodic, Bloch theorem can be applied, i.e., the EM field propagating in this structure can be expressed by

$$\boldsymbol{E}(\boldsymbol{r}+\boldsymbol{a}) = \mathrm{e}^{i\boldsymbol{k}\cdot\boldsymbol{a}}\boldsymbol{E}(\boldsymbol{r}) \tag{6.68}$$

here $\boldsymbol{k}$ is the corresponding Bloch wave vector, and $\boldsymbol{a}$ is the periodic lattice constant.[58, 59] In the FDTD scheme, this means that for a periodic system along the z direction,

$$E_x(x, y, z + n\delta z) = \mathrm{e}^{ik_z \times n\delta z} E_x(x, y, z) \tag{6.69}$$

here k_z is the Bloch wave vector and $n\delta z$ is the periodic lattice along the z direction.

Auxiliary differential equation (ADE) method: In order to implement FDTD models for dispersive materials such as metals, time-domain ADEs are often used to link the polarization field and the electric flux density.[60–62] These equations are time-stepped in synchronism with Maxwell's differential equations, yielding a composite self-consistent numerical system. For example, the frequency-domain Maxwell equation about $\boldsymbol{H}(\omega)$ can be written as

$$\nabla \times \boldsymbol{H}(\omega) = \boldsymbol{J}(\omega) - i\omega\epsilon_\infty \boldsymbol{E}(\omega) \tag{6.70}$$

where ϵ_∞ is the permittivity at the infinite frequency and $\boldsymbol{J}(\omega)$ is the current density. The corresponding time-domain analog equation is given by

$$\nabla \times \boldsymbol{H}(t) = \boldsymbol{J}(t) + \epsilon_\infty \frac{\partial}{\partial t}\boldsymbol{E}(t) \tag{6.71}$$

so that the time-domain difference equation can be written as

$$\frac{\epsilon_\infty}{\delta t}(\boldsymbol{E}^{n+1/2} - \boldsymbol{E}^{n-1/2}) = \nabla \times \boldsymbol{H}^n - \boldsymbol{J}^n \tag{6.72}$$

Incident wave sources: There are many EM sources for different practical situations: (1) hard-source $\boldsymbol{E}$ and $\boldsymbol{H}$ fields in one- and two-dimensional grids, (2) $\boldsymbol{J}$ and $\boldsymbol{M}$ current sources in three-dimensional lattices, (3) the total-field/scattered-field (TF/SF) formulation for plane-wave excitation in one, two, and three dimensions,[63, 64] (4) the waveguide sources. Among these sources, the TF/SF technique is very popular, which permits numerical FDTD modeling of a long-time-duration pulsed or sinusoidal illumination at an arbitrary plane-wave propagation direction.

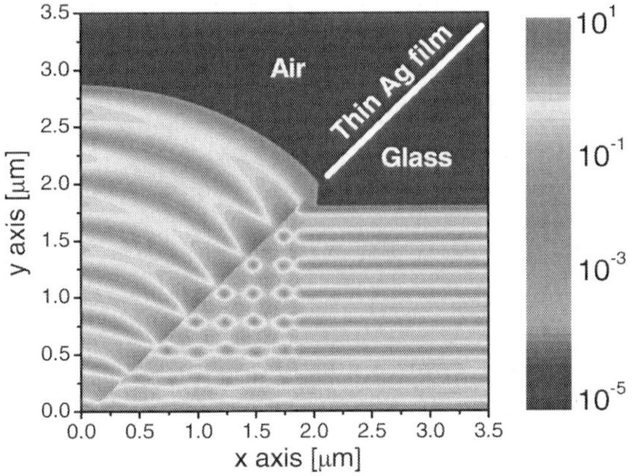

Fig. 6.4 Excitation of the SPP in the Kretschmann configuration of Fig. 6.1a. The vacuum wavelength of the incident EM plane wave is 800 nm. The plot shows the spatial distribution of the amplitude of the electric field at a time of 9.63 fs after the injection of the electromagnetic plane wave at $y = 0$ along the y direction. See also Color Insert.

As a brief demonstration, we simulate the SPP excitation in the Kretschmann configuration of a thin silver metal film (thickness 50 nm), which is illuminated through a dielectric prism (glass with a relative permittivity of 2.25) at an angular of incidence greater than the critical angle of total internal reflection by using a two-dimensional FDTD approach. In the numerical calculation, the TF/SF technique is used to excite a sinusoidal plane wave along the y direction at $y = 0$, the ADE technique is used to simulate the dispersive silver metal, and the PML is used to terminate the numerical spatial regime. The corresponding result of the electric field intensity, i.e., $\sqrt{E_x^2 + E_y^2}$ is plotted in Fig. 6.4 at a time of 9.63 fs after the starting injection of the EM plane wave with a vacuum wavelength of 800 nm at $t = 0$ and $y = 0$ along the y direction. The propagation of the EM field along the y direction is clearly visualized in Fig. 6.4, showing a traveling distance of about 1.9 μm after 9.63 fs in the glass. SPP has been generated in the metal–air surface, which causes the reflection of the EM field from the metal film back to the dielectric prism. Diffusion of the EM field into the air is also clearly observed, whereas the EM field distribution shows the strong spatial confinement and the wave-like motion fashion of the SPP in the vicinity of the air/metal interface.

References

[1] R. Zia, J. A. Schuller, A. Chandran, and M. L. Brongersma, "Plasmonics: the next chip-scale technology," Mater. today, **9**, 20–27, 2006.

[2] L. P. Sempere, "Surface plasmon polaritons (SPP) and their use in subwavelength optics," Term paper for physics of Nanostructures.

[3] A. V. Zayats, I. I. Smolyaninov, and A. A. Maradudin, "Nano-optics of surface plasmon polaritons," Phys. Rep., **408**, 131–314, 2005.

[4] A. N. Sommerfeld, "Propagation of waves in wireless telegraphy," Ann. Phy., **28**, 665–736, 1909.

[5] U. Fano, "The theory of anomalous diffraction gratings and of quasi-stationary waves on metallic surfaces (Sommerfeld's waves)," J. Opt. Soc. Am., **31**, 213–222, 1941.

[6] R. H. Ritchie, "Plasma losses by fast electrons in thin films," Phys. Rev., **106**, 874–881, 1957.

[7] E. A. Stern and R. A. Ferrell, "Surface plasma oscillations of a degenerate electron gas," Phys. Rev., **120**, 130–136, 1960.

[8] C. J. Powell and J. B. Swan, "Origin of the characteristic electron energy losses in aluminum," Phys. Rev., **115**, 869–875, 1959.

[9] E. Kretschmann and H. Raether, "Radiative decay of nonradiative surface plasmon excited by light," Z. Naturf. A, **23**, 2135–2136, 1968.

[10] A. Otto, "Excitation of nonradiative surface plasma waves in silver by the method of frustrated total reflection," Z. Phy., **216**, 398–410, 1968.

[11] K. J. Webb and J. Li, "Waveguide cavity surface-enhanced Raman scattering," Phy. Rev. B, **73**, 073404(4), 2006.

[12] K. J. Webb and J. Li, "Resonant waveguide field enhancement in dimers," Opt. Lett., **31**, 3348(3), 2006.

[13] H. J. Simon, D. E. Mitchell, and J. G. Watson, "Optical second-harmonic generation with surface plasmons in silver films ," Phys. Rev. Lett., **33**, 1531–1534, 1974.

[14] C. K. Chen, A. R. B. de Castro, and Y. R. Shen, "Surface-enhanced second-harmonic generation," Phys. Rev. Lett., **46**, 145–148, 1981.

[15] H. Kano and S. Kawata, "Two-photon-excited fluorescence enhanced by a surface plasmon," Opt. Lett., **21**, 1848–1850, 1996.

[16] T. Neumann, M.-L. Johansson, D. Kambhampati, W. Knoll, "Surface-plasmon fluorescence spectroscopy," Adv. Funct. Mater. **12**, 575–586, 2002.

[17] T. W. Ebbesen, H. J. Lezec, H. F. Ghaemi, T. Thio, and P. A. Wolff, "Extraordinary optical transmission through sub-wavelength hole arrays," Nature, **391**, 667–669, 1998.

[18] H. F. Ghaemi, T. Thio, D. E. Grupp, T. W. Ebbesen, and H. J. Lezec, "Surface plasmons enhance optical transmission through subwavelength holes," Phys. Rev. B, **58**, 6779–6782, 1998.

[19] L. Allen and J. H. Eberly, *Optical resonance and two-level atoms.* John Wiley New York 1975.

[20] R. A. Ferrell, "Predicted radiation of plasma oscillations in metal films," Phys. Rev., **111**, 1214–1222, 1958.

[21] W. Steinmann, "Experimental verification of radiation of plasma oscillations in thin silver films," Phys. Rev. Lett., **5**, 470–472, 1960.

[22] R. W. Brown, P. Wessel, and E. P. Trounson, "Plasmon re-radiation from silver films," Phys. Rev. Lett., **5**, 472–473, 1960.

[23] E. T. Arakawa, R. J. Herickhoff, and R. D. Birkhoff, "Detection of plasma radiation from electron-bombarded Al and Mg foils," Phys. Rev. Lett., **12**, 319–320, 1964.

[24] J. Brambring and H. Raether, "Plasma radiation from thin silver foils excited by light," Phys. Rev. Lett., **15**, 882–883, 1965.

[25] E. A. Stern, "Plasma radiation by rough surfaces," Phys. Rev. Lett., **19**, 1321–1324, 1967.

[26] E. Kröger and E. Kretschmann, "Scattering of light by slightly rough surfaces or thin films including plasma resonance emission," Z. Physik, 237, 1–15, 1970.

[27] E. Kretschmann and E. Köger, "Reflection and transmission of light by a rough surface, including results for surface-plasmon effects," J. Opt. Soc. Am., **65**, 150–154, 1975.

[28] A. V. Shchegrov, I. V. Novikov, and A. A. Maradudin, "Scattering of surface plasmon polaritons by a circularly symmetric surface defect," Phys. Rev. Lett., **78**, 4269–4272, 1997.

[29] U. Schröter and D. Heitmann, "Surface-plasmon-enhanced transmission through metallic gratings," Phys. Rev. B, **58**, 15419–15421, 1998.

[30] J. A. Sanchez-Gil and A. A. Maradudin, "Near-field and far-field scattering of surface plasmon polaritons by one-dimensional surface defects," Phys. Rev. B, **60**, 8359–8367, 1999.

[31] L. Salomon, F. Grillot, A. V. Zayats, F. de Fornel, "Near-field distribution of optical transmission of periodic subwavelength holes in a metal film," Phys. Rev. Lett., **86**, 1110–1113, 2001.

[32] L. Salomon, G. Bassou, H. Aourag, J. P. Dufour, F. de Fornel, F. Carcenac, and A. V. Zayats, "Local excitation of surface plasmon polaritons at discontinuities of a metal film: theoretical analysis and optical near-field measurements," Phys. Rev. B, **65**, 125409(5), 2002.

[33] S. Suto, K.-D. Tsuei, E. W. Plummer, and E. Burstein, "Surface-plasmon energy and dispersion on Ag single crystals," Phys. Rev. Lett., **63**, 2590–2593, 1989.

[34] D. Bohm and E. P. Gross, "Theory of plasma oscillations. A. Origin of medium-like behavior," Phys. Rev., **75**, 1851–1864, 1949.

[35] S. E. Irvine and A. Y. Elezzabi, "Surface-plasmon-based electron acceleration," Phys. Rev. A, **73**, 013815(8), 2006.

[36] R. A. Innes and J. R. Sambles, "Optical characterisation of gold using surface plasmon-polaritons," J. Phys. F: Met. Phys., **17**, 277–287, 1987.

[37] J. B. Pendry, "Negative refraction makes a perfect lens," Phys. Rev. Lett., **85**, 3966–3969, 2000.

[38] E. Hao, S. Li, R. C. Bailey, S. Zou, G. C. Schatz, and J. T. Hupp, "Optical properties of metal nanoshells," J. Phys. Chem. B, **108**, 1224–1229, 2004.

[39] J. A. Stratton, *Electromagnetic Theory*, McGraw-Hill Book Company, Inc., New York, 1941, pp. 563–573.

[40] A. L. Aden and M. Kerker, "Scattering of electromagnetic waves from two concentric spheres" J. Appl. Phys., **22**, 1242–1246, 1951.

[41] E. Prodan and P. Nordlander, "Exchange and correlations effects in small metallic nanoshells," Chem. Phys. Lett., **349**, 153–160, 2001.

[42] B. T. Draine and P. J. Flatau, "Discrete-dipole approximation for scattering calculations," J. Opt. Soc. Am. A, **11**, 1491–1499, 1994.

[43] A. Taflove and S. C. Hagness, *Computational Electrodynamics: The Finite-Difference Time-Domain Method*, 2nd edn, Artech House, Boston, 2000.

[44] L. Li, "New formulation of the Fourier modal method for crossed surface-relief gratings," J. Opt. Soc. Am. A, **14**, 2758–2767, 1997.

[45] Z. Li and L. Lin, "Photonic band structures solved by a plane-wave-based transfer-matrix method," Phys. Rev. E, **67**, 046607(11), 2003.

[46] Y. Zeng, Y. Fu, X. Chen, W. Lu, and H. Ågren, "Optical properties of two-dimensional negative-phase-velocity-medium photonic crystals," Phys. Rev. E, **73**, 066625(5), 2006.

[47] S. Fan, P. R. Villeneuve, and J. D. Joannopoulos, "Large omnidirectional band gaps in metallodielectric photonic crystals," Phys. Rev. B, **54**, p.11245–11251, 1996.

[48] X. Zhang, B. Sun, R. H. Friend, H. Guo, D. Nau, and H. Giessen, "Metallic photonic crystals based on solution-processible gold nanoparticles," Nano Lett., **6**, 651–655, 2006.

[49] B. Lamprecht, G. Schider, R. T. Lechner, H. Ditlbacher, J. R. Krenn, A. Leitner, and F. R. Aussenegg, "Metal nanoparticle gratings: Influence of dipolar particle interaction on the plasmon resonance," Phys. Rev. Lett., **84**, 4721–4724, 2000.

[50] S. Linden, J. Kuhl, and H. Giessen, "Controlling the interaction between light and gold nanoparticles: Selective suppression of extinction," Phys. Rev. Lett., **86**, 4688–91, 2001.

[51] A. Christ, S. G. Tikhodeev, N. A. Gippius, J. Kuhl, and H. Giessen, "Waveguide-plasmon polaritons: Strong coupling of photonic and electronic resonances in a metallic photonic crystal slab," Phys. Rev. Lett., **91**, 183901(4), 2003.

[52] W. Y. Zhang, X. Y. Lei, Z. L. Wang, D. G. Zheng, W. Y. Tam, C. T. Chan, and P. Sheng, "Robust photonic band gap from tunable scatterers," Phys. Rev. Lett., **84**, 2853–2856, 2000.

[53] E. Yablonovitch, "Inhibited spontaneous emission in solid-state physics and electronics," Phys. Rev. Lett., **58**, 2059–2062, 1987.

[54] S. John, "Strong localization of photons in certain disordered dielectric superlattices," Phys. Rev. Lett., **58**, 2486–2489, 1987.

[55] K. Sakoda, *Optical Properties of Photonic Crystals*, Springer, Berlin, 2001.

[56] K. S. Yee, "Numerical solution of initial boundary value problems involving Maxwell's equations in isotropic media," IEEE Trans. Antennas Propag., **14**, 302–307, 1966.

[57] J. P. Berenger, "A perfectly matched layer for the absorption of electromagnetic waves," J. Comput. Phys., **114**, 185–200, 1994.

[58] W. J. Tsay and D. M. Pozar, "Application of the FDTD technique to periodic problems in scattering and radiation," IEEE Micro. Guid. Wave Lett., **3**, 250–252, 1993.

[59] E. A, Navarro, B. Gimeno, and J. L. Cruz, "Modeling of periodic structures using the finite-difference time domain method combined with the Floquet theorem," Electron. Lett., **29**, 446–447, 1993.

[60] R. J. Luebbers, D. Steich, and K. K. Kunz, "FDTD calculation of scattering from frequency-dependent materials," IEEE Trans. Antennas Propag., **41**, 1249–1257, 1993.

[61] P. M. Goorjian and A. Taflove, "Direct time integration of Maxwell's equations in non-linear dispersive media for propagation and scattering of femtosecond electromagnetic solitons," Opt. Lett., **17**, 180–182, 1992.

[62] T. Kashiwa and I. Fukai, "A treatment by FDTD method of dispersive characteristic associated with electronic polarization," Microw. Opt. Technol. Lett., **3**, 203–205, 1990.

[63] G. Mur, "Absorbing boundary ocnditions for the finite-difference approximation of the time-domain electromagnetic field equation," IEEE Trans. Electromagn. Compatibility, **23**, 377–382, 1981.

[64] K. R. Umashankar and A. Taflove, "A novel method to analyze electromagnetic scattering of complex objects," IEEE Trans. Electromag. Compatibility, **24**, 397–405, 1982.

Chapter 7

Surface Plasmonics Devices

Microchips and contemporary optical data storage media are currently being produced at the limit of the conventional optical possibilities, whereas future photonics will require increased integration of photonic devices of up to 1000×1000 input and output channels on a single substrate for increased data transfer rates and for improved photolithography with feature lengths well below one wavelength.[1] Similar to electronic circuits where the transport and storage of electrons can be controlled, the concept of surface plasmon polariton (SPP)-based circuits has been proposed.[2] Many plasmonics devices have been designed and fabricated. In this chapter, we discuss a few principal plasmonics devices.

7.1 PASSIVE SURFACE PLASMONICS DEVICES

7.1.1 *Plasmonic Lens*

The function of plasmonic lens is to focus the EM energy by exciting SPPs. Two mechanisms are often used; one is the (periodic) surface defects to reflect SPPs.[3] The other utilizes the sharp edge of a circular slit etched through a metal film to directionally excite the SPPs toward the focus point.[4–6] For the first case, the reflected energy is normally quite low because of the low reflection coefficient of SPPs.

In 2004, Sun *et al.* reported a special setup of nanostructures, which, in combination with SPP, allowed them to focus light through a subwavelength-scale nanoslit array.[7] Their first experiment was to shoot a planar wave on a single nanoslit, as shown in Fig. 7.1. The planar wave seemed to propagate through the slot and emitted photons radial with a uniform power distribution. They predicted that this effect was due to SPP generation on the bottom of the film, which traveled through the slit and emitted the detected photons. Their calculations show that the transmittance and phase are only depending on the slit depth/metal thickness if SPPs do not couple to each other. Upon the conclusion of their experiments it was shown that it was possible to focus or collimate photons by using only nanoslits (Fig. 7.2). On top of the film, they built a convex metal bump with nanoslits and shot it with a planar wave as they had done before. To obtain the special effect of

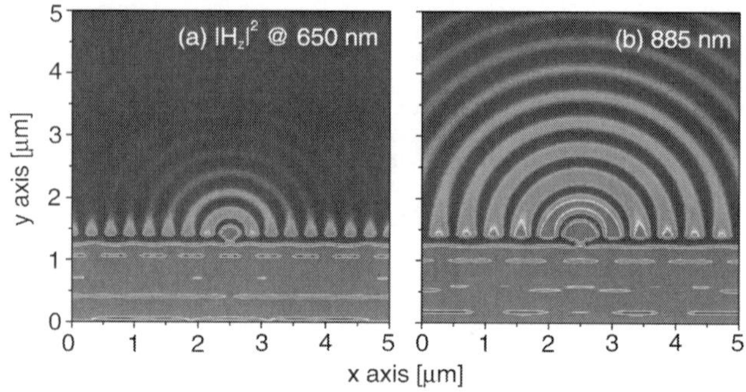

Fig. 7.1 A TM polarized plane wave, (a) λ = 650 nm, (b) λ = 885 nm, is incident to the slit (width 80 nm) on a metal film of 200 nm thickness from the bottom. A SPP is generated at the bottom of the dark-blue metal. It travels through the slit and reconverts into a radial light wave after the slit.[7] See also Color Insert.

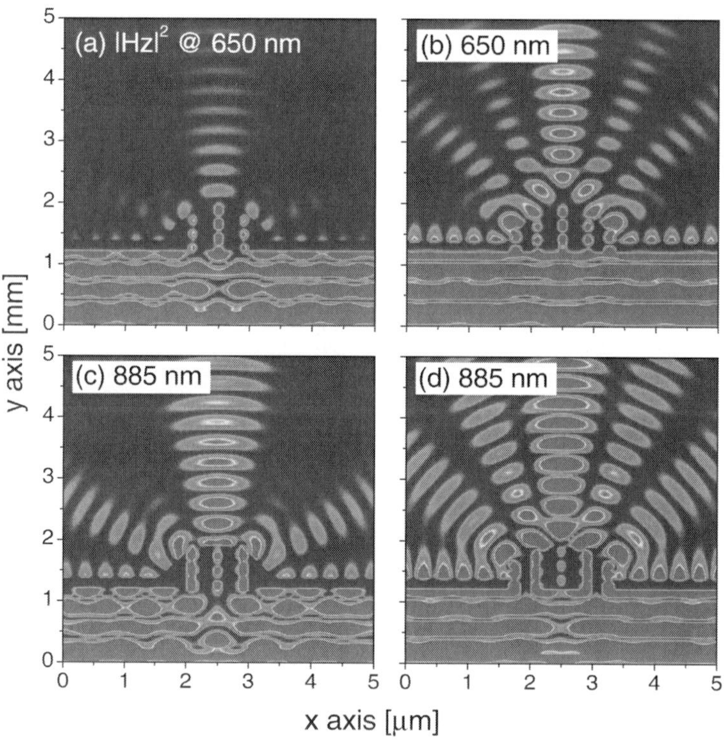

Fig. 7.2 A 2 μm wide convex bump and a planar wave (λ = 650 and 885 nm, respectively) hitting the metal film from the bottom (a,c) 80 nm-wide slits with 400 nm slit spacing and 700, 750, and 700 nm depth. (b,d) 80 nm-wide slits with 400 nm slit spacing and 450, 700, 750, 700, and 450nm depth.[7] See also Color Insert.

focusing, they calculated the required slit depth to receive the desired phase shift. They tried the same experiment with three and five slits and extrapolated from the results that the focal length of the nano-lens increases with the slit amount.

If we compare the wavelength of the laser ($\lambda = 650\,\text{nm}$) with the slit width (80 nm), we can say that we used a real subwavelength "optical" lens to focus our light. As opposed to the conventional lenses we now have the advantage of controlling each phase shift separately by changing the corresponding slit depth.

Another example of the second kind of plasmonic lens is shown in Fig. 7.3 including the geometry shape and the near-field pattern of the focusing points. When the slit width is smaller than half of the wavelength of the incident light, no propagating modes are allowed so that a majority of the light is diffracted. According to the momentum conservation requirement, the sharp edge of the slit can be treated as a line of SPP point sources, and an SPP can be excited by a portion of the diffracted light whose wave vector is determined by the metallic surface according to the dispersion relationship. In addition, the direction of the wave vector, which determines the energy propagation direction, will be normal to the slit when the incident light is normal to the metal film. Consequentially, the energy will be guided toward the focal point of the lens. Because a wide band of SPPs can be excited by the slit edge, different wavelengths of light may be used to excite SPPs in the same structure. This is an advantage over periodic surface defects that can only reflect efficiently those SPPs having a specified wavelength.[4, 5]

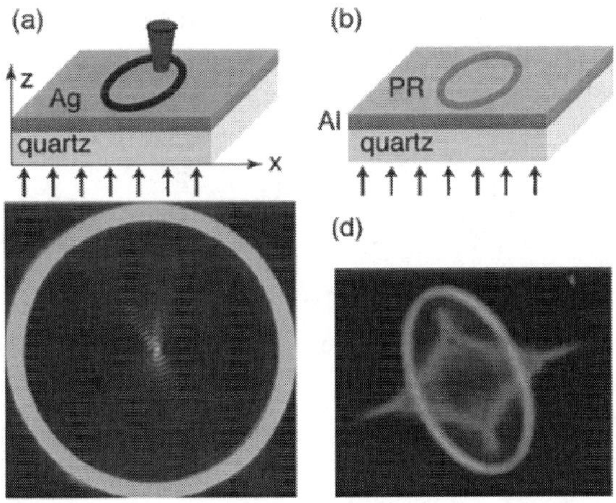

Fig. 7.3 Experimental setup for (a) near-field scanning microscopy (NSOM) and (b) plasmonic lithography measurements for recording the near-field pattern for plasmonic lenses. (c) Near-field pattern for a 14 μm-diameter circle cut into a 150 nm-thick silver film recorded with NSOM. Polarization of incident light is indicated with an arrow. (d) Near-field pattern for an ellipse with a long axis of approximately 4 μm and a short axis of 2.5 μm cut into a 70 nm-thick aluminum film recorded with plasmonic lithography. Z. Liu, J. M. Steele, W. Srituravanich, Y. Pikus, C. Sun, and X. Zhang, Focusing surface plasmons with a plasmonic lens, Nano Lett. vol. 5, p. 1726–9, 2005. Copyright (2005) by American Chemical Society. See also Color Insert.

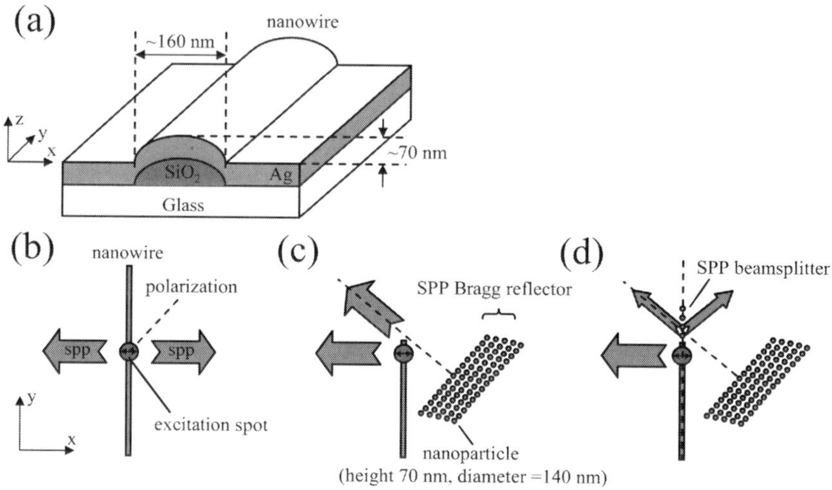

Fig. 7.4 (a) Schematic nanoscale surface structure. Nanowire is shown, while nanoparticle is obtained by modulations along the y axis. (b) Single nanowire: excitation light spot and polarization, and SPP generation and propagation. (c) SPP propagation in a single nanowire and SFF Bragg reflector from nanoparticles. (d) SPP beamsplitter.

7.1.2 *Nano Optic Beam Splitter*

In 2002 Ditlbacher et al. submitted a paper which described a setup to control SPP in a way we do control light waves with mirrors and a beam splitter.[8] This control over SPPs, the ability to travel several 100 nm to some μm,[9] and the ability to reconvert into photons allow us to approach our aim of subwavelength optics.

Fig. 7.4a shows the schematic drawing of the nanowire along the y axis, while nanoparticle can be obtained similarly with modulations along the y axis. Fig. 7.4b shows the generation and propagation of SPPs from a single nanowire. When five lines of nanoparticles are added (the distance between the lines is 350 nm to form the first-order Bragg reflection), SPP reflection is achieved, as shown in Fig. 7.4c. Finally, an extra line of nanoparticles over the nanowire is added to achiev beam splitting, see Fig. 7.4d.

7.1.3 *Surface Plasmonics Waveguide*

The function of the SPP waveguide is to guide EM energy of SPP at optical frequencies below the diffraction limit.[10] Due to the hybrid nature of the SPP, i.e., light waves coupled to free electron oscillations in a metal, the SPP can be laterally confined below the diffraction limit using subwavelength metal structures. However, it is very difficult to realize simultaneously a strong confinement as well as a propagation loss sufficiently low for practical applications, which is usually dictated by loss (Ohmic losses) in the metal.[9]

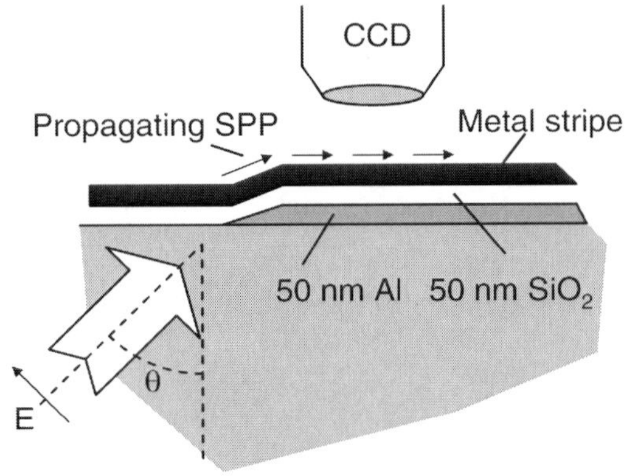

Fig. 7.5 Spatially confined SP excitation in the prism-coupling arrangement is provided by an opaque Al screen. Scattered light from the propagating SP is observed with a CCD camera. See also Color Insert.

In order to achieve efficient guiding, three different approaches are currently being investigated: (1) SPP mode propagation along finite-width thin metal stripe (MS).[10–15] When buried in a homogeneous dielectric medium, MS can guide so-called "long-range" SPP modes, whose field is more confined in the metal over several millimeters at telecommunication frequencies; Fig. 7.5 shows the schematics of microscale MS.[12] (2) Channel SPP mode propagation along finite-width grooves cut into planar metal surface.[16–18] For subwavelength groove widths, the channel SPP mode field in the groove is nearly constant and close to its maximum, thereby rendering the propagation losses relatively low.[18] Most recently, channel SPP based subwavelength waveguide components operating at telecom wavelengths have been designed and fabricated, including Y-splitters, Mach-Zehnder interferometers, and waveguide-ring resonators;[18] see Fig. 7.6 as an example. (3) Through electrodynamic interparticle coupling, SPP propagation along lines of defects created into a so-called "plasmonic crystal",[19] or linear chains of metal nanoparticles.[20–24] The optical properties of such a structure depend strongly on the individual nanoparticle behavior as well as on the separation distance and on the configuration of the nanoparticles relative to the illumination direction.

Figure 7.7 shows a waveguide design using a closely spaced array of metallic nanoparticles.[25]

7.1.4 *Corner Plasmonic Waveguides*

Unlike guided by an all-dielectric waveguide, SPP can be guided along a two-dimensional metal corner.[26–29] This property, on one hand, allows SPP waveguides to have a simple structure as primitive as a mono-angular corner. On the other hand, it also suggests that SPP waveguides fabricated with traditional methods are not

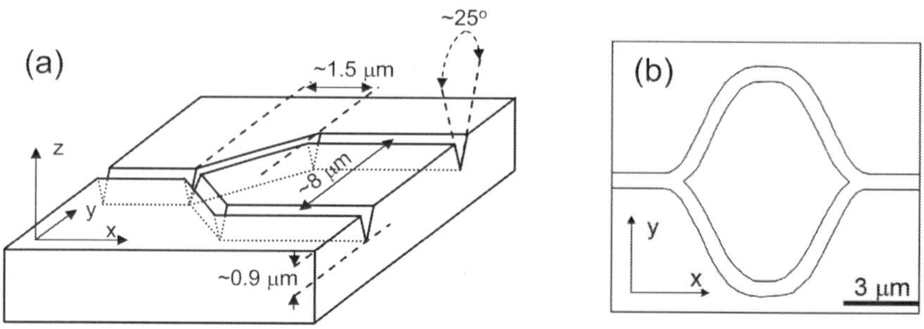

Fig. 7.6 Schematic drawings of plasmonic Y-splitter (a) and Mach-Zehnder (MZ) interferometer (b) studied in Ref.[18]

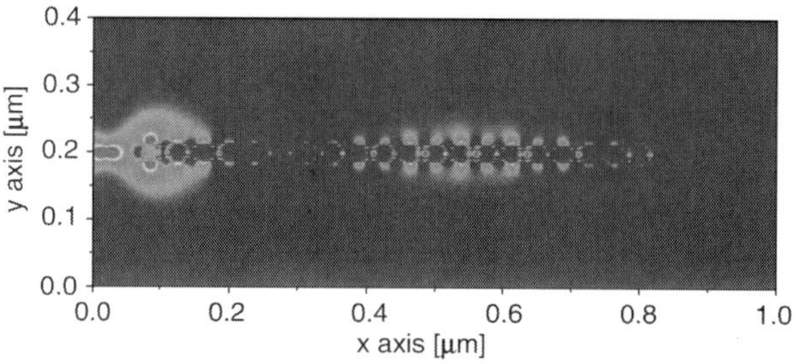

Fig. 7.7 $|H_y|$ for a 20 Ag nanoparticle chain waveguide. Excitation is at the nanoparticle on the left. See also Color Insert.

as simple as their all-dielectric counterparts in terms of mode properties. In fact, most plasmonic waveguides reported so far include several corners in their cross-sections, as a direct consequence of the current lithography technology. Examples in reported literature include the MS waveguide,[16, 30–32] V-channel waveguide,[27, 33] metal-slot waveguide,[34–37] U-channel waveguide,[38] and Λ-wedge waveguide.[39] When the metal corners are in close proximity, the modal property of the overall waveguide can be very problematic to comprehend without in-depth knowledge of a single-corner waveguide. It is the objective of this section to illustrate how the modal properties of a single-corner SPP waveguide vary according to its geometry and the operating wavelength. Our discussion encompasses both the V-channel and Λ-wedge waveguides. Their performances are closely compared.

The cross-sections of V-channel and Λ-wedge waveguides are shown schematically in Fig. 7.8a and c, respectively. The flat surface waveguide (Fig. 7.8b) is a special case in our study, which has a corner angle at 180°. A mono-angular metal-corner waveguide has two long enough side walls, so the guided mode is solely attributed to this single corner, not influenced by any adjacent corner or curvature changes. We refer to modes guided by these 2D metal corners in general as corner plasmon polaritons, i.e., CPPs.

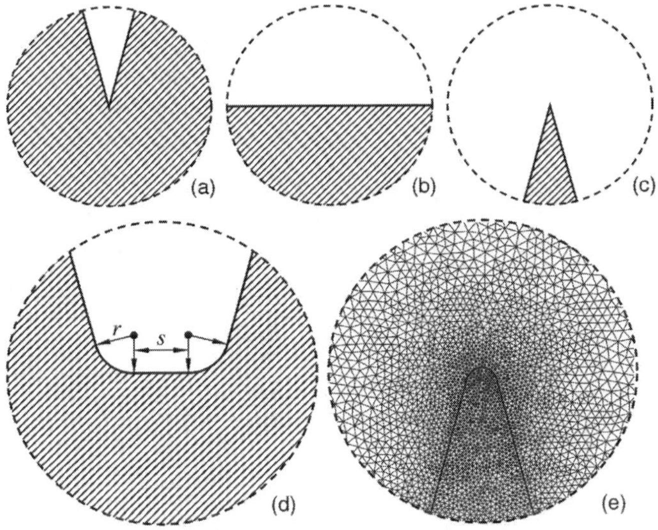

Fig. 7.8 (a)–(c) show schematic diagrams of V-channel, 1D surface, and Λ-wedge metal waveguides, respectively. Hatched region is metal, and white region is dielectric material. (d) Geometric parameters for a generalized metal-corner waveguide. (e) Sample mesh of a circular region with a 200 nm diameter. The corner has a 30° angle, and is rounded with an arc of 10 nm in radius.

We deploy the finite element method (FEM) as our numerical tool for the following investigations. FEM has the advantage of defining the material interfaces accurately, straight or curved. In addition, FEM uses an adaptive mesh resolution. Due to these reasons, FEM is expected to achieve convergence faster than finite-difference-based numerical methods. For most structures concerned in this section, triangle edge size as small as 1.5 nm is used at the metal corner tip. A sample finite-element mesh is shown in Fig. 7.8e, which has a domain diameter of 200 nm and the tip arc has a radius of 10 nm. Owing to the EM field singularity at the corner of a penetrateble material wedge,[40–43] all corners are rounded with an appropriate arc. In fact, we argue that localization of the guided mode in such interface waveguides is largely owing to the variation in local surface curvature. Given a certain wavelength, smaller curvature tends to attract more light and gives rise to a larger propagation constant. We will demonstrate how sensitively the mode changes with the corner sharpness later.

The schematic CPP waveguide shown in Fig. 7.8d gives a general picture of metal-corner waveguides to be studied. Note that s can be zero, in which case the corner tip is a single arc (Fig. 7.8e). It should be kept in mind that actual corners fabricated in experiments may be far more complex than what have been shown in the figure. It is certainly not possible to exhaust all possible structural variations. However, we believe the results presented in the following give a general idea how the corner waveguides behave, even when the corners are at a slightly higher complexity.

Without loss of generality, we choose silver (Ag) as the metal, and air as the dielectric material. The dielectric constant of silver is described in Drude model as

$$\epsilon = \epsilon_\infty - \frac{(\epsilon_0 - \epsilon_\infty)\omega_p^2}{\omega(\omega + i\gamma)} \tag{7.1}$$

with $\epsilon_\infty = 4.017$, $\epsilon_0 = 4.896$, $\omega_p = 1.419 \times 10^{16}$ rad/s, and $\gamma = 1.117 \times 10^{14}$ rad/s. This Drude model is fitted according to the measured data from Palik's handbook.[44] Note that in Ref.,[44] there are two experimental datasets presented at the near-infrared wavelength regime. We have in particular used the data from the Ref. [33] as cited in that book chapter.

7.1.5 *General Properties of Corner Modes*

Refer to Figs. 7.8a–c; a mono-angular CPP waveguide has reflection symmetry about the vertical line passing through its vertex. It is possible to make use of this symmetry to reduce the computational domain size by half, with a proper symmetry condition defined on the reflection axis. A preliminary symmetry analysis tells that both perfect electric conductor (PEC), i.e.,

$$E_\parallel = 0 \ , \quad \frac{\mathrm{d}E_\perp}{\mathrm{d}n} = 0 \ , \quad \frac{\mathrm{d}H_\parallel}{\mathrm{d}n} = 0 \ , \quad H_\perp = 0$$

where n is the surface normal direction, and perfect magnetic conductor (PMC), i.e.,

$$H_\parallel = 0 \ , \quad \frac{\mathrm{d}H_\perp}{\mathrm{d}n} = 0 \ , \quad \frac{\mathrm{d}E_\parallel}{\mathrm{d}n} = 0 \ , \quad E_\perp = 0$$

symmetry conditions are allowed at the reflection axis.[45] However, after numerous calculations using full structures, we concluded that the V-channel waveguide only guides mode with a PEC symmetry, and the Λ-wedge waveguide only guides mode with a PMC symmetry.

The transverse electric and magnetic field of the allowed modes supported by the V-channel and Λ-wedge waveguides are schematically shown in Fig. 7.9a, f, respectively. Modes in Fig. 7.9c, d, though not confined, are shown according to their predicted symmetries. Here we offer a graphical understanding on why these two modes are not guided. Modes shown in the upper three panels in Fig. 7.9 are of the same gene. That is, they all have the PEC symmetry condition. The channel mode in Fig. 7.9a is well-bounded, especially when the channel angle is far smaller than 180°. In other words, we can say $\beta > k_0 n_{\mathrm{SPP}}$, where β is propagation constant, k_0 is free-space wave number, and n_{SPP} is the effective index of the SPP mode guided by a flat Ag–air interface. As the angle increases to 180° (Fig. 7.9b), the mode becomes a folded SPP (or second-order SPP), and is infinitely extended. Under the circumstance, we have, equivalently, $\beta = k_0 n_{\mathrm{SPP}}$.

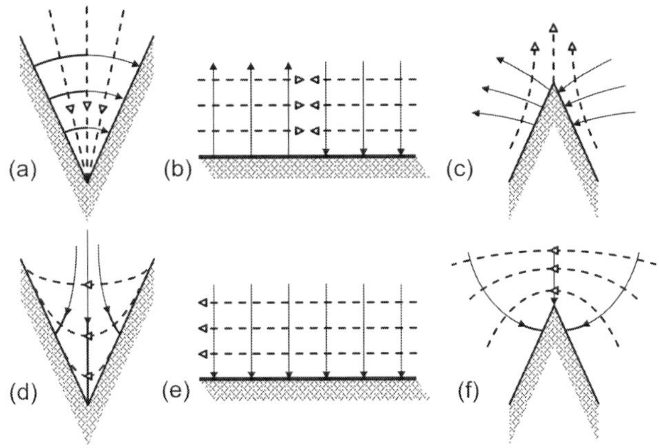

Fig. 7.9 Schematic mode field patterns suggested by symmetry analysis. Only transverse electric (solid lines with arrow heads) and magnetic (dashed lines with arrow heads) fields in air region are shown. Domain below the thick solid line is metal. (a) Mode in a V-channel waveguide with PEC symmetry. (b) Mode on a flat surface with PEC symmetry. (c) Mode on a Λ wedge with PEC symmetry. (d) Mode in a V-channel waveguide with PMC symmetry. (e) Mode on a flat surface with PMC symmetry. (f) Mode on a Λ wedge with PMC symmetry.

When the angle increases further (channel becomes wedge, Fig. 7.9c), by imaging that the process is physically continuous, we come to know that $\beta < k_0 n_{\mathrm{SPP}}$. That is, the mode now spectrally stays inside the mode continuum region of a single Ag–air interface, and therefore will radiate away along the side walls. Similarly, modes in the lower three panels in Fig. 7.9 all have PMC symmetry.

The mode in Fig. 7.9f is bounded ($\beta > k_0 n_{\mathrm{SPP}}$). When the wedge angle increases to 180° (Fig. 7.9e), the mode becomes the well-known SPP mode ($\beta = k_0 n_{\mathrm{SPP}}$). When the angle increases further (wedge becomes channel, Fig. 7.9d), we will expect $\beta < k_0 n_{\mathrm{SPP}}$. That is, the mode will no longer be bounded. Here, once again, we stress that both the V-channel and Λ-wedge waveguides have their two side interfaces extend to a very far distance at which the corner-bounded mode has decayed significantly in its strength. In actual numerical executions, we ensure that the domain size is big enough such that the field at the domain boundary has decayed at least by three orders from its maximum value. It should, however, be noticed that, a realistic CPP waveguide, whose side walls are of a finite length, normally consists of both V and Λ corners in close proximity. Such a composite waveguide might support modes of both types of symmetries.

From our numerical simulation, the Λ-wedge waveguide is found to guide only a single mode, regardless of wavelength and geometric parameters. The V-channel waveguide is multi-mode, especially when its angle is small and at a small operating wavelength. However, here we particularly pay our attention to the fundamental mode, which should be most useful in practical applications. V channels normally support multiple modes, but excitation and manipulation of the higher order modes

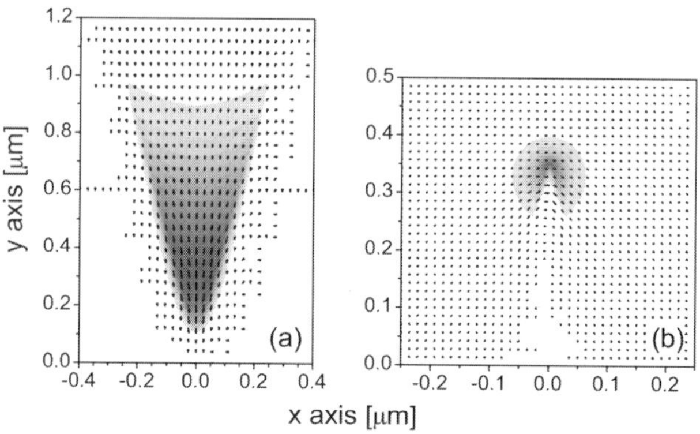

Fig. 7.10 Fundamental mode guided by two types of corner waveguides. (a) V-channel waveguide; (b) Λ-wedge waveguide. Color map is for $|\boldsymbol{H}_t|^2$, and quiver is for $\boldsymbol{H}_t$. Wavelength is at 0.633 μm. Corner angle is 30°. Corner tip is rounded with an arc with 10 nm in radius. See also Color Insert.

are not easy since the waveguides are supposed to guide light in a sub-micron scale. The calculated fundamental mode patterns for both V-channel and Λ-wedge waveguides are presented in Fig. 7.10. For the Λ-type waveguide, the maximum field values are found to be almost at the corner tip. For the V-channel waveguide, although the maximum field is not exactly at the corner tip, it is however very close to the tip. The distribution of the mode field in a V channel suggests that the coupling between two side walls (each of them is a surface waveguide) is also a contributing factor of the channel mode, whereas in a Λ wedge, such coupling is somewhat suppressed by the opaqueness of the wedge body. For both types of waveguides, fields are noticed to stretch longer along the two side interfaces as compared to in silver or air.

It is meaningful to give a quantitative definition of mode field size (MFS) of such corner waveguides. In all-dielectric optical waveguides, e.g., optical fibers, the confinement of an EM field is due to refractive index elevation in a region. Therefore, the mode tends to be of a 2D Gaussian shape, which leads to easy definition of the mode size. CPPs are confined by a material interface. Their MFS needs a slightly different interpretation. The transverse decay constant of a CPP mode field along the interface and in the air can be written as

$$\begin{aligned} k_t^s &= k_0\sqrt{n_{\text{eff}}^2 - n_{\text{SPP}}^2} \\ k_t^a &= k_0\sqrt{n_{\text{eff}}^2 - n_{\text{air}}^2} \end{aligned} \tag{7.2}$$

respectively. The effective mode index, n_{eff}, can be derived from the propagation constant as $n_{\text{eff}} = \beta/k_0$. As n_{SPP} is always larger than n_{air}, field tends to stretch longer along the two side interfaces compared to in the air. For this reason the MFS can be estimated by only considering its extension along the Ag–air interface. From Fig. 7.10, we notice that the maximum field values are located very close to the corner tip for both types of waveguides. Therefore, MFS of a CPP can be fairly accurately defined as

$$\text{MFS} = \frac{1}{k_t^s} \tag{7.3}$$

which denotes the distance from the position where field amplitude is at its maximum to where the amplitude is decayed to $1/e$ of the maximum. It is very important to keep in mind that the field of a V-channel waveguide is well insulated laterally by two side metal walls, although it may extend very long in the vertical (channel-opening) direction. An example can be a V-channel waveguide with a 10° angle working at 2 μm wavelength. In this case such waveguides can still be densely packed shoulder by shoulder without inducing any cross talk, but not in the vertical direction. The mode field of Λ-wedge waveguide is almost like in the open air, and decay of the field in air is only slightly faster than that along the side interfaces (Fig. 7.10d). Therefore, the package density of such waveguides is almost solely limited by the MFS computed using Eq. 7.3.

7.1.6 *The Effect of Geometry*

In this section, we discuss the effect of changing geometric parameters on the modal properties of both V-channel and Λ-wedge waveguides. We will focus on the corner waveguides with $s = 0$, as shown by Fig. 7.8d.

First, the guided mode in a V channel is computed as the channel angle increases. The V bottom is rounded with a 10 nm radius arc. The geometrical dispersion curves, at both 0.633 and 1.55 μm wavelengths, are shown in Fig. 7.11a. At either wavelength, the effective mode index drops quickly as the V angle increases. Such tendency is especially evident for long wavelength. The effective mode index asymptotically approaches to that of the SPP mode as the angle approaches to 180°. Figure 7.11b shows the variation of MFS as a function of the V angle. As suggested by the n_{eff} curves, the mode size increases very quickly as the angle increases. At $\lambda = 0.633$ μm, the MFS exceeds the wavelength when the angle reaches ~47.5°. At $\lambda = 1.55$ μm, the MFS exceeds the wavelength when the angle reaches ~22.5°. The loss values are shown in Fig. 7.11c. At both wavelengths, loss initially decreases almost linearly in logarithmic scale with respect to the channel angle. Such linear loss decrease persists up to a V angle of ~40° for $\lambda = 0.633$ μm, and ~15° for $\lambda = 1.55$ μm. As the angle increases further, the loss approaches to that of the SPP mode at a slower rate. Loss at $\lambda = 1.55$ μm is smaller than that at $\lambda = 0.633$ μm, which is mainly caused by the drastic increase (negatively) of the real part of silver's permittivity value. In other words, the metal appears more perfect electromagnetically at longer wavelengths.

Let us now discuss similarly the geometric dispersion, MFS, and loss curves for the Λ-wedge type waveguide, also at 0.633 and 1.55 μm wavelengths, which are imposed in the respective panels in Fig. 7.11. The lower bounds of the n_{eff}

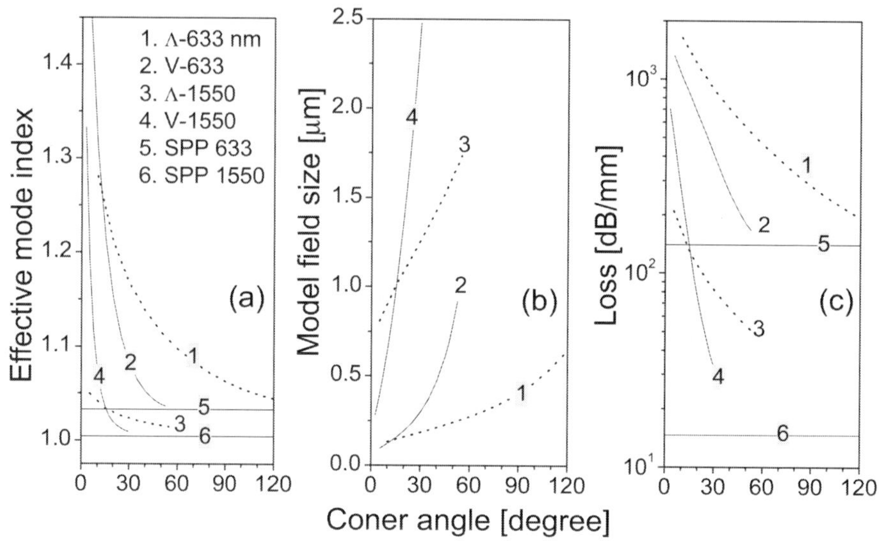

Fig. 7.11 Dispersion (a), mode field size (b) and loss (c) curves of the guided CPP mode as the corner angle changes.

and loss curves are still the values associated with the SPP mode. The n_{eff}-angle relationships at both wavelengths in general have the same trend as those for the V-channel waveguide. However, despite their common tendency, some differences exist. First of all, from structural point of view, the Λ waveguide confines light better at large corner angles; whereas the V waveguide confines light better at small corner angles. Second, from wavelength point of view, at a *small* corner angle, confinement of a Λ wedge is far more sensitive to wavelength. That is, the mode guided by a Λ-wedge with an angle of, say 5°, will expand more quickly in size compared to a V channel with the same angle as the wavelength increases. As a consequence of these two differences, at a long operating wavelength, a V-channel waveguide will be especially superior as compared to a Λ-wedge in terms of confinement.

In general, from Fig. 7.11b, c, we see by examining the V- and Λ-waveguide categories separately or even in a whole that better confinement is achieved at the expense of higher loss. However, there is a small exception. At 0.663 μm wavelength and a small angle ($< 15°$), it is noticed that the V-channel waveguide has a better confinement, but with lower loss, as compared to the Λ-wedge waveguide. Therefore, we are better-off to use the V-channel type waveguide if we want to achieve a tiny mode size at $\lambda = 0.633$ μm in consideration of the propagation loss.

It should be kept in mind that one condition for our above comparison is that the V channel and Λ wedge have the same tip curvature. The effect of the corner tip curvature is discussed below.

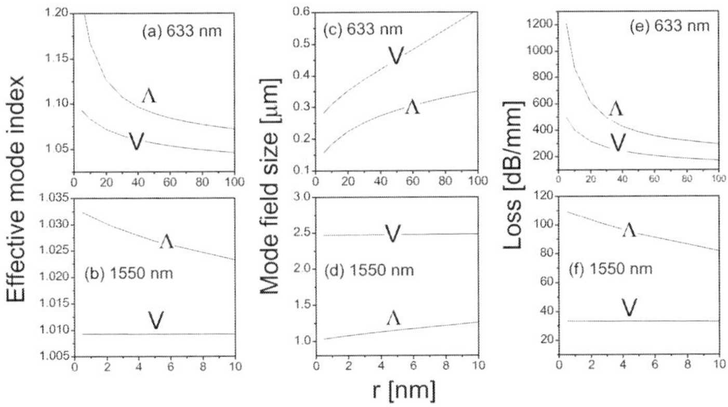

Fig. 7.12 Dispersion (a), mode field size (b), and loss (c) curves of the guided CPP mode as the curvature of the corner tip increases. All waveguides have a corner angle at 30°.

In Fig. 7.12, we show the geometric dispersion, MFS, and loss curves of modes guided by V-channel and Λ-wedge waveguide when tip sharpness, or r parameter, varies. Both waveguides have a 30° corner angle. First we let wavelength be at 0.633 μm, and change the tip curvature from 5 to 100 nm. The increase of the corner curvature leads to decreases of both the n_{eff} and the loss values, and correspondingly an increase in MFS. It can be imagined that a corner with an infinite r would make photons there think they are guided by a flat interface. Therefore, both modes asympototically approach to a SPP. From all three panels in Fig. 7.12, we see that the curves are affected more heavily when r, the tip sharpness, is small (< 30 nm). This is quite different from all-dielectric waveguides, in which such a minor structural variation usually does not perturb the mode as much. This also suggests that irregularities, especially *small* ones, might affect performance of a plasmon polariton waveguide considerably. When r is large (> 30 nm), its variation induces steady changes in all three curves, as shown in Fig. 7.12. Among the two types of waveguides, the Λ-wedge waveguide is observed to be more sensitive to the r value, especially when r is very small.

We then change the wavelength to 1.55 μm, and vary the tip curvature from 0.5 to 10 nm. We would like to know whether a very sharp tip would result in a small MFS even when the wavelength is large, and in turn we could achieve high packaging density even at this relatively long wavelength. Numerical resolution issue is specially reminded here when studying the sharp tip, which can be easily resolved by making use of FEM's adaptive meshing advantage. The results are imposed in the respective panels in Fig. 7.12. It is noticed that the V-channel waveguide improves only slightly in confinement as the corner is getting sharper. Its MFS decreases only from 2.482 μm at $r = 10$ nm to 2.469 μm at $r = 0.5$ nm. This confirms what has been noticed in Ref.[46] Therefore, at increasing wavelength, the mode in a realistic V-channel waveguide will expand in size and inevitably cause modal couplings with the wedge corners. In contrast, mode guided by the Λ-wedge waveguide shows relatively more sensitive changes with decreasing r value. In fact the modal change of the Λ-wedge waveguide due to variation in r is almost comparable to that induced by variation of the corner angle (Fig. 7.11) at both wavelengths. Again, from Fig. 7.12, all improvement of confinement is at the cost of higher propagation loss.

7.1.7 *The Chromatic Properties*

In this section, we present the chromatic modal properties (effect of wavelength) of the corner waveguides. Among all chromatic properties, the group velocity dispersion (GVD) is especially important, as its presence will limit the data rate that can be transmitted over the waveguide. Although the waveguides discussed here are meant to propagate light in the millimeter range (e.g., for integrated optics applications), severe optical pulse broadening due to GVD can still deteriorate the optical signal and render the signal unintelligible. Therefore, it is critical to understand quantitatively the GVD property of such waveguides. Besides, the variations of the n_{eff} and MFS values as a function of wavelength provide hint on how a composite waveguide, which consists of both V and Λ corners in close proximity, behaves as the wavelength increases.

We show the chromatic properties of two particular corner waveguides in Fig. 7.13. Both the V-channel and Λ-wedge waveguides have an angle of 20°, and their tips are rounded with an arc of 10 nm in radius. The 20° angle is chosen here so that the mode size is reasonably small even at relatively long wavelengths, and in turn derivation of the modes can be numerically manageable. From Fig. 7.13a, the Λ-wedge waveguide always has a larger n_{eff} value than the V-channel waveguide over

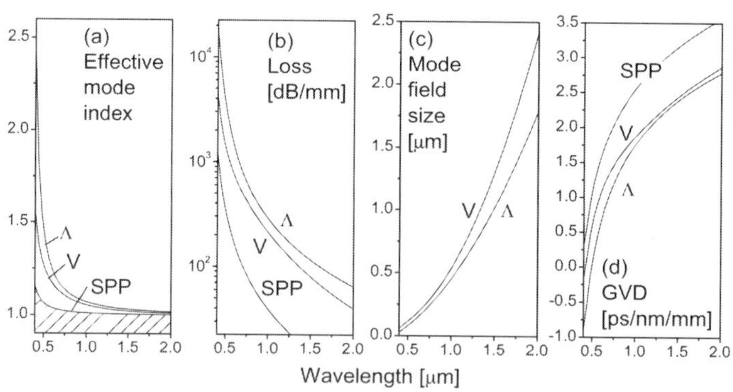

Fig. 7.13 Chromatic properties of the corner waveguides. The corner angle is 20°, and the tip is rounded with an arc of 10 nm in radius. (a) Effective mode index; (b) loss; (c) mode field size (MFS) and (d) group velocity dispersion (GVD). The gray region below the black line in (a) denotes the mode continuum region of a flat Ag–air interface.

the whole 0.4–2.0 μm wavelength span. Correspondingly, the Λ-wedge waveguide has a larger loss (Fig. 7.13b) and a smaller MFS (Fig. 7.13c). The GVDs of the two waveguides are shown in Fig. 7.13d. Note we have used the unit of ps/nm/mm, rather than the ps/nm/km commonly adopted in optical fiber communication systems. The curves for the SPP mode are also shown wherever possible. All three waveguides (including the SPP mode) have a normal dispersion value at the examined wavelength range, which suggests that the higher-frequency components travel slower. The two corner waveguides have comparable dispersion values, both of which decrease (more close to zero) as wavelength increases. Except at the extremely short wavelength, i.e., near to 0.4 μm, the GVD value is relatively small as compared to that (in the unit of ps/nm/km) in a standard modern single-mode optical fiber.[47] Therefore, we expect that both channel- and wedge-type waveguides have a communication channel capacity at least as large as that in an optical fiber link. Of course, the concerned distance with the metal-corner waveguides is calculated in millimeters, and the loss issue is not taken into consideration. By further calculating the GVD values of a V-channel and a Λ-wedge waveguides, both at a 30° angle (not shown) and comparing to the 20° cases, it is noticed that GVD of the 30° V-channel waveguide changes little at short wavelengths (<∼0.55 μm), and the value becomes even smaller (closer to zero) when the wavelength increases. Very similar variation in GVD value is noticed for the wedge-type waveguide as its angle increases from 20 to 30°.

A realistic V-channel waveguide, which is etched from a flat metal surface,[27, 28] has a finite depth. And the whole waveguide structure consists of both V- and Λ-type corners. When operating wavelength increases to a certain value (which depends on how deeply the V channel is etched), mode guided by the channel bottom will interact or couple with modes guided by two side wedges. This interaction is inevitable due to expansion of the channel mode size. A quantitative knowledge of the "critical" wavelength can be obtained by examining the MFS of the mode in an infinitely deep V channel. For instance, from Fig. 7.13c, the MFS would be as large as 1 μm when wavelength is close to 1.4 μm. Therefore, strong coupling of the two types of modes are expected if the V channel has side walls at 1 μm long. Note that the consequence of increasing wavelength is more than just modal mixing; the channel mode can go completely cutoff. To explain such cutoff needs a bit more work. We dedicate the next sub-section to this problem, in which we also mention the effect of increasing wavelength on the mode guided by a realistic Λ-wedge waveguide.

7.1.8 *Realistic Metal-Corner Waveguides*

In practice, a V channel can be realized by etching a groove from a flat metal film.[28] Similarly a Λ wedge can be sitting on a flat metal substrate. In other words, real corner waveguides have finite side walls. In addition, there will be some extra corners at the terminations of the side walls. At a small enough wavelength, the mode can be so well confined that it may not be aware of neighboring corners. However, when wavelength increases, the mode will expand and interact with neighboring corners and waveguiding structures further beyond.

In fact, the mode guided by a V channel of finite height will experience a cut-off when wavelength increases to a certain value. The explanation was partially mentioned in Ref.[46] Here we examine the phenomenon from the viewpoint of supermodes, with special emphasis on the mode symmetry. In simple words, at a long wavelength, a realistic V-channel waveguide is rather a super-guide formed by three potential corner waveguides. Two types of modes are possible: one is of PEC symmetry, and the other is of PMC symmetry. All symmetries are about its reflection axis. Note that the V-channel bottom, if regarded as a single waveguide, only guides modes with PEC symmetry (Fig. 7.9a). That is, the channel bottom only contributes to supermodes with the PEC symmetry. Reversely, supermodes with PMC symmetry are anti-resonant with the channel bottom for the symmetry reason, and its field is guided solely by two side wedges. Unfortunate to the channel bottom, modes with the PMC symmetry are favored at a large wavelength, since it can eventually evolve into the SPP mode (also of PMC symmetry, Fig. 7.9e) as wavelength increases. In contrast, supermodes with PEC symmetry are likely to be cutoff at large wavelengths. And when the groove appears so small to the operating wavelength, the channel, or rather the channel bottom, can be completely overlooked by the EM wave.

We examined two types of modes guided by a realistic V channel with a 20° angle and a 0.36 µm opening (depth is about 1 μm) in the previous sub-section. Corners are rounded by an arc with a 10 nm radius. Their $n_{\rm eff}$ values are shown in Fig. 7.14. The $n_{\rm eff}$ curve of the SPP mode is also superimposed. We can see that the supermode with PEC symmetry experiences cutoff at λ ~1.47 µm. The

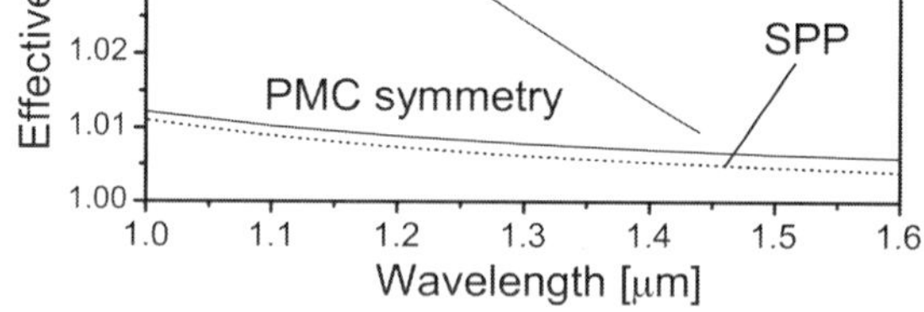

Fig. 7.14 Variations of $n_{\rm eff}$ values of two supermodes supported by a realistic V-channel waveguide.

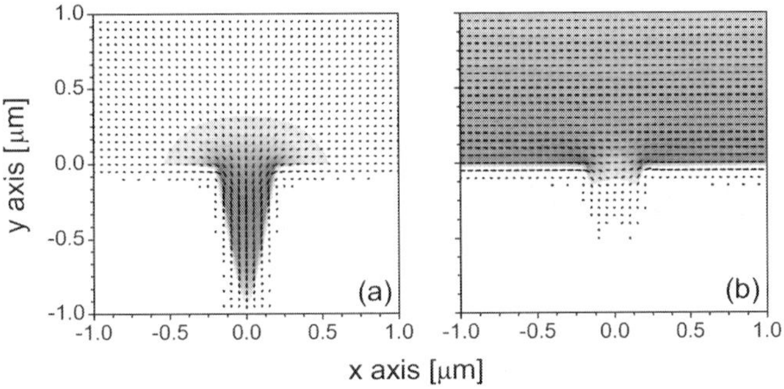

Fig. 7.15 $|\boldsymbol{H}_t|^2$ (colormap) and $\boldsymbol{H}_t$ (quiver) plots of two supermodes guided by a realistic V-channel waveguide at $\lambda = 1.4$ μm. (a) Mode with PEC symmetry; (b) Mode with PMC symmetry. See also Color Insert.

supermode with PMC symmetry has its n_{eff} value only slightly larger than that of the SPP mode at the examined wavelength region, which indicates that its mode field should extend very far along the Ag–air interface. Figure 7.15 shows the modes in two symmetries at $\lambda = 1.4$ μm. It is evident that the mode with PEC symmetry (Fig. 7.15a) is inter-played by three individual corner waveguides, whereas the mode with PMC symmetry (Fig. 7.15b) only takes efforts from two side wedges.

Let us examine a Λ-wedge waveguide with the same geometrical parameters. Note now that we have two V-shaped corners with a 100° (always larger than 90°) angle at two sides. From Fig. 7.11, V channels with such a large angle will have difficulties in guiding light. Therefore, their contribution to the waveguiding ability of the super-guide is subtle. Due to the fact that a single Λ wedge guides a mode with PMC symmetry, which coincides with the SPP mode symmetry, the wedge mode in a realistic structure will not experience a cutoff. And the mode will evolve into a SPP mode as wavelength increases. We observe in Fig. 7.16 the variation of n_{eff} value of the supermode guided by the realistic Λ waveguide. The mode field at $\lambda = 2$ μm is plotted in Fig. 7.17. Note colormap now is for the absolute value (not squared) of the transverse magnetic field. As expected, at such a wavelength the mode has a quite large field size. And its coupling with the Ag–air interfaces adjacent to the wedge is also negligible. The two side V channels seem to contribute little to the waveguiding.

7.1.9 *Plasmonic Resonator Channel Drop Filters*

Channel drop filters (CDFs), which single out one channel from wavelength division multiplexed signals and pass through other channels undisturbed, are useful and essential elements for photonic integrated circuits and dense wavelength division multiplexing optical communication systems. In 2006, Xiao *et al.* theoretically

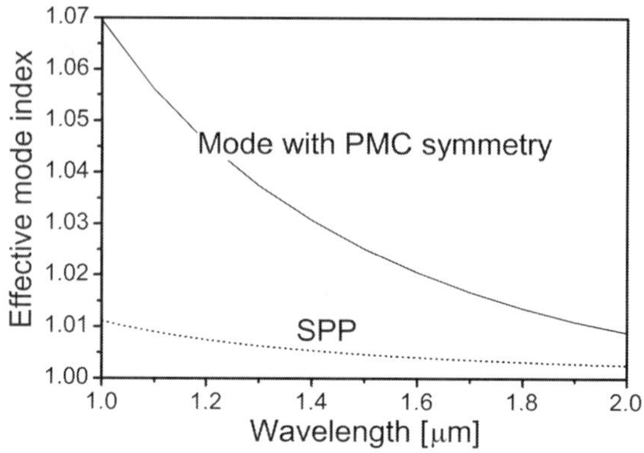

Fig. 7.16 Variation of n_{eff} value of the mode guided by a realistic Λ-wedge waveguide.

investigated disk/ring resonator CDFs realized in a 2D plasmonic structure using the FDTD method, together with the coupled mode theory.[48]

Consider firstly a symmetrical resonant CDF system with an air circular disk resonator in a plasmon–polariton metal, where the waveguide and the resonator are introduced by removing the metal with a specific shape. The corresponding structure is shown in Fig. 7.18a, where d, L, and R are the waveguide width, distance between the boundaries of the air disk and the waveguide, and the radius of the air disk resonator, respectively. The radius of the air disk is chosen as 1 μm, and the dielectric function of the metal (silver) is described by the lossy Drude model (Eq. 7.1.) The waveguide width, chosen as 60 nm, is much smaller than

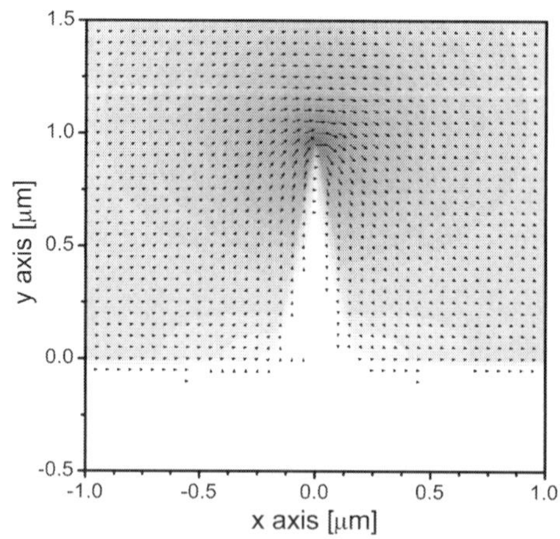

Fig. 7.17 $|\boldsymbol{H}_t|$ (colormap) and $\boldsymbol{H}_t$ (quiver) of the mode guided by a realistic Λ-wedge waveguide at $\lambda = 2$ μm. Axis unit: μm. See also Color Insert.

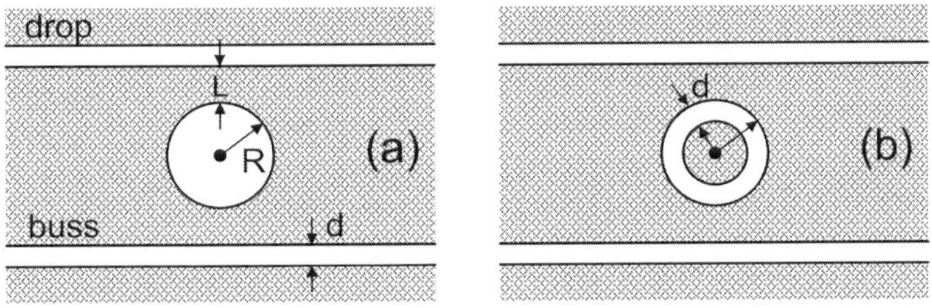

Fig. 7.18 Channel drop filter (CDF) structures realized in a plasmon–polaritons metal with (a) an air disk resonator; (b) a ring resonator.

the wavelength so that only the fundamental transverse magnetic (TM) waveguide mode is excited.

The coupled mode theory is used to analyze theoretically the interaction of a cavity resonator with a waveguide system.[49] According to the coupled mode theory, we can easily obtain the filter response of the symmetrical system,[49]

$$|R|^2 = \left| \frac{\sqrt{1/(\tau_{e1}\tau_{e2})}}{j(\omega - \omega_0) + 1/\tau_o + 2/\tau_e} \right|^2 \tag{7.4}$$

$$|T|^2 = \left| 1 - \frac{1/\tau_{e1}}{j(\omega - \omega_0) + 1/\tau_o + 2/\tau_e} \right|^2 \tag{7.5}$$

$$|D_L|^2 = \left| \frac{\sqrt{1/(\tau_{e1}\tau_{e4})}}{j(\omega - \omega_0) + 1/\tau_o + 2/\tau_e} \right|^2 \tag{7.6}$$

$$|D_R|^2 = \left| \frac{\sqrt{1/(\tau_{e1}\tau_{e3})}}{j(\omega - \omega_0) + 1/\tau_o + 2/\tau_e} \right|^2 \tag{7.7}$$

where ω_0 is the resonant frequency, $1/\tau_o$ and $1/\tau_e$ are the decay rate due to loss and the rate of decay into the bus/drop waveguide, respectively, R is the reflection from the input port, T is the transmission through the bus, and D_L and D_R represent the transmission into the left and right ports of the receiver, respectively. $1/\tau_{e1,3}$ and $1/\tau_{e2,4}$, defined as the decay rates in the forward and backward direction, satisfy

$$\frac{1}{\tau_{e1}} + \frac{1}{\tau_{e2}} = \frac{1}{\tau_{e3}} + \frac{1}{\tau_{e4}} = \frac{2}{\tau_e} \tag{7.8}$$

For a traveling-wave resonant mode, the power flows continuously in only one direction in the resonator. It can be easily obtained that the incident power in the bus in the forward direction is partially transferred to the receiver in the backward direction, limited only by the loss for the unloaded resonator, i.e., for $1/\tau_0 = 0$ the channel with resonant frequency ω_0, can be transferred completely into the drop waveguide. For a pure standing-wave resonant mode, the resonant mode decays equally into the forward and backward propagating waveguide mode. From Eqs. 7.4–7.7, one knows that the maximum drop efficiency is

$$|D_L|^2 = |D_R|^2 = 0.25\left(1 - \frac{\tau_e}{\tau_o}\right)$$

i.e., half the input power at resonant frequency can be dropped into the drop waveguide if the loss for the unloaded resonator is ignored.

Look back to the channel drop filter system, as shown in Fig. 7.18a. Not only surface traveling resonant modes but also standing wave modes in the air disk resonator will be excited. Numerical calculations show that the intrinsic quality factor Q_i for the unload resonator is quite large and reaches 10^{10} if we ignore the loss of the metal, and the quality factor ($Q_{\|}$) for the filter system is about 600 with the resonant frequency $f = 378.3\,\text{THz}$ when the distance between the waveguide and resonator is $L = 25\,\text{nm}$. Basically, $Q_{\|}$ can increase if we enlarge L. Based on the coupled mode analysis above, almost 100% power can be dropped by the surface-traveling mode, while the drop efficiency at one output based on standing waves is quite small, at most 25%. The spectral response after 500, 000 time steps, for $L = 25\,\text{nm}$, is shown in Fig. 7.19a, where the solid and dashed lines represent the transmissions at the forward bus and backward drop waveguide, respectively. It can be seen from Fig. 7.19a that the light can be efficiently dropped for some frequencies. For some resonant frequencies, the drop efficiency is quite high, close to 100%, due to the surface traveling mode. However, the drop efficiency, governed by

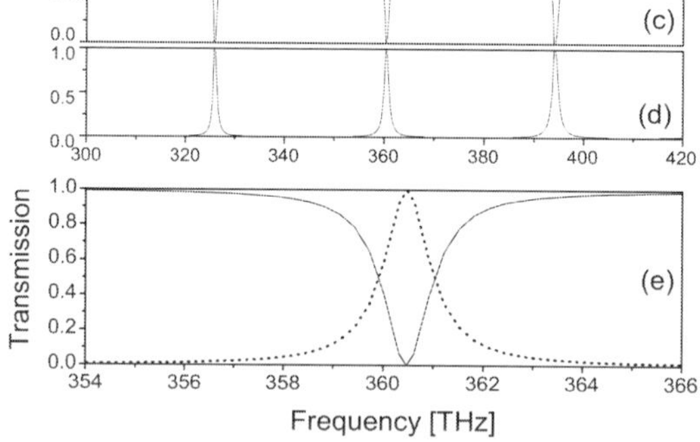

Fig. 7.19 Spectral response of the channel drop filter (CDF) at the forward bus (a,c) and backward drop waveguide (b,d) with (a) the air disk cavity; (b) the ring resonator cavity. The solid/dotted lines in (d) represent the spectral responses obtained from the coupled mode theory and results from (c) and (d) in the vicinity of 360.3 THz.

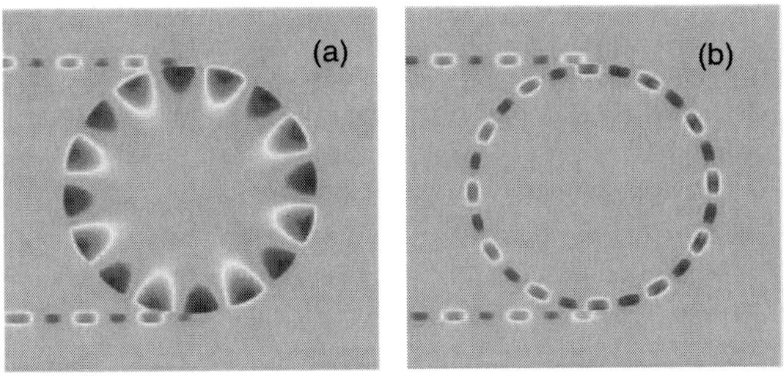

Fig. 7.20 Stead-state H_y field oscillation for a resonant frequency of (a) $f = 421.2\,\mathrm{THz}$; (b) $f = 427.8\,\mathrm{THz}$. Corresponding structures are shown in Fig. 7.20. See also Color Insert.

the standing wave, is below 25%, which is in agreement with results from coupled mode theory. Figure 7.20a shows the oscillation of the steady-state field distribution at a resonant frequency with $f = 421.2\,\mathrm{THz}$, where almost all energy is dropped by a surface traveling resonant wave.

One important factor characterizing a CDF is its channel isolation (CI), defined as

$$\mathrm{CI} = 10 \log \left(\frac{P_1}{P_2} \right)$$

where P_1 and P_2 are the power of the selected channel transferred to the drop waveguide and the power remaining in the bus, respectively. The CI should be as large as possible to avoid the cross talk. For the resonator with a surface traveling mode mentioned above, CI is quite large and theoretically infinite if ignored the loss of the metal. However, CI for a single-mode standing wave resonator is quite small. It is possible to get a high CI for a single-mode traveling-wave resonant filter using two standing-wave modes. Here to avoid exciting the standing-wave modes in the resonator, let us investigate a CDF with a ring resonator, as shown in Fig. 7.18b. The width of the ring waveguide is the same as the bus/drop waveguide and the outer radius of the ring resonator is also chosen as 1 μm. the mechanism in this system is quite similar to a conventional ring resonator. Coupled mode theory tells us that a single-mode traveling wave resonator side-coupled to the bus and the receiver can fully transfer a channel at the resonant frequency from the bus to the receiver, only limited by $Q_{\|}/Q_i$.

The spectral response at both the forward bus (solid line) and backward drop waveguide (dashed line) is shown in Fig. 7.20b when $L = 25\,\mathrm{nm}$. It can be seen from Fig. 7.19b that light can be almost completely dropped for resonant frequencies. The results for the resonant frequency $f = 360.3\,\mathrm{THz}$, obtained by coupled mode theory, are shown in Fig. 7.20c by circled and asterisk lines. It can be seen from Fig. 7.20c that the result obtained from coupled mode theory are consistent with the results obtained by FDTD method. Combined with the resonant conditions

$$N \times 2\pi = 2\pi R n_{\mathrm{eff}} k_0$$

the effective index (n_{eff}) for the ring waveguide with the width of 60 nm is about 1.339 for the resonant frequency 427.8 THz, where N is also obtained from the FDTD results. However, the effective index for the straight waveguide with the same width is 1.501 for the same resonant frequency. There is still no good analytical approximate method to calculate the effective index of the ring waveguide when the working wavelength is close to the radius of the ring. Figure 7.21b shows the oscillation of the steady-state field distribution at a resonant frequency of $f = 427.8$ THz, where almost 100% of the field energy is transferred to the backward drop waveguide. However, the quality factor for the filter is quite slow, only about 250 for the resonant frequency $f = 427.8$ THz. Since the intrinsic quality factor for the unload resonator is much high, the quality factor of the CDF can be increased if we enlarge the distance between waveguide and ring resonator, which also keeps a high drop efficiency. $Q_{\|}$ reaches 4×10^3 when $L = 50$ nm.

A metal ($\epsilon = \epsilon' + i\epsilon''$) is always a loss material especially in the visible and infrared frequency region. Here, we consider a CDF with a ring resonator in a lossy plasmon–polaritons metal. If we take into account the loss of the metal, the quality factor for the unloaded resonant cavity decreases significantly. Q_i becomes about 60 for the same ring resonator cavity previously studied. To increase the drop efficiency, we shorten the distance between the waveguide and the cavity in order to enlarge the coupling. We set the source and two detectors with a distance of 0.5 μm with the center of the ring resonator in the propagation direction. For the channel drop filter system with $L = 10$ nm, the quality factor of the CDF is only 30, which leads to weak drop efficiency. The spectral response at both the forward bus (solid line) and backward drop waveguide (dashed line) are shown in Fig. 7.21. It can be seen from Fig. 7.21 that the drop efficiency is quite slow, as well as the quality

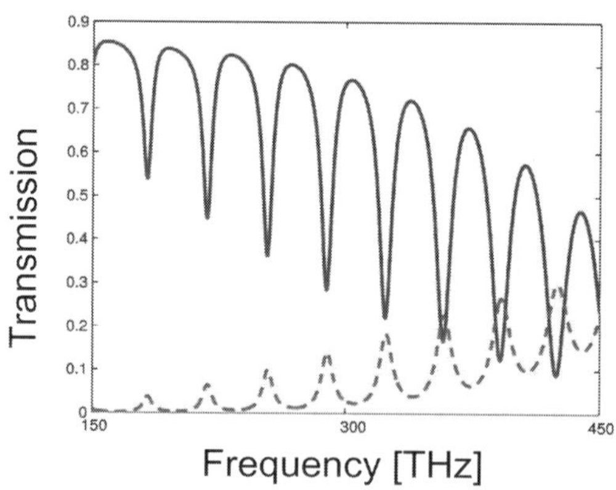

Fig. 7.21 Spectral responses of the channel drop filter (CDF) at the forward bus (solid line) and backward drop waveguide (dashed line) in a loss metal. See also Color Insert.

factor. The drop efficiency becomes larger with increasing resonant frequency. Since the quality factor for the filter becomes smaller with increasing resonant frequency, based on the coupled mode theory, the drop efficiency will naturally increase.

7.1.10 *Broadband Plasmonic Coupler*

One fundamental issue of plasmonics is how to efficiently couple light from conventional optical waveguides (e.g., fibers, or silicon wire waveguides) into plasmonic waveguides. There are two general schemes that are used to couple light into plasmonic structures.[50] The most widely used one is through momentum matching. The external light usually has largely different momentum than those of plasmonic waves. An efficient free-space coupling method must provide the momentum difference. The matching can be obtained by a number of methods, such as prism coupling,[51, 52] grating coupling,[53] and nanoparticle scattering.[54] Those methods always require large-scale (lager than the wavelength) devices, such as the grating coupler with more than 10 μm dimensions, and the transmission bandwidth is usually limited.

The other scheme for coupling is through field matching across the interface, i.e., end-fire coupling.[55, 56] The method could be highly efficient and broadband. A modulated-index metal-slot waveguide transfer with lateral mode confinement and a squeezing light coupler with vertical mode confinement have been investigated theoretically.[57, 58] The calculated coupling efficiency can be as high as 95%. However, only a few experimental demonstrations of light coupling from a dielectric waveguide into plasmonic tapers have been shown,[59] mostly due to the fact that the footprint of a plasmonic waveguide is much smaller than that of a dielectric waveguide; thus, efficient coupling is difficult to achieve.

In 2009, Tian *et al.* experimentally demonstrate a planar plasmonic coupler coupling light between a silicon wire waveguide and a metal-dielectric-metal plasmonic gap waveguide.[60] The experiment results give a broadband coupling efficiency as high as 35%. The schematic picture of the plasmonic coupler structure is shown in Fig. 7.22a. The whole structure is based on the silicon-on-insulator platform. The coupler is composed of two tapered silicon strip waveguides and a plasmonic gap waveguide that extends symmetrical bevel edge that are parallel with the silicon taper edges. The strip silicon waveguide has a 450 nm-wide by a 250 nm-high rectangular cross-section, which are typical parameters for single-mode silicon wire waveguides. The thickness of the metal material slabs for forming the plasmonic gap waveguide is 250 nm, the same as that of the silicon tapered waveguide. Figure 7.22(b) indicates the taper coupler parameters from the top view. The light mode coupling between the silicon taper and the gold film and the light propagation in the plasmonic gap are shown in Fig. 7.22c. Fig. 7.22d is the optical microscope image of the structure. A scanning electron micrograph of a gold plasmonic coupler is shown in Fig. 7.22e.

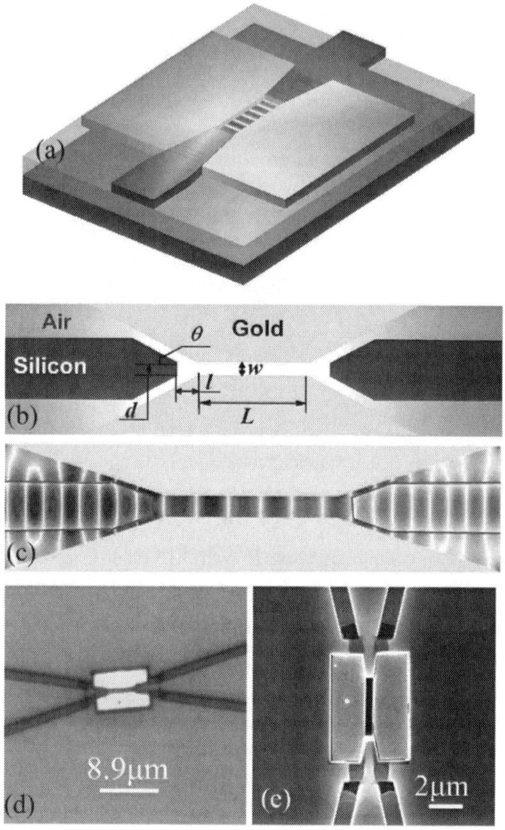

Fig. 7.22 Two tapered silicon strip waveguides coupled with a plasmonic gap waveguide. (a) The schematic picture of the silicon-metal plasmonic coupler. (b) The top-view with the taper parameters: θ is the angle between the taper and the center axis of silicon waveguide (the plasmonic taper edge is set to be parallel to the silicon taper), d is the width of the silicon taper end, l is the distance between the end of silicon waveguide and the start of the plasmonic waveguide. w and L are the width and length of the waveguide gap, respectively. (c) The H_z field distribution at the wavelength 1.5 μm simulated by 3D FDTD. (d) Optical microscope image of a fabricated silicon-gold plasmonic coupler ($\theta = 10°$, $d = 200$ nm, $l = 0.25$ μm, $w = 200$ nm, $L = 3$ μm). (d) The scanning electronic micrograph of the fabricated taper. J. Tian, S. Q. Yu, W. Yan, and M. Qiu, Broadband high-efficiency surface-plasmon-polariton coupler with silicon-metal interface, Appl. Phys. Lett. vol. 95, p. 013504, 2009. Copyright (2009) by The American Institute of Physics. See also Color Insert.

Figure 7.23 shows the normalized transmission before and after the formation of the gold plasmonic gap waveguide. The gold waveguide is $L = 5\,\mu\text{m}$ long. The transmission spectrum is normalized by a reference single-mode silicon waveguide with $450 \times 250\,\text{nm}^2$ core size. From Fig. 7.23, it can be seen that the average

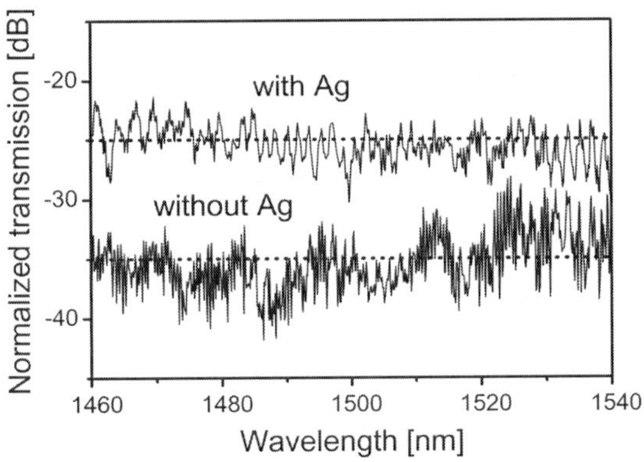

Fig. 7.23 Normalized transmission spectrum before and after the formation of the gold plasmonic gap waveguide.

transmission power before and after the gold formation is −35 and −25 dB, respectively. The power enhancement is proximate 10 dB by comparing two reference lines. To determine the coupling efficiency and the propagation loss, gold waveguides with different length L but with the same tapered coupler structure were fabricated and measured. Figure 7.24a plots the measured transmission for various waveguide lengths. From the plot, one can estimate the values of the propagation loss and coupling efficiency, and they are 2.9 dB/μm and 5.5 dB (28%)/facet at wavelength around 1.5 μm, respectively. Figure 7.24b shows the measured coupling efficiency per facet as the function of the wavelength. From Fig. 7.24b, one can find out that the average coupling efficiency is about 4.5 dB (35%). The simulation result

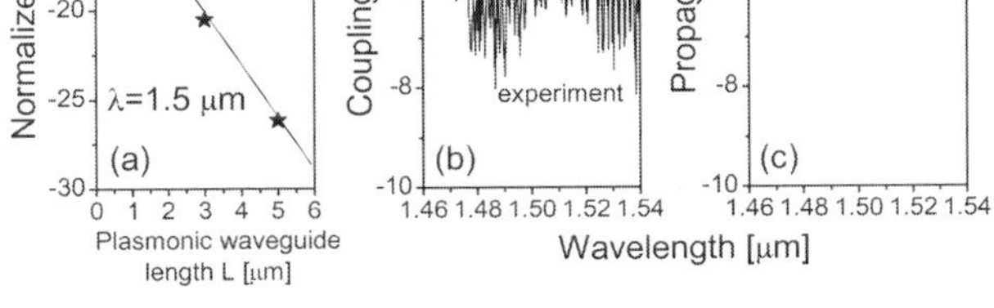

Fig. 7.24 The theoretical and experiment results of the coupling efficiency and the propagation loss. (a) Normalized transmission versus the length of the slot waveguide at wavelength 1.5 μm. (b) Coupling efficiency per facet as the function of the wavelength. (c) Propagation loss in the slot waveguide as the function of the wavelength.

of the structure is added for comparison. The discrepancy between the simulation and experimental results in terms of efficiency is caused by the gold pattern defects and roughness and also due to the fact that the gold waveguide is not perfectly aligned to the silicon taper. Figure 7.24c shows the propagation loss in the gap waveguide versus the wavelength. The average loss is about $-2.5\,\text{dB}/\mu\text{m}$, compared with the simulation result of $-1.5\,\text{dB}/\mu\text{m}$, demonstrating a relatively good agreement between theory and reality. This silicon-metal coupling structure enables the potential application in the hybrid integration of semiconductor devices and plasmonics.

7.2 DEEP-SUBWAVELENGTH PLASMONIC WAVEGUIDE

7.2.1 *General Consideration*

Owing to the fact that SPP does not experience a diffraction limit, plasmonic waveguides are expected by many to be promising for sub-wavelength waveguiding, and in turn for dense optical integration. However, it should be noticed that not all SPP waveguides reported so far have subwavelength MFS. For example, most long-range SPP waveguides (e.g., Refs.[61,62]) achieve relatively low propagation loss because their modal energy is largely distributed in the low-loss dielectric material. Such waveguides usually have a very large MFS. In order to deploy SPP waveguides as integrated optical circuits, one has to design true subwavelength waveguides. To date, there are several types of SPP waveguides known for achieving the goal. The first type is derived from metal-insulator-metal (MIM) structure. In 1D slab limit, two vertically placed parallel metal plates can be used to confine light in the middle dielectric layer. The even TM gap mode is free of geometrical cutoff and its effective mode index (n_{eff}) diverges as the gap distance shrinks.[63] A horizontal slice of the tri-layered structure can, in principle, squeeze light into subwavelength scale in 2D.[36,64,65] Another known method for realizing subwavelength SPP waveguiding is to use a 2D metal corner, be it either a V channel or a Λ wedge.[26,39,66] It has been shown that the smallest MFS of most practical corner waveguides made of Ag and air is limited to ~100 nm at 633 nm wavelength.[26] For an even smaller MFS, a corner angle at less than 5° and (or) a very sharp corner with a less-than-5 nm tip curvature are necessary, which would impose great challenge on our current nanofabrication technology. Note that the subwavelength field confinement capability of metal-corner waveguides, especially the Λ-wedge waveguide, roots in the tiny surface curvature at their corner tips. From this viewpoint, such Λ-wedge waveguides can be regarded as derivatives from the metallic fiber-shaped waveguides. The singular behavior of the SPP mode guided by a metallic fiber as its radius decreases has been reported in Ref.[67]

The aforementioned subwavelength SPP waveguides are all based on geometrical tailoring. In fact, the peculiar guidance principle of SPP waveguides allows their mode field to be confined in a subwavelength fashion without resorting to geometrical tailoring. For a 1D metal–dielectric interface, the guided SPP mode has its n_{eff} value defined as

$$n_{\text{eff}} = \sqrt{\frac{\epsilon_+ \epsilon_-}{\epsilon_+ + \epsilon_-}} \tag{7.9}$$

where ϵ_+ and ϵ_- ($|\epsilon_-| > \epsilon_+$) are permittivities of the dielectric material and metal, respectively. The materials are assumed to be nonpermeable. From Eq. 7.9, it is noticed that n_{eff} can be arbitrarily large, depending on how close $(\epsilon_+ + \epsilon_-)$ is to zero. It follows that the transverse field decay constant in the cladding

$$k_t = k_0 \sqrt{n_{\text{eff}}^2 - \epsilon_{\text{clad}}}$$

(ϵ_{clad} is either ϵ_+ or ϵ_-) can also be made arbitrarily large. This gives rise to the possibility of tightly confining the EM field at the interface. A section of the interface can potentially confine light in nanodimensions in 2D. Such a subwavelength waveguide has the obvious advantage of being structurally very simple. A primary reason for the lack of proper study on such waveguide probably is that, in addition to the divergent propagation constant, the propagation loss will also tend to infinity as the operation is near to the $\epsilon_+ = -\epsilon_-$ resonance condition. In fact, the trade-off between confinement and loss for SPP waveguides has been observed for a wide variety of waveguiding structures (e.g., Refs.[26, 34]). In view of many published results on SPP waveguides, it has generally been accepted that some loss reduction technique (rather than merely geometric optimization) has to be deployed in order to make functioning integrated optical circuits based on SPP.

We argue that the near-resonant SPP waveguides deserve as much attention as other types of SPP waveguides in realizing subwavelength light channeling.[68] From the perspective of integrated photonic circuit, here we specifically look into a realistic waveguide design based on a finite section of near-resonant silver–silicon interface. In particular, we show that the material loss for both silver and silicon has to be reduced by ~1000 times in order for the waveguide to achieve practical propagation length. The urgency of material loss reduction is also evident for geometry-tailored SPP waveguides in many published works. Decreasing environment temperature[69] and using QD-based metamaterials[70] could be two viable ways to achieving the goal.

7.2.2 *Realistic Design and Simulation*

Based on the above discussion, we pay particular attention to a realistic SPP waveguide design that relies on a section of near-resonant metal-dielectric interface for achieving tight field confinement. We purposely choose $|\epsilon_+| \approx |\epsilon_-|$, i.e., the near-resonant material parameters, in order to realize deep subwavelength MFS. Two materials with close but opposite permittivity values (in their real part ϵ') at certain wavelengths do exist in nature, but not without loss. One example

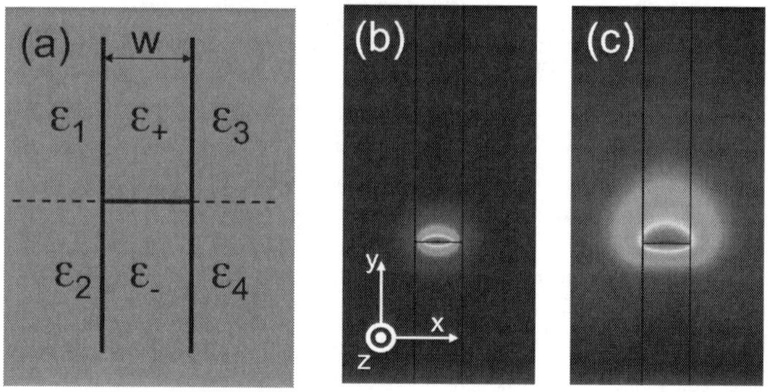

Fig. 7.25 (a) Schematic diagram of a general near-resonant SPP waveguide. (b) Mode field (H_x) supported by a sample near-resonant SPP waveguide ($\epsilon_+ = 2.1$, $\epsilon_- = -2.3$, $\epsilon_{1,2,3,4} = 1$, $w = 50\,\text{nm}$, $\lambda = 600\,\text{nm}$). (c) Field in the same waveguide but far from the singular condition ($\epsilon_- = -15$). See also Color Insert.

is silver (Ag) and silicon (Si). An examination of their dispersion curves tells that their permittivity values meet our requirement around the free-space wavelength of 600 nm, at which $\epsilon_{\text{Ag}} = -16.08 + 0.4434i$ and $\epsilon_{\text{Si}} = 15.58 + 0.2004i$.[44] The imaginary part of the permittivity values (denoted as ϵ'') is directly responsible for attenuation of the guided surface mode. A single surface mode formed by the two materials at $\lambda = 600\,\text{nm}$ has a loss value as large as $690.7\,\text{dB}/\mu\text{m}$, rendering almost any waveguide built upon such a surface impractical. One of our objectives is to investigate how small ϵ'' of Ag and Si should be for practical applications.

A general waveguide cross-section is shown in Fig. 7.25a. Composition choices of the left and the right cladding regions are many. In the simplest case, the two claddings can be the same homogeneous material ($\epsilon_1 = \epsilon_2 = \epsilon_3 = \epsilon_4$). Among other variations, the whole structure can share the same metal substrate ($\epsilon_2 = \epsilon_4 = \epsilon_-$), or the same dielectric superstrate ($\epsilon_1 = \epsilon_3 = \epsilon_+$); each cladding can also have a different metal–dielectric interface. However, the underlying principle for achieving subwavelength light guiding is the same, i.e., relying upon the near-resonant metal-dielectric interface in the center. As the waveguiding interface has finite lateral size (w), the structure can be used to achieve high-density photonic integration in 2D. In Fig. 7.25b, c, two similar SPP waveguides working under two different conditions are compared. Advantage of near-resonant operation is obvious in terms of confinement. Several SPP waveguides that are similar to that sketched in Fig. 7.25a have been reported (e.g., Refs.[69, 71]). However, few of the studies have paid particular attention to achieving subwavelength guidance. Here, by starting from naturally available materials, we analyze the feasibility and challenges of such a nanoscaled near-resonant SPP waveguide. In the following, without loss of generality we take a representative configuration in which $\epsilon_2 = \epsilon_4 = \epsilon_-$ (i.e., common metal substrate). In addition, we assume the cladding materials in the upper half

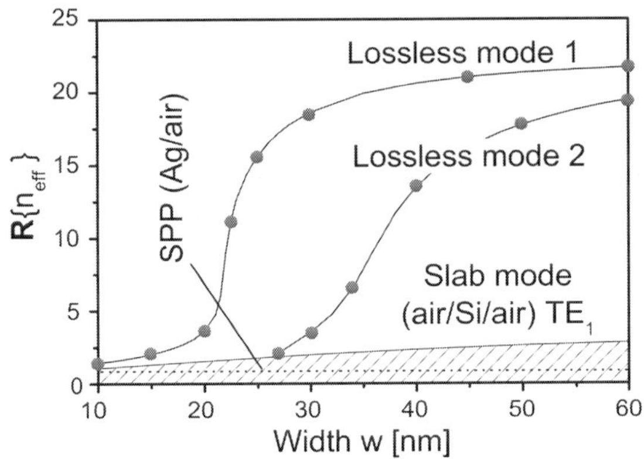

Fig. 7.26 Geometric dispersions of first two modes of the waveguide with respect to w. Loss is assumed to be zero. Red dots: the n_{eff} values when ϵ'' values of both Ag and Si are reduced to their 1%.

plane (ϵ_1 and ϵ_3) as air. Note that the near-resonant operation ensures a tight field confinement even when ϵ_1 and ϵ_3 are larger than ϵ_+ (see e.g., Ref.[69]).

First, to make sure the waveguide is single mode, we analyze the geometric dispersion as a function of the core width w at $\lambda = 600\,\text{nm}$ (Fig. 7.26). Mode derivation is done in COMSOL with an electric-field- and edge-element-based FEM. The solid curves (with dots) are the first two modes derived with $\epsilon'' = 0$ for both Si and Ag materials. The red dots are calculated with ϵ'' values reduced to 1% (compared to their natural values). The mode index changes little when the ϵ'' values change from 0 to 1%. We will see later that when losses are higher, the waveguide is too lossy to be useful. From Fig. 7.26, it is seen that the waveguide is single mode when $w < 27\,\text{nm}$. We hence take $w = 25\,\text{nm}$ in our following analyses. The n_{eff} value is about 15.6 at $w = 25\,\text{nm}$, which ensures the mode field is highly evanescent in the cladding regions.

Mode supported by the waveguide is depicted in Fig. 7.27. The field does not change appreciably when the material losses vary from 0 to 0.1. In the cladding regions, the mode field decreases to its $1/e$ over a distance of approximately 6 nm. Therefore, its MFS is approximated to be $37 \times 12\,\text{nm}^2$. The mode field has a major polarization along the y direction. The z-component of the Poynting vector (S_z) shown in Fig. 7.27c confirms the highly confined energy flow in the waveguide. Note that, although S_z in Ag region is negative, the net energy flow is positive.

We now analyze the loss of the waveguide with $w = 25\,\text{nm}$ as ϵ'' values of both Ag and Si are varied, which is shown in a contour map in Fig. 7.28. ϵ'' values of both materials are increased from 10^{-6} to 10^{-1}, in fractions of their natural values. It is observed that the waveguide loss is almost equally sensitive to variations in each of the two ϵ'' values. In practice, the requirement of propagation length depends on the application. Here, given such a tiny circuit cross-section, a loss level of 1 dB/μm

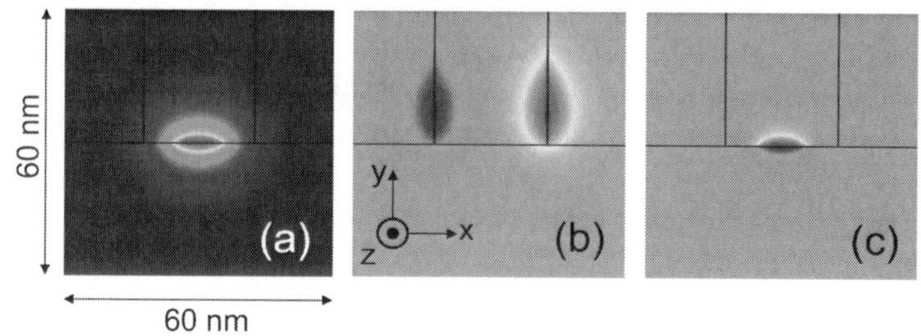

Fig. 7.27 Field plots of the SPP-based waveguide with a 25 nm-sized core. Three panels share the same color scale. (a) H_x field (min:0, max:1.27); (b) H_y field (min:-5.02e-2, max:5.02e-2); and (c) z-component Poynting vector S_z (min:-6.0e2, max:6.2e2). Axis unit: nm. See also Color Insert.

(corresponding to a propagation length of a few micrometers) could be suitable for a wide range of purposes. A circuit with over 100 length-to-cross section aspect ratio permits necessary waveguide bends for forming basic components (coupler, interferometer, etc.) and inter-connecting various ports in a high-density fashion. Fig. 7.28 shows it is shown that both ϵ'' values (or equivalently, conductivities of the two materials, see sub-section 6.1.2) have to be decreased by $\sim$1000 times in order to reach 1 dB/μm propagation loss. It should be noted that keeping the desired negative ϵ' and decreasing ϵ'' will, as dictated by the Kramers–Krönig relations, require either other (meta)materials than the materials employed here, or possibly low-temperature operation.

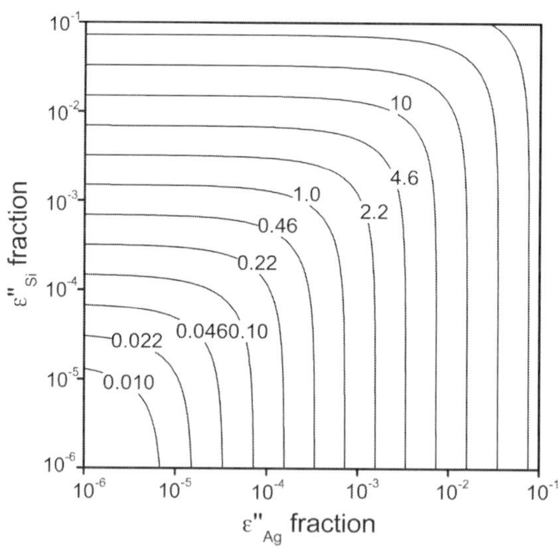

Fig. 7.28 Contour plot of the loss values in dB/μm when the ϵ'' values of both Ag and Si are varied in fractions of their natural values at room temperature.

Table 7.1 Group velocity (GV) and group velocity dispersion (GVD).

λ [nm]	600	610	620	630
GV [c]	5.7712×10^{-4}	0.0090	0.0285	0.0560
GVD [ps/nm/mm]	−459.2	−67.20	−7.632	−4.969

It is understood that near-resonant 1D SPP exhibits an interesting property of slow group velocity (GV).[69] However, explicit studies on GV and also the group velocity dispersion (or GVD, which is responsible for pulse broadening in digital communication links) for 2D SPP waveguides are often ignored in most published works. GV and GVD of the particular waveguide depicted in Fig. 7.27 are shown in Table 7.1. Frequency-dependent ϵ_{Ag} and ϵ_{Si} values are taken from Refs.[72] and,[44] respectively, except that the imaginary parts are kept at their 1%. The result shows that the GV in this particular waveguide can be slowed down by over 1700 times at the near-resonant operation condition. The huge negative GVD value (notice the propagation length unit is in mm) at 600 nm suggests that such waveguide may be promising for dispersion compensation applications. Further GV and GVD tailoring can be realized by using a multilayer dielectric material, in replacement of the homogeneous region denoted by ϵ_+ in Fig. 7.25.[69]

7.3 HYBRID PLASMONIC AND INDEX-GUIDING WAVEGUIDE

So far, most SPP waveguides proposed have their propagation lengths limited to only several micrometers once light is squeezed beyond the diffraction limit.[26, 35, 73] Such a high loss value ultimately prevents the waveguides from being deployed as integrated photonic circuits. In this section, we look into a type of waveguides that guide light using both the dielectric index guiding mechanism and the SPP effect. Such hybrid waveguides should inherit both the low-loss feature of all-dielectric waveguides as well as the superior mode confinement possessed by SPP waveguides. Compared to SPP waveguides, the proposed waveguides are expected to be able to achieve a comparable mode field confinement and lower propagation loss.

The waveguide to be examined, in a 1D slab limit, has a high-index dielectric core layer surrounded by low-index dielectric cladding layers, which is then further jacketed by metal claddings (see Fig. 7.29a). When the distance between two metals (or metal gap size) is large enough, both TE and TM modes can be guided by the central high-index dielectric core. As the gap distance gets smaller, the effect of metal claddings becomes more evident. In fact, if the gap distance is close to the

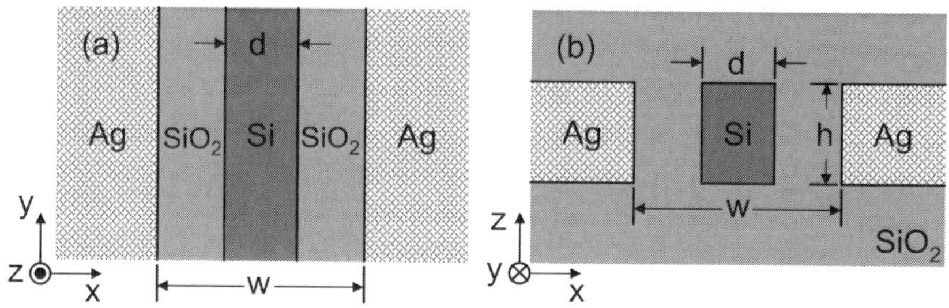

Fig. 7.29 Schematic diagrams of the hybrid optical waveguide cross-sections. (a) 1D; (b) 2D.

wavelength, the dielectric materials between metal claddings can be treated as an effective dielectric medium. Among the two modes, the TE mode will be cut off if the metal gap is as short as half the light wavelength (measured in the effective dielectric medium), while the TM mode is an SPP in nature and does not suffer from a diffraction limit. In practice, the waveguide in Fig. 7.29a should be fabricated with a finite height. Figure 7.29b depicts the schematic cross-section of a corresponding planar 2D waveguides. Modes guided by the waveguide shown in Fig. 7.29b are all hybrid. The first two modes have their major electric field polarized along either x or y direction. These two modes, therefore, can be referred to as quasi-TE or quasi-TM mode. Both index-guiding and SPP-guiding mechanisms, in general, contribute to the mode field confinement in such a waveguide. Without the central high-index core, the waveguides shown in Fig. 7.29a, b correspond to the previously reported MIM[74] and metal gap waveguides,[34–37] respectively. We will see later that introduction of a high-index inner core will reduce metal absorption loss considerably for the TM mode (or quasi-TM mode for a 2D waveguide). At the same time the TE mode (or quasi-TE mode) can be saved from cutoff, due to an increase in the effective index of the dielectric filling between two metal claddings. The optimum inner core thickness will be determined for achieving relatively low-loss quasi-TE or quasi-TM operation around 1550 nm wavelength, given a fixed metal cladding separation.

7.3.1 *1D Analysis*

We choose commonly available materials, namely silver (Ag), silica (SiO_2), and silicon (Si), for the current theoretical study. Throughout the section, we have explicitly considered the material dispersion. The dielectric constant of silver is described in Drude model by Eq. 7.1. The dielectric constant of silica is assumed to adhere to the well-recognized Sellmeier equation.[75] And similarly, a Sellmeier equation for silicon[44] is adopted. The operating wavelength is chosen to be within the range 1.2–2.0 μm, in which both silica and silicon are transparent.

We now employ a transfer matrix method (TMM) for deriving modes propagating in the 1D waveguide. A time dependence of $e^{i\omega t}$ is assumed throughout this study for harmonic EM field. The guided modes in a 1D waveguide can be separated into TE and TM two groups. We describe briefly the TMM method for deriving TM modes. TE probelms can be handled accordingly. As a mode field in any single layer fulfils the wave equation in homogeneous material, its general solution, for the tangential field H_y, can be written as

$$H_{yi} = A\,e^{ik_{xi}x} + B\,e^{-ik_{xi}x} \tag{7.10}$$

where

$$k_{xi}^2 = k_0^2 \mu_{ri}\epsilon_{ri} - \beta^2$$

Care should be taken for choosing the sign of the square root when deducing k_{xi}. k_0 is the wave vector. β is the propagation constant, which is in general complex. The effective mode index is defined as $n_{\text{eff}} = \beta/k_0$. In Eq. 7.10, the first term denotes the left-propagating wave, and the second denotes the right-propagating wave. Another tangential field E_z can be derived as

$$E_{zi} = A\frac{k_{xi}}{\omega\epsilon_i}\,e^{ik_{xi}x} - B\frac{k_{xi}}{\omega\epsilon_i}\,e^{-ik_{xi}x} \tag{7.11}$$

Equations 7.10 and 7.11 can be written compactly as

$$\begin{bmatrix} H_{yi} \\ E_{zi} \end{bmatrix} = \mathbf{M}_i \begin{bmatrix} A \\ B \end{bmatrix} \tag{7.12}$$

As tangential fields are continuous across an material interface, by setting $x = x_i$ with x_i being the ith interface position, we can relate fields in two adjacent layers by

$$\begin{bmatrix} H_{yi} \\ E_{zi} \end{bmatrix} = \mathbf{M}_{xi} \begin{bmatrix} H_{y,i+1} \\ E_{z,i+1} \end{bmatrix} \tag{7.13}$$

where $\mathbf{M}_{xi} = \mathbf{M}_i^{-1}\mathbf{M}_{i+1}$ is the transfer matrix. In this manner, the field in the leftmost layer can be related to that in the rightmost layer. By considering that there is no right-propagating wave in the leftmost layer, and that there is no left-propagating wave in the rightmost layer, we can get a characteristic equation. β values can be determined by solving for the roots of the equation.

Figure 7.30a, b shows, respectively, the variations of the TE and TM modes in their n_{eff} and loss values as the Si layer thickness d is changed from 0 to 500 nm while the distance between two metals w is kept at 500 nm. Wavelength is 1550 nm. For the first TE mode, its propagation loss decreases sharply by more than one order and reaches its minimum at 25.7 dB/mm around d = 220 nm. The loss increases gently as d is further increased. In fact the TE_0 mode is close to cutoff at d = 0, and introduction of the Si layer pushes the cutoff to longer wavelength, which explains the initial sharp decrease of loss. When d reaches 275 nm, the second TE mode (TE_1) begins to be guided. We can easily avoid this TE_1 mode by letting

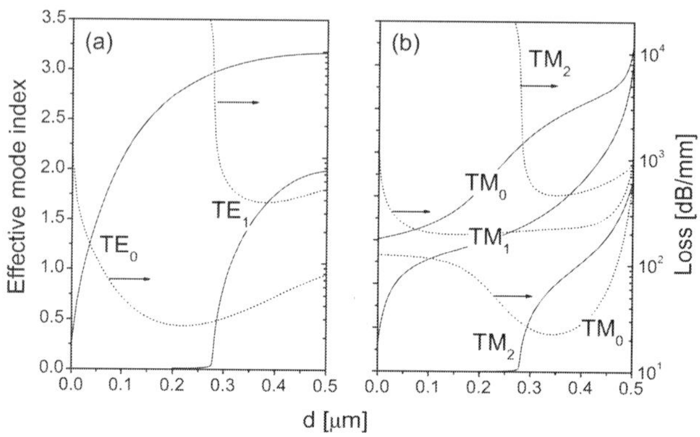

Fig. 7.30 The variations of n_{eff} (left vertical axis) and loss (right vertical axis) values for the TE (a) and TM (b) modes with respect to the Si layer thickness.

$d < 275$ nm. The TM modes in Fig. 7.30b in general exhibit the same trend. The first two modes, TM_0 and TM_1, are the Fano mode pair in a conventional MIM slab waveguide. Among them, the TM_0 mode does not experience cutoff even if w decreases to a near-zero value. Propagation loss of this fundamental mode decreases as d increases. The minimum loss of 22.6 dB/mm occurs when $d = 340$ nm. This drop in loss is attributed to the trapping of light in the Si layer, and therefore the interaction of mode with metal surfaces is reduced. Note that all-Si filling results in a higher loss than that induced by all-SiO_2 filling. The TM_1 mode is close to cutoff at $d = 0$, and it becomes better confined as the Si layer is introduced. However, the improvement is not as significant as that for the fundamental mode. The is due to the fact that there is a nodal line for the dominant field of the TM_1 mode, which effectively expels light out of the Si layer. The TM_2 mode appears at $d = 275$ nm, which experiences similar dispersion and loss behaviors as the TE_1 mode. The Fano mode pair is seen to always exist regardless of different d values. However, we will see in the next section that the second Fano mode can be stripped off if a proper substrate and superstrate are deployed for a realistic 2D waveguide. Here $w = 500$ nm has been chosen for this case analysis. Although not shown here, we point out that an extra Si layer only helps to reduce propagation loss when w is larger than ~250 nm. When $w = 200$ nm, no TE mode is supported, and loss of the first TM mode increases monotonically with d.

In Fig. 7.31a, we plot the fundamental TE mode at $d = 200$ nm, and in Fig. 7.31b the fundamental TM mode at $d = 340$ nm. It is evident that a major part of the light is trapped in the Si layer, which reduces the interaction of the modes with the metal claddings. The field penetration into metal claddings is limited to only tens of nanometers (so-called skin depth). Therefore, it is expected that such waveguides can be placed very close to each other without suffering from heavy cross talks, which is key for achieving high-density optical integration.

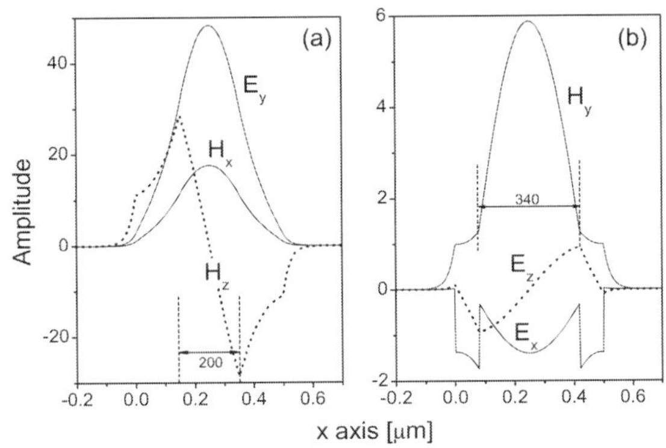

Fig. 7.31 Modes supported by 1D waveguides. (a) Field components of the TE_0 mode guided by a slab waveguide with $d = 200\,\text{nm}$. (b) Field components of the TM_0 mode guided in a slab waveguide with $d = 340\,\text{nm}$. In both cases, the z field component is the minor component.

7.3.2 *2D Analysis*

In 2D scenario, the waveguide in Fig. 7.29a will have a finite height h. A substrate and a superstrate will be present below and above the waveguide, respectively, as shown in Fig. 7.29b. The guided mode is affected by not only the height of the planar waveguide, but also the permittivities of substrate and superstrate. Most importantly, the high-order modes that appear in 1D waveguides as shown in Fig. 7.30 can be trimmed off for some adequate substrate or superstrate. For example, it is relatively simple to achieve low-loss single-mode quasi-TE operation at $\lambda = 1550\,\text{nm}$, since a substrate and a superstrate made of SiO_2 ($n \approx 1.444$) would be sufficient to get rid of the TE_1 mode at $d = 200\,\text{nm}$ (see Fig. 7.30a). For low-loss single-mode quasi-TM operation a substrate and a superstrate of index greater than $\sim$2.0 (but less than 2.75) can be used to get rid of the odd Fano mode (TM_1) while d can be kept around 340 nm (see Fig. 7.30b). In the following, we will introduce two separate waveguides that are, respectively, intended for low-loss quasi-TE and quasi-TM operations.

7.3.2.1 *Waveguide for quasi-TE operation*

For low-loss quasi-TE operation, we let both substrate and superstrate be silica. d is fixed at 200 nm. In 1D limit, the loss of the TE-mode will be as low as 26.1 dB/mm at 1550 nm wavelength (Fig. 7.30a). A further complication is on choosing an appropriate height h to prevent nodal line from appearing in the vertical direction. Although a vector FEM can be used to decide h for single-mode operation, it is found that the EIM is much more efficient way to achieve the objective.

In EIM, a 2D planar waveguide is simplified to a horizontally placed three-layer slab waveguide. For the particular waveguide shown in Fig. 7.29b, the upper and bottom layers of the corresponding simplified waveguide are simply SiO_2. The central layer has a refractive index that is the $n_{\rm eff}$ value of the mode guided by the waveguide shown in Fig. 7.29a. The EIM is noticed to have a reasonable accuracy, especially when wavelength is small. Figure 7.32a shows the geometric dispersion curves for the first four TE modes as h is varied at $\lambda = 1550\,\text{nm}$. Only the TE_0 mode is presented in Fig. 7.30a. It is seen that all high-order modes approach cutoff when h decreases. When h is less than ~400 nm, only one mode, i.e., TE_{00}, is supported, see Fig. 7.32a. Note that the first subscript in the mode designation corresponds to the number of nodal lines along the x direction, while the second subscript denotes the number of nodal lines along the y direction. Although the modes are all hybrid, for simplicity we still use the abbreviations "TE" and "TM."

In the same manner, we examine the geometric dispersions for the TM modes, as shown in Fig. 7.32b. And it is noticed that all high-order modes have been cut off for h <~500 nm. Again, only TM_0 mode in Fig. 7.30b is studied here. Although the second mode in Fig. 7.30b (TM_1) has a $n_{\rm eff} = 1.472768$, which is slightly larger than the refractive index of the substrate/superstrate (1.444024), the effective index of all modes in 1D case will be reduced as the height of the waveguide is reduced. It will be shown later that $TM_{1,n}$ modes will not be supported in our designed 2D waveguide by using an appropriate h. Here n is an integer.

Next, let us adopt the a vector FEM method to investigate explicitly a particular 2D planar waveguide that is optimized for low-loss quasi-TE operation. The advantage of FEM includes its accurate definition of material boundaries as well as its adaptive meshing of the numerical domain. The waveguide under study has a height of 450 nm. The dispersion and loss curves for the first four modes are plotted in Fig. 7.33a, 7.33b. It is seen that the high-order TM_{01} and TE_{01} modes approach their cutoffs at $\lambda \approx 1380$ nm and $\lambda \approx 1430$ nm, respectively. Hence the waveguide operates in single mode for each polarization at wavelength beyond ~1430 nm. At $\lambda = 1550\,\text{nm}$, the waveguide is highly birefringent, with birefringence value (defined

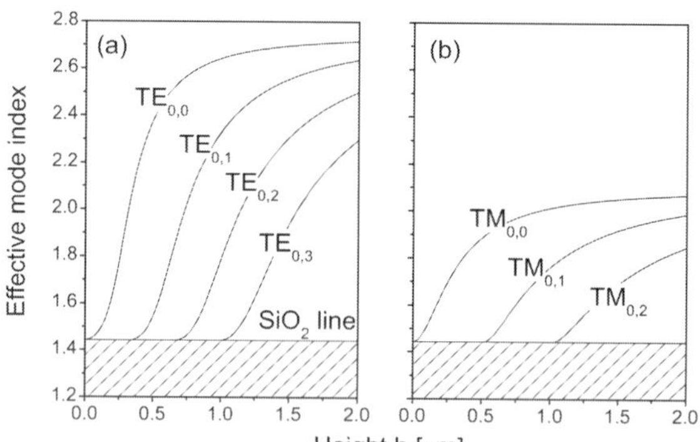

Fig. 7.32 Dispersion of the TE (a) and TM (b) modes as the waveguide height is changed. $d = 200\,\text{nm}$.

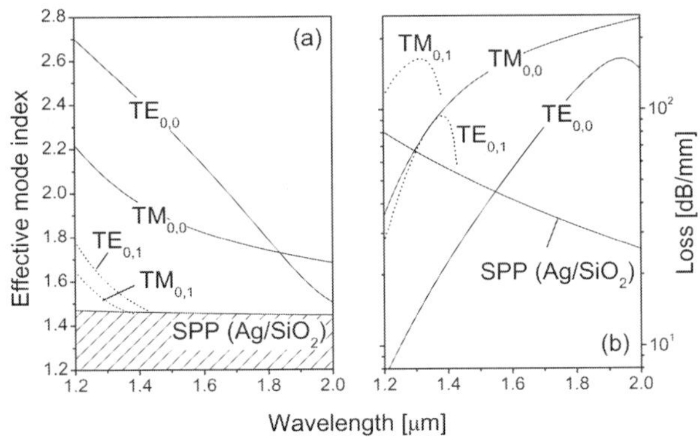

Fig. 7.33 Dispersion (a) and loss (b) values for the first four modes guided by a 2D waveguide with $d = 200\,\text{nm}$ and $h = 450\,\text{nm}$.

as $|n^x_{\text{eff}} - n^y_{\text{eff}}|$) as high as 0.3569. This indicates that the $TM_{0,0}$ and $TE_{0,0}$ modes, though both supported by the waveguide, will hardly interact with each other. A careful excitation by a linearly polarized light source will ensure only one of the modes will be propagating down the waveguide. Generally speaking, the propagation loss of the waveguide increases as h is decreased. The propagation losses of both $TE_{0,0}$ and $TM_{0,0}$ modes are in general higher as compared to the pure 1D modes. The propagation loss of the $TE_{0,0}$ mode at 1550 nm is 44.9 dB/mm, and that for $TM_{0,1}$-like mode it is 150.1 dB/mm. The loss of the TE_{01}-like mode is almost equal to that of the SPP mode supported by a single Ag–SiO_2 surface (Fig. 7.33b), which is *in theory* impossible to be achieved by the coupled SPPs in MIM waveguides.

To know how strong two waveguides placed in parallel interact with each other, one needs to examine the beat length L_p for the coupled supermodes, which is defined as

$$L_p = \frac{2\pi}{|\beta_e - \beta_o|}$$

where β_e and β_o are propagation constants of the even and odd supermodes, respectively. The longer the L_p, the weaker the two waveguides interact. In Fig. 7.34, we show the variation of beat length as a function of the waveguide separation S (defined as the distance between core center to core center). The same curve is also derived for an identical waveguide but without the metal claddings. Both curves are approximately linear with the beat length represented in logarithmic scale. It is observed that the beat length is increased 3.5 times when metal claddings are present. With the same cross-talk requirement, waveguides with metal claddings can be put 23% closer as compared to without metal cladding. According to the coupled-mode theory, field overlapping of two waveguide modes (examined for each individual waveguide in separation) has direct effect on the coupling strength between two waveguides. From the field distribution of the mode supported by the

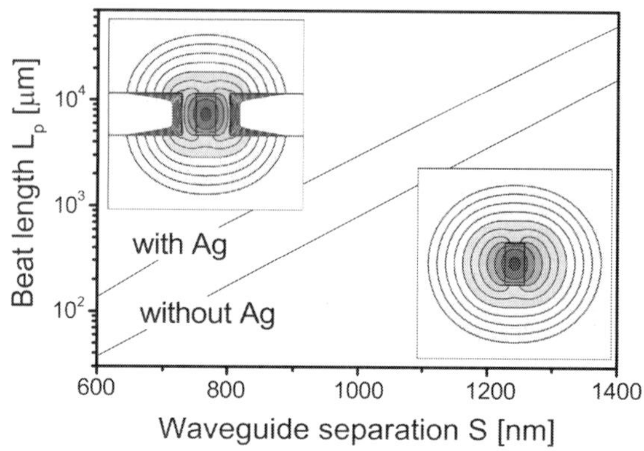

Fig. 7.34 Beat length variation as a function of waveguide separation for the quasi-TE mode at $\lambda = 1550\,\text{nm}$. The inset in the upper-left corner gives the $TE_{0,0}$ mode supported by an individual waveguide with metal cladding, while that in the bottom-right corner gives the same mode supported by a waveguide without metal cladding. The field contour lines are in 3 dB separation. The first contour line is an exception, which is at 90% of the maximum value. Domain size of the field plots: $2 \times 2\,\mu\text{m}^2$. See also Color Insert.

hybrid waveguide (upper-left inset in Fig. 7.34), it is not difficult to understand that the field overlapping of two such modes will be smaller, as compared to waveguides without metal claddings.

7.3.2.2 *Waveguide for quasi-TM operation*

In this sub-section, we look into waveguides that can operate with a relatively low-loss $TM_{0,0}$ mode. We follow the previous sub-section by setting $w = 500\,\text{nm}$. To get rid of the odd Fano mode shown in Fig. 7.30b, we can choose silicon nitride (Si_3N_4, whose index is assumed to be 2.05 for $\lambda = 1.2$–$2\,\mu\text{m}$), as both the substrate and the superstrate. The Si_3N_4 layers are not placed in direct contact with the metals in order to avoid the high loss induced by Ag–Si_3N_4 SPP. And the Si_3N_4 materials are placed 200 nm away from the metal surfaces. See Fig. 7.36 insets for the waveguide cross-section.

First, according to Fig. 7.30b, we choose $d = 340\,\text{nm}$ in order to minimize the absorption loss. It can be estimated, for example by using EIM that the height h of the core layer should be kept less than 500 nm. Let us select $h = 480\,\text{nm}$. In Fig. 7.35, variations of the n_{eff} and loss values are presented with respect to wavelength. Similar to previously designed waveguide, this waveguide supports both $TE_{0,0}$ and $TM_{0,0}$ modes at $\lambda = 1.55\,\mu\text{m}$. Losses of the two modes are comparable, at 45.3 and 42.0 dB/mm, respectively. The birefringence at $\lambda = 1.55\,\mu\text{m}$ is noticed to be 0.052, which is smaller as compared to the value possessed by the waveguide shown in the previous sub-section. Still, the birefringence is significantly larger than those in commercial HI-BI fibers, which are normally in the order of 10^{-4}.

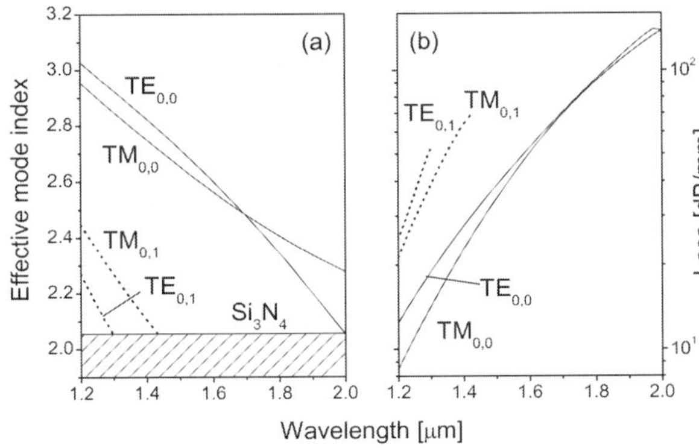

Fig. 7.35 Dispersion (a) and loss (b) curves for guided modes in a waveguide with $d = 340\,\text{nm}$ and $h = 480\,\text{nm}$. $\lambda = 1550\,\text{nm}$.

Therefore, a carefully launched polarization state is likely to be preserved at the output end of the waveguide.

To examine the coupling strength between two such waveguides placed in parallel, we show the beat length L_p as a function of their separation S in Fig. 7.36. The same curve for an identical waveguide but without the metal cladding is also superimposed. With the metal cladding, the beat length is 5.0 times at $S = 600\,\text{nm}$ and 10.7 times at $S = 1400\,\text{nm}$. At the same level of cross-talk requirement, the

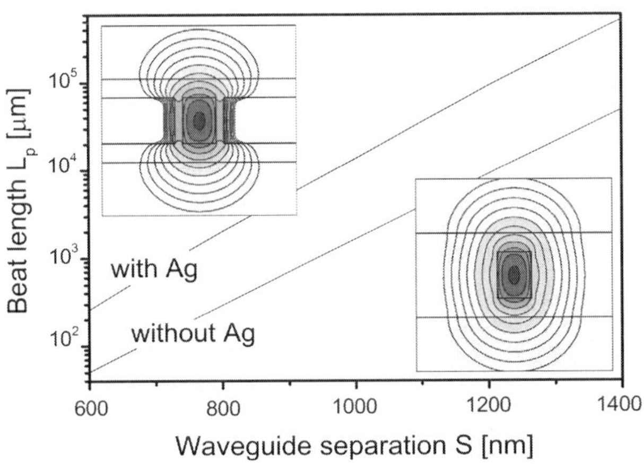

Fig. 7.36 Beat length variation as a function of waveguide separation for the quasi-TE mode at $\lambda = 1550\,\text{nm}$. The inset in the upper-left corner gives the $\text{TM}_{0,0}$ mode supported by an individual waveguide with metal cladding, while that in the bottom-right corner gives the mode supported by a waveguide without metal cladding. The field contour lines are in 3 dB separation. The first contour line is an exception, which is at 90% of the maximum value. Domain size of the field plots is $2\times2\,\mu\text{m}^2$. See also Color Insert.

proposed hybrid waveguide can be placed up to 25% closer as compared to the waveguides without metal claddings (examined around $S = 700\,\text{nm}$). The major magnetic field H_y supported by an individual waveguide, for both with and without metal claddings, are shown as insets in Fig. 7.36. Comparing the two fields, it can be seen that the mode guided by the hybrid waveguide is better confined laterally due to the presence of metal claddings. In fact, the major field components H_y and E_x are anti-resonant with the horizontal Ag-SiO_2 interfaces; therefore, a quicker field decay along the surfaces is resulted.

7.3.3 *Further Considerations*

The waveguides thus far discussed have confirmed that it is possible to obtain both reasonably low-loss propagation as well as good field confinement by combining both index-guiding and SPP-guiding mechanisms. There are several ways to achieve even lower loss, probably at the expense of field confinement. First, the propagation loss can be reduced by using some dielectric material with a lower permittivity value (e.g., air), in replacement of the SiO_2 material. A drop in propagation loss is expected because the SPP guided by a Ag–air interface has a significantly lower loss than that guided by a Ag–SiO_2 interface. However, planar waveguides with their core exposed in air are relatively more vulnerable to environmental changes. Notably, contamination of water molecules will affect the guided modes significantly. Second, propagation loss can be drastically reduced if the gap size is allowed to increase to some higher value. And finally, waveguides with lower loss may be possible by applying the same design methodology to other waveguide structures, other than based on the MIM structure.

7.4 ACTIVE SURFACE PLASMONICS DEVICES

7.4.1 *Plasmonic Light Sources*

It is well known by quantum mechanism that the spontaneous emission process of an excited molecule (can also be an excited semiconductor QD, for example) is totally determined by the local density of EM states (LDOS) in the vicinity of the molecule.[76] Increasing the LDOS of the EM field means an increased possibilities of coupling between the emission photon and the allowed EM modes.[77]

Since evanescent modes, the occurrence, and the increase of local density of SPP waves, surely tend to enhance the LDOS at the close proximity of nanostructures, they will also accelerate the spontaneous emission process of the light-emitting active material. Based on this mechanism, surface-plasmon-enhanced light sources have been designed, fabricated, and reported.[78–82] Enhanced luminescence enhancements from QDs were discussed previously. Another example is shown in

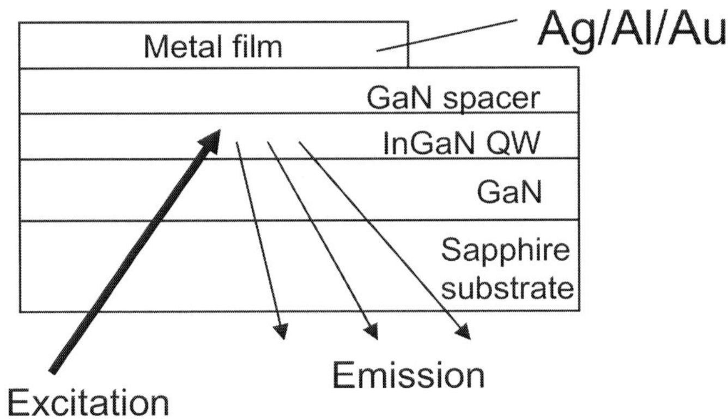

Fig. 7.37 Sample structure and excitation/emission configuration of the PL measurement studied in Ref.[82] The distance between the metal layer and QW was 10 nm. The intensity of PL spectrum of InGaN/GaN QWs is enhanced when a metal layer is coated, and the enhancement increases mostly when Ag film is coated, the next is the Al layer.

Fig. 7.37, where the light-emitting active InGaN/GaN QW coated with a nanometer metal film was fabricated, and a 14-fold enhancement in the peak PL intensity was recorded. The appearance of SPP increases the LDOS of the EM field. This leads to the enhancement of energy transfer between the QW and SPPs, which results in the large enhancement of internal quantum efficiency.[2, 82, 83] Similar promising results were also obtained for organic light-emitting diodes.[80]

7.4.2 *Active Plasmonic Devices*

To fabricate active plasmonic devices such as modulators and switches, the efficient techniques to manipulate actively the plasmonic signals are to be found. Based on the most fundamental dependence and sensitivity of SPP properties to the metal surface to which the SPP are bounded, Krasavin *et al.* have shown that SPP signals in metal-on-dielectric waveguides containing a Ga section a few microns long can be effectively controlled by switching the crystalline Ga section of the waveguide from one structural phase to another.[84] Furthermore, the switching may be achieved by changing both the waveguide temperature and external optical excitation. The signal modulation depth can exceed 80%, and switching times are expected to be in the picosecond time scale. Most recently, their results have received much experimental[85, 86] and theoretical attention.[87–89]

On the other hand, an active plasmonic device can also be formulated by combining thin polymer films containing molecular chromophores with a thin metal film; see Fig. 7.38 for more details.[90, 91] This molecular plasmonic device consists of two polymer layers, one containing donor chromophore molecules and the other containing acceptor fluorophore molecules. These layers are deposited on either side

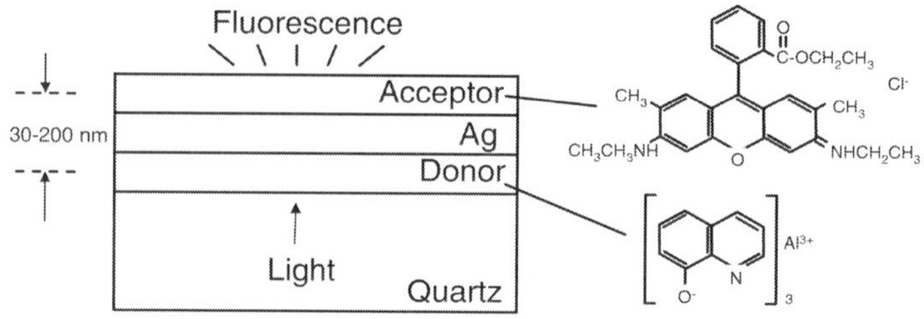

Fig. 7.38 Schematic representation of the plasmonic device created by Duyne.[90] Light is absorbed by the donor molecules, transferred across the silver film with the help of the SPPs. The acceptor molecules in the top layer absorb the SPP and emit their characteristic fluorescence.

of a thin metal film to form a sandwich structure. The donors absorb incident light and transfer this excitation energy by dipole–dipole interactions to the acceptors. The latter then emit their characteristic fluorescence. The chromophores are the molecular equivalent of a Wi-Fi transmitter and receiver, transferring energy and hence information across the metal film with the help of SPPs.

7.5 NANOSTRUCTURE SURFACE PLASMONICS

Once the surface modes at one interface were understood, research moved on to thin metal slabs[92–96] and it was found that if the slab was made thin enough, a special class of coupled SPPs emerged with longer propagation distances. This led to a number of device applications such as modulators,[97–99] sensors,[100] and polarizing devices.[101] SPP has been associated with a number of other novel ideas, from optical transistors[102] to plasmon lasers[103, 104] and light emitters.[105] The broad term plasmonics is now used popularly to describe plasmon-based technologies.

In the last few years, the rapid improvement and the better availability of nanofabrication techniques such as electron-beam lithography, ion-beam milling, and self-assembling, together with modern nano characterization techniques such as dark-field and near-field optical microscopies, and the emergence of a growing number of specific theories, models, and quantitative numerical simulation tools, have led to a resurgence of interest in this field. Many different dielectric and metallic nanostructures have been designed and fabricated including photonic band gap waveguides,[106] plasmons in nanoparticle chains,[107] channel plasmons,[108] and nano-wires.[109] Finite-width SPP waveguides were predicted[110, 111] and experimentally realized.[112] These structures have the potential applications for creating subwavelength optical devices enabling the miniaturization of optical components to size dimensions of their electronic counterparts in the sub-100 nm-size regime as well as the direct integration of the optical components with micro- and nanoscale electronic systems.[2, 113–116]

7.5.1 *Surface Plasmon Polaritons in Periodic Nano Hole Arrays*

In 1998, Ebbesen et al.[117] published an experiment in which he shone light through nano holes in a silver (Ag) film having a thickness of h. The holes had a diameter d between 150 nm and 1.0 μm and a spacing L about 0.6 $\sim$ 1.8 μm. More interesting are the high transmission peaks when $\lambda > L$, which are transmissions of light waves through subwavelength nanostructures. The maximal transmitted intensity occurs at 1370 nm, almost 10 times the diameter of an individual hole in the array. The standard aperture theory predicts a very small light transmission for these small holes[118] with a transmission efficiency of as small as 0.01,[119] whereas the experimental spectrum shows much higher transmissions and transmission efficiencies. Since the fraction of exposed area divided by the fraction of the hole area is larger than 2, twice as much light is transmitted as the illuminated hole area together. When searching for the reason of this phenomenon Ebbesen et al. analyzed different setups by experimenting with different materials, propagation directions of the probing EM field, hole sizes, hole shapes, and the periodicities of the hole arrays. The experiments showed that the enhanced transmission occurs on all metal films they have tested except the Ge films. By including the strong propagation-direction dependency, which is an important characteristic of the coupling between the SPP and metallic gratings, it was concluded that the enhanced transmission is induced by the SPP. Furthermore, it was pointed out that the periodicity of the hole array determines the positions of the transmission peaks in the long wavelength regime. The width of the transmission peak can be controlled by modifying the h/d ratio.

The Ebbesen's experimental report has received much intensive experimental[119–123] and theoretical[18, 124–129] attention. In attempts to understand the underlying physical mechanisms, many aspects of the structure have been studied including the arrangement of the subwavelength holes,[17] thickness of the metal film,[119] polarization of the incident light,[120] hole shape,[120] and the symmetry of the whole structure.[130] Many explanations have been presented such as the appearance of localized waveguide resonances[125, 131] and the dependence of the resonances on the hole geometric shape (shape resonance).[120] In addition, various complicated hole array structures have been proposed to further enhance the light transmission efficiency.[132, 133]

It is generally agreed that the observed transmission enhancement is largely attributed to the SPP excitations on the top and/or bottom metal–air interfaces of the gratings, which enhances the evanescent field in the hole and thus contributes to the tunneling of the incident EM wave through the hole.[128] The SPP waves can be excited in the spatial periodic hole array in a metal film, with wave vectors given by Eq. 6.19. Low-index ($(m, n) = (\pm 1, 0)$ or $(0, \pm 1)$) SPP Bloch waves, which have been shown to play very important roles in the extraordinary optical transmission.[119, 129]

Figure 7.39 shows the well-studied free-standing Ag film perforated by an array of holes. The thickness of the metal film is denoted as h, the hole has a square

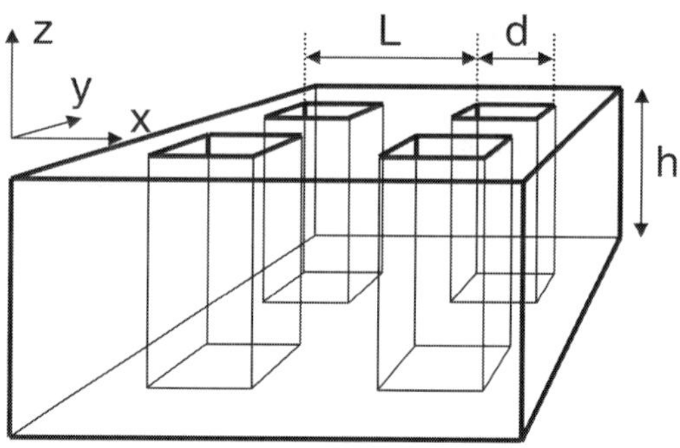

Fig. 7.39 Schematic drawing of a free-standing Ag film perforated with a square array of square air holes. h is the thickness of the metal film, the hole has a square cross-section of $d \times d$, and the square hole array has a period of L.

cross-section of $d \times d$ and $d = 280\,\text{nm}$, and the square array has a period of $L = 750\,\text{nm}$. The relative dielectric constant of silver is given by the Drude model of Eq. 7.1.

The phenomena of the enhanced transmission through subwavelength holes in a metallic film are usually characterized by the normalized transmission, which is defined as

$$T_{\text{norm}} = \frac{P_{\text{out}}}{P_{\text{in}}} \frac{S_{\text{cell}}}{S_{\text{hole}}} \tag{7.14}$$

where $P_{\text{in(out)}}$ is the power flux through into (out) the metal film, S_{hole} is the area of the hole, and S_{cell} is the area of a unit cell in the periodic array. $S_{\text{hole}} = d^2$ and $S_{\text{cell}} = L^2$ for the structure of Fig. 7.39. The enhanced transmission happens when T_{norm} becomes larger than unity.

Numerical calculation based on the Fourier modal method shows the dependence of the normalized transmission of the hole arrays in the metal film with a thickness h in Fig. 7.40. One transmission peak is observed when h is rather large ($h = 600\,\text{nm}$) and the peak wavelength is 802 nm, which is the SPP wavelength of the thick Ag film perforated with holes. Note that the wavelength of the SPP is about 466 nm at the surface between the air and the half-infinite Ag bulk. Figure 7.40 shows that there always exists wavelength region(s) within which T_{norm} is larger than unity for all the h values and the transmission peaks depend strongly on the metal film thickness.[119] A two-level characteristics can be observed concerning the transmission peaks as functions of the film thickness. When the metal film becomes thin enough, the SPP on one of the two metal–air surfaces will interact strongly with the other one on the other metal–air surface so that a "repulsion of the levels" occurs.[3] One evidence of such a repulsion is that the sum of the energies of the photons of two transmission peaks for the same h is always close to twice of the energy of 802 nm (the SPP wavelength when the Ag film is thick under which circumstance the two SPPs on the two metal–air surfaces become dissociated from each other).

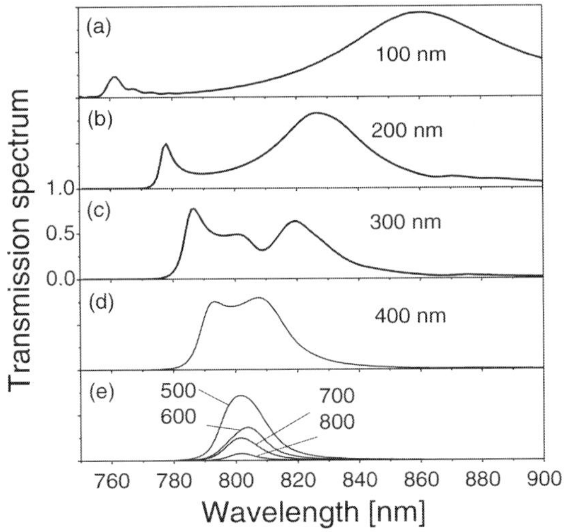

Fig. 7.40 Normalized transmission spectra of a silver film perforated with a square array of square holes. The metal film thickness h increases from 100 to 600 nm.

In Fig. 7.41, the transmission, reflection, and absorption spectra for $h = 700\,\text{nm}$ are plotted, showing the coincidence of the three peaks. This was explained by the increase of the fields associated with the evanescent waves via the SPP modes, which enhances the field transmission.[121]

The existence of SPP modes in the extraordinary transmission can be directly observed through the near-field distributions of the EM field components.[126] In Fig. 7.42, the near-field intensity distributions of three resonant wavelengths are

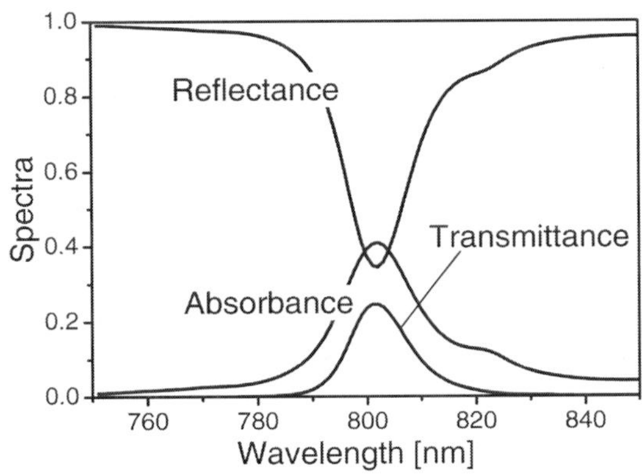

Fig. 7.41 Normal-incident field transmission, reflection and absorption spectra of an Ag film perforated with a square array of square holes. $L = 750\,\text{nm}$, $d = 280\,\text{nm}$. The metal film thickness $h = 700\,\text{nm}$.

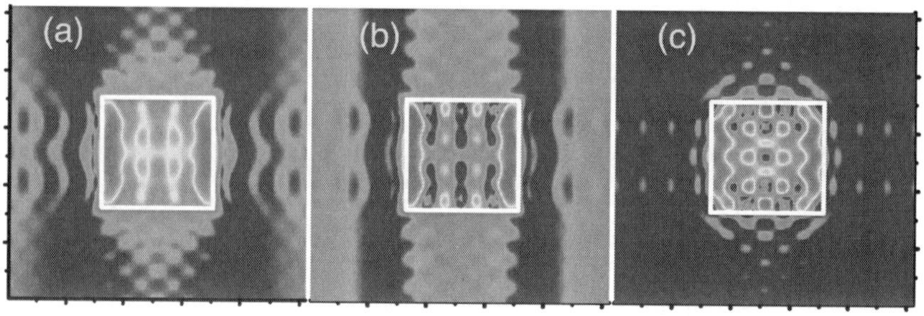

Fig. 7.42 EM field intensity distributions across a unit cell of the periodic hole array at a wavelength of (a) 802 nm at the transmission peak of $h = 600$ nm, (b) 767 nm when $h = 100$ nm, and (c) 853 nm when $h = 100$ nm. The observation plane is at a distance of 1.0 nm from the metal–air surface. The incident light is polarized along the horizontal axis of the images. The square white frames mark the spatial positions of the hole. See also Color Insert.

plotted where important feature properties can be noticed: the EM field at the corresponding peak transmission is strongly localized in the vicinity of the hole. High symmetry is presented in every intensity distribution, which reflects the boundary conditions of the Brillouin zone of the periodic hole array.[126]

7.5.2 *Localized Plasmon Resonances in Metal Nanoparticles*

The optical properties of metal nanoparticles show striking differences relative to the ones of metal bulk and thin film because of the three-dimensional localization of the SPP modes. One of the major differences is that the resonant wavelength of a metal nanoparticle can be turned through the changing of its geometry shape. This property was adopted by ancient people to fabricate the color glass windows and ceramic potteries. The Lycurgus cup (Byzantine empire, the fourth century AD), shown in Fig. 7.43,[134] is one of the most-cited masterpieces. Because of the small gold nanoparticles embedded in the glass, the cup displays a striking red color when being viewed in transmitted light, while it shows green in the reflection configuration.

The fundamental physical mechanism of resonant EM behavior of metal nanoparticles is the confinement of the conduction electrons to the small three-dimensional volume of the nanoparticle. For a nanoparticle with a diameter d, which is far less than the wavelength λ of the incident EM field, the conduction electrons inside the nanoparticle move all in phase upon plane-wave excitation, leading to the build-up of polarization charges on the nanoparticle surface. These charges act as an effective restoring force, allowing for a resonance to occur at a specific frequency— the dipole plasmon frequency of the nanoparticle, where the response of the electrons shows a $\pi/2$ phase lag with respect to the external driving field. Thus, a resonantly enhanced field builds up inside the nanoparticle, which in the small particle limit is homogeneous throughout its volume,

Fig. 7.43 The Lycurgus glass cup shows the bright red color of gold nanoparticles via transmitted light (the British museum).[134] See also Color Insert.

producing a dipolar field outside the nanoparticle. This leads to an enhanced absorption and scattering cross-sections for EM waves, as well as to a much enhanced near field in the close vicinity of the nanoparticle surface. It is this resonantly enhanced near field from which most of the promising applications of metal nanoparticles stem.[115]

There has been a tremendous progress in the fabrication and functionlization of solid metal nanoparticles, and many kinds of nanoscale structures have been fabricated including triangular prisms,[135] cubes,[136] disks,[137] rods,[138] wires,[139] and nanoshell.[140, 141] Among them, metallic nanoshell is one of the most interesting and useful structures.[38, 142–149] The metal nanoshell consists of a spherical dielectric core surrounded by an ultra-thin conducting metallic shell, see Fig. 7.44. More recently, this geometry has been extended into multilayered concentric nanoshells that show far more complex hybridized resonances whose spectral profiles span from the infrared spectrum regime to thermal wavelengths.[149]

The EM response of the nanoshell structures obeys quantitatively the classical Mie scattering theory,[40] with an additional contribution to the plasmon linewidth from the electron scattering at the inner and outer interfaces of the shell layer.[143]

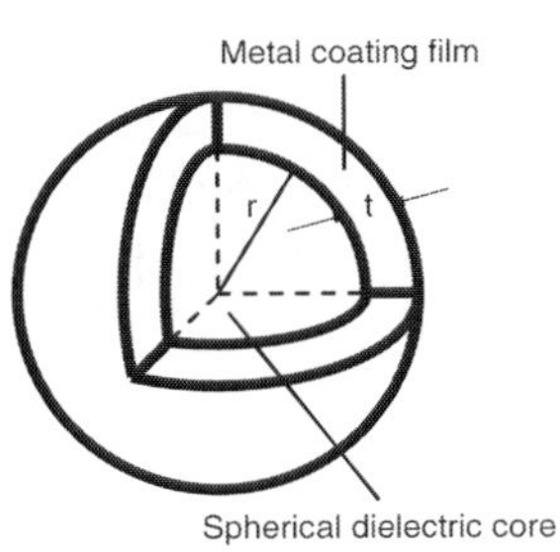

Fig. 7.44 Schematic nanoshell geometry.

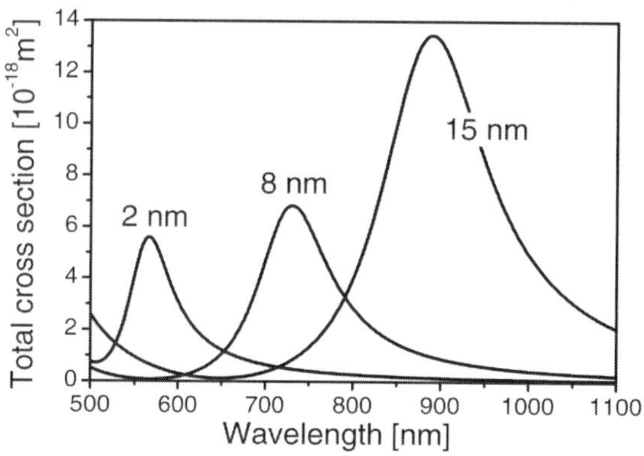

Fig. 7.45 The relationship between the total scattering cross section and the wavelength of an Au (gold) nanoshell having a 2-nm-thick shell. The core dielectric material is Au_2S with a dielectric constant 5.44. The radius of the dielectric core is 2, 8, and 15 nm, respectively.

By varying the composition and dimensions of the layers of the nanoparticles, nanoshells can be designed and fabricated with plasmon resonances from the visible to infrared regions in the optical spectrum, and the magnitude of the local EM field at the surface can also be properly manipulated. Figure 7.45 shows the relationships between the total scattering cross-section and the wavelength of three types of Au (gold) nanoshells. Moreover, the plasmon resonance of the nanoparticle is very sensitive to the local chemical environment of the surface, leading to the possibility of sensitive optical chemical detectors. Therefore, this structure is important in surface-enhanced Raman scattering (SERS) and surface nonlinear process.[150] In fact, it has demonstrated that a strong SERS— at least by a factor of 10^{10}, is very possible.[144] Because metal nanoshells can be easily tuned to have strong scattering or absorption properties in the near infrared regions where light penetration through tissue is optimal, there have already been many biomedical applications such as confocal imaging, laser iridotomy, and photothermal coagulation.[142, 151]

Figure 7.46 shows the temporal development of the integrated electromagnetic optical power inside a metal-thin-film-coated semiconductor QD. The QD is a spherical CdS QD having a radius of 3.7 nm and its energy bandgap is 2.67 eV. Since the QD shows a multiphoton spectrum in the useful region for bioimaging between 500 and 1300 nm,[152] a two-photon excitation EM field having a photon energy of 1.34 eV (half of the QD bandgap) is applied. The Au metal thin film is added in order to enhance the EM field inside the QD to improve its bioimaging properties. The numerical analysis of Fig. 7.46 is carried out by the three-dimensional FDTD algorithm, which shows that with the increase of the coated metal film thickness, the integrated optical power inside the QD increases due to the generation of the SPP at the nanoshell. For $\hbar\omega = 1.34$ eV (which is in the two-photon regime, since the QD bandgap is 2.67 eV), the enhancement of the integrated optical power reaches

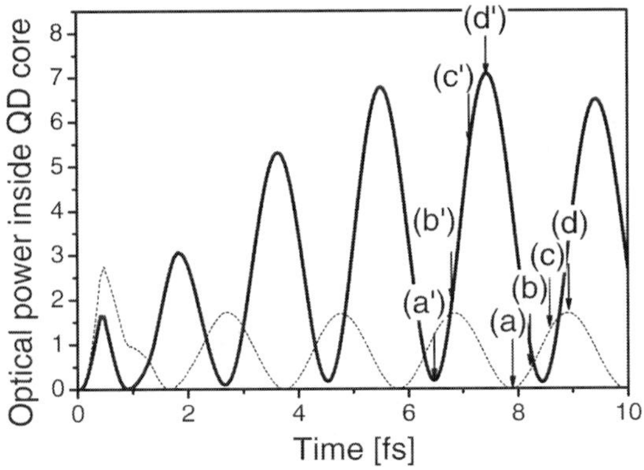

Fig. 7.46 Integrated optical power inside the CdS QD. The QD has an energy bandgap of 2.67 eV and the photon energy of the two-photon excitation electromagnetic field is 1.34 eV. Solid line: the thickness of the metal film is 0.37 nm. Dashed line: bare QD.

its maximal value of about 10 when the metal film thickness is about 0.37 nm when compared with bare CdS QD.

Using a linear scale (from 0.0 to 3.5×10^{-2}) for the field intensity, Fig. 7.47 shows motion pictures of the optical power distribution around the bare CdS QD and the metal-coated CdS QD. The times in the cases of (a)–(d) in Fig. 7.47 are marked by arrows in Fig. 7.46. We observe here the sequential development of the EM field where the field distribution around the bare CdS QD follows the incident EM field passively; the peaks and valleys in Fig. 7.47a coincide exactly with the ones of the incident EM wave. One can notice here the accumulation of the EM power in the vicinity of the upper and lower poles of the bare CdS QD. At the same time, depletion is also observed around the left and right poles. For the metal-coated CdS QD, the high field enhancement, i.e., the localized SPP, is clearly observed, as can be expected. The field distribution around the metal-coated CdS QD, both the field strength and the peak-to-valley relationship in Fig. 7.46, is greatly modified by the generation of the localized SPP. Moreover, it is observed that the localized SPP around the coated CdS QD is in the form of a dipole. Experimental near-field PL images of an elliptical cluster of Au nanoparticles reveal similar dipole nature of the EM field surrounding the nanoparticles.[153]

Systematic PL measurements and theoretical studies of metal-coated semiconductor nanoparticles have been reported concerning the optical enhancement and quenching effects from both the fundamental research and the practical application points of view. In addition, the mixture of the QDs and the metal nanoparticles has also been investigated showing a 5-fold enhancement of luminescence intensity and a blue shift of the emission peak as compared to unconjugated semiconductor QDs.[154] Theoretical work was also performed to confirm and support the conclusion that it is the confinement and repulsion of the SPP field at the surfaces of

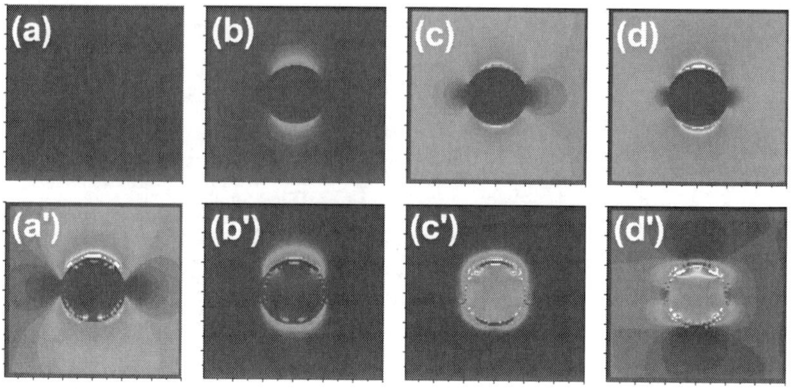

Fig. 7.47 Spatial distribution of the optical power around the CdS QD. The QD has an energy bandgap of 2.67 eV, and the photon energy of the two-photon excitation electromagnetic field is 1.34 eV. (a)+(b)+(c)+(d): bare CdS QD. (a')+(b')+(c')+(d'): 0.37 nm metal-coated CdS QD. The times are marked by arrows in Fig. 7.46. See also Color Insert.

metal nanoparticles that causes the enhancement and quenching of the PL signal from QDs. A much more detailed theoretical work was further reported showing that the suppression of nanoparticle PL was caused by the energy transfer from semiconductor QD to metal nanoparticles.[155, 156]

7.6 BARRIERS AND FRONTIERS

SPP have brought many important solutions for the old problems of optics in the subwavelength regime. The hole array, for example, is used for signal enhancement on near-field microscopy. Discovering this effect birthed many new research areas in SPP and subwavelength optics. In[157] a similar effect on single nano holes using SPP enhancing through corrugations was described. The possibility of manipulating SPP-like light beams, but in two dimension, will provide many new possibilities. Numerous sensors based on SPP have been invented. The knowledge about Bragg reflectors and beam splitters will enhance the flexibility of such devices. The achievement of the nano lens is a good indication that physicists have accomplished a real nano optics device that manipulates light beams in certain controllable ways. In future, we will be able to use subwavelength lithography and terabyte optical data storage.

Other areas also profit from the properties of SPP. Many biosensors use the evanescent field change that takes place when a molecule docks on the metal surface,[158, 159] and diodes become more efficient using the strong EM fields of SPP in nanoparticles.[160]

Because plasmonic components can be crafted from the same materials chipmakers currently use, it is hopeful that all the devices can be made to route light around a processor or other kind of chips. These would include plasmon sources, detectors and wires, as well as splitters, and even transistors. While an all-plasmonic chip might someday be possible, in the near term plasmonic wires will act as high-traffic freeways on chips with otherwise conventional electronics. Local arrays of electronic transistors would carry out the switching necessary for computation, but when a large amount of data need an express lane to travel from one section of a chip to another one, electronic bits could be converted to plasmon waves, sent along a plasmonic wire, and converted back to electronic bits at their destination.

As shown via the subwavelength imaging with a multilayer silver film structure, loss is the major limitation.[161] The potential of plasmonics right now is mainly limited by the fact that plasmons typically can travel only several millimeters before they peter out. Chips, meanwhile, are typically about a centimeter across, so plasmons cannot yet go the whole distance. The distance a plasmon can travel before dying out is a function of several aspects of the metal.[116] For optimal transfer through a wire of any metal, the surface of contact with surrounding materials must be as smooth as possible and the metal should not have impurities. There are other forms of losses, e.g., the loss from intrinsic higher-energy multipole modes and the bend loss in a bended metal line, which was shown to increase approximately quadratically with the bend angle.[162]

For most wavelengths of visible light, aluminum allows plasmons to travel farther than other metals such as gold, silver, and copper. It is somewhat ironic that aluminum is the best metal to use because the semiconductor industry recently dumped aluminum in favor of copper— the better electrical conductor— as its wiring. Of course, it may turn out that some kind of alloy will have even better plasmonic properties than either aluminum or copper.

Another classic semiconductor industry issue that researchers will have to address is heat. Chipmakers are constantly battling to ensure that their electronic chips donot run too hot. Plasmonics also will likely generate some heat, but exactly how much is not yet known. Even if plasmonics run as hot as electronics, they will still have the advantage of having a higher data capacity in the same space.

References

[1] M. Ohtsu, K. Kobayashi, T. Kawazoe, S. Sangu, and T. Yatsui, "Nanophotonics: design, fabrication, and operation of nanometric devices using optical near fields," IEEE Sel. Top. Quantum Electron., **8**, 839–862, 2002.

[2] E. Ozbay, "Plasmonics: merging photonics and electronics at nanoscale dimensions," Science, **311**, 189–193, 2006.

[3] A. Drezet, A. L. Stepanov, H. Ditlbacher, A. Hohenau, B. Steinberger, F. R. Aussenegg, A. Leitner, and J. R. Krenn, "Surface plasmon propagation in an elliptical corral," Appl. Phys. Lett., **86**, 074104(3), 2005.

[4] Z. Liu, J. M. Steele, W. Srituravanich, Y. Pikus, C. Sun, and X. Zhang, "Focusing surface plasmons with a plasmonic lens," Nano Lett., **5**, 1726–1729, 2005.

[5] Z. Liu, J. M. Steele, H. Lee, and X. Zhang, "Tuning the focus of a plasmonic lens by the incident angle," Appl. Phys. Lett., **88**, 171108(3), 2006.

[6] J. M. Steele, Z. Liu, Y. Wang, and X. Zhang, "Resonant and non-resonant generation and focusing of surface plasmons with circular gratings," Opt. Exp., **14**, 5664–5670, 2006.

[7] Z. Sun and H. K. Kim, "Refractive transmission of light and beam shaping with metallic nano-optic lenses," Appl. Phys. Lett., **85**, 642–644, 2004.

[8] H. Ditlbacher, J. R. Krenn, G. Schider, A. Leitner, and F. R. Aussenegg, "Two-dimensional optics with surface plasmon polaritons," Appl. Phys. Lett., **81**, 1762–1764, 2002.

[9] W. L. Barnes, A. Dereux, and T. W. Ebbesen, "Surface plasmon subwavelength optics," Nature, **424**, 824–830, 2003.

[10] P. Berini, "Plasmon-polariton modes guided by a metal film of finite width," Opt. Lett., **24**, 1011–1013, 1999.

[11] J. C. Weeber, A. Dereux, C. Girard, J. R. Krenn, and J. P. Goudonnet, "Plasmon polaritons of metallic nanowires for controlling submicron propagation of light," Phys. Rev. B, **60**, 9061–9068, 1999.

[12] B. Lamprecht, J. R. Krenn, G. Schider, H. Ditlbacher, M. Salerno, N. Felidj, A. Leitner, F. R. Aussenegg, and J. C. Weeber, "Surface plasmon propagation in microscale metal stripes," Appl. Phys. Lett., **79**, 51(3), 2001.

[13] J. C. Weeber, Y. Lacroute, and A. Dereux, "Optical near-field distributions of surface plasmon waveguide modes," Phys. Rev. B, **68**, 115401(10), 2003.

[14] R. Zia, M. D. Selker, and M. L. Brongersma, "Leaky and bound modes of surface plasmon waveguides," Phys. Rev. B, **71**, 165431(9), 2005.

[15] J. A. Dionne, L. A. Sweatlock, H. A. Atwater, and A. Polman, "Planar metal plasmon waveguides: frequency-dependent dispersion, propagation, localization, and loss beyond the free electron model," Phys. Rev. B, **72**, 075405(11), 2005.

[16] I. V. Novikov and A. A. Maradudin, "Channel polaritons," Phys. Rev. B, **66**, 035403(13), 2002.

[17] S. I. Bozhevolnyi, V. S. Volkov, E. Devaux, and T. W. Ebbesen, "Channel plasmon-polariton guiding by subwavelength metal grooves," Phys. Rev. Lett., **95**, 046802(4), 2005.

[18] S. I. Bozhevolnyi, V. S. Volkov, E. Devaux, J. Y. Laluet, and T. W. Ebbesen, "Channel plasmon subwavelength waveguide components including interferometers and ring resonators," Nature, **440**, 508–511, 2006.

[19] S. I. Bozhevolnyi, J. Erland, K. Leosson, P. M. W. Skovgaard, and J. M. Hvam, "Waveguiding in surface plasmon polariton band gap structures," Phys. Rev. Lett., **86**, 3008–3011, 2001.

[20] M. Quinten, A. Leitner, J. R. Krenn, and F. R. Aussenegg, "Electromagnetic energy transport via linear chains of silver nanoparticles," Opt. Lett., **23**, 1331–1333, 1998.

[21] J. Takahara, S. Yamagishi, H. Taki, A. Morimoto, and T. Kobayashi, "Guiding of a one-dimensional optical beam with nanometer diameter," Opt. Lett., **22**, 475–477, 1997.

[22] A. Mcmillan, C. D. Paavola, J. Howard, S. L. Chan, N. J. Zaluzec, and J. D. Trent, "Ordered nanoparticle arrays formed on engineered chaperonin protein templates," Nat. Mater., **1**, 247–252, 2002.

[23] S. A. Maier, P. G. Kik, H. A. Atwater, S. Meltzer, E. Harel, B. E. Koel, and A. A. G. Requicha, "Local detection of electromagnetic energy transport below the diffraction limit in metal nanoparticle plasmon waveguides," Nat. Mater., **2**, 229–232, 2003.

[24] J. R. Krenn, A. Dereux, J. C. Weeber, E. Bourillot, Y. Lacroute, J. P. Goudonnet, G. Schider, W. Gotschy, A. Leitner, F. R. Aussenegg, and C. Girard, "Squeezing the optical near-field zone by plasmon coupling of metallic nanoparticles," Phys. Rev. Lett., **82**, 2590–2593, 1999.

[25] K. J. Webb and J. Li, "Resonant slot optical guiding in metallic nanoparticle chains," Phys. Rev. B, **72**, 201402R(4), 2005.

[26] M. Yan and M. Qiu, "Guided plasmon polariton at 2D metal corners," J. Opt. Soc. Am. B, **24**, 2333–2342, 2007.

[27] S. I. Bozhevolnyi, V. S. Volkov, E. Devaux, and T. W. Ebbesen, "Channel plasmon-polariton guiding by subwavelength metal grooves," Phys. Rev. Lett., **95**, 046802, 2005.

[28] S.I. Bozhevolnyi, V. S. Volkov, E. Devaux, J.-Y. Laluet, and T. W. Ebbesen, "Channel plasmon subwavelength waveguide components including interferometers and ring resonators," Nature, **440**, 508–511, 2006.

[29] D. F. P. Pile, T. Ogawa, D. K. Gramotnev, T. Okamoto, M. Haraguchi, M. Fukui, and S. Matsuo, "Theoretical and experimental investigation of strongly localized plasmons on triangular metal wedges for subwavelength waveguiding," Appl. Phys. Lett., **87**, 061106, 2005.

[30] A. Boltasseva, T. Nikolajsen, K. Leosson, K. Kjaer, M. S. Larsen, and S. I. Bozhevolnyi, "Integrated optical components utilizing long-range surface plasmon polaritons," J. Light. Technol., **23**, 413–422, 2005.

[31] P. Berini, "Plasmon-polariton waves guided by thin lossy metal films of finite width: bound modes of symmetric structures," Phys. Rev. B **61**, 10484–10503, 2000.

[32] P. Berini, "Plasmon-polariton waves guided by thin lossy metal films of finite width: bound modes of asymmetric structures," Phys. Rev. B, **63**, 125417, 2001.

[33] D. K. Gramotnev and D. F. P. Pile, "Single-mode subwavelength waveguide with channel plasmon-polaritons in triangular grooves on a metal surface," Appl. Phys. Lett., **85**, 6323–6325, 2004.

[34] G. Veronis and S. Fan, "Guided subwavelength plasmonic mode supported by a slot in a thin metal film," Opt. Lett., **30**, 3359–3361, 2005.

[35] L. Liu, Z. Han, and S. He, "Novel surface plasmon waveguide for high integration," Opt. Exp., **13**, 6645–6650, 2005.

[36] D. F. P. Pile, T. Ogawa, D. K. Gramotnev, Y. Matsuzaki, K. C. Vernon, K. Yamaguchi, T. OKamoto, M. Haraguchi, and M. Fukui, "Two-dimensionally localized modes of a nanoscale gap plasmon waveguide," Appl. Phy. Lett., **87**, 261114, 2005.

[37] L. Chen, J. Shakya, and M. Lipson, "Subwavelength confinement in an integrated metal slot waveguide on silicon," Opt. Lett., **31**, 2133–2135, 2006.

[38] S. I. Bozhevolnyi, "Effective-index modeling of channel plasmon polaritons," Opt. Exp., **14**, 9467–9476, 2006.

[39] D. F. P. Pile, T. Ogawa, D. K. Gramotnev, T. Okamoto, M. Haraguchi, M. Fukui, and S. Matsuo, "Theoretical and experimental investigation of strongly localized plasmons on triangular metal wedges for subwavelength waveguiding," Appl. Phy. Lett., **87**, 061106, 2005.

[40] J. Meixner, "The behavior of electromagnetic fields at edges," IEEE Trans. Antennas Propag., **20**, 442–446, 1972.

[41] C. J. Bouwkamp, "A note on singularities occurring at sharp edges in electromagnetic diffraction theory," Physica, **12**, 467–474, 1946.

[42] J. Andersen and V. Solodukhov, "Field behavior near a dielectric wedge," IEEE Trans. Antennas Propag., **26**, 598–602, 1978.

[43] A. S. Sudbø, "Why are accurate computations of mode fields in rectangular dielectric waveguides difficult?," J. Lightw. Technol., **10**, 418–419, 1992.

[44] *Handbook of Optical Constants of Solids*, ed. E. D. Palik, Academic, Orlando, 1985.

[45] P. McIsaac, "Symmetry-induced modal characteristics of uniform waveguides — I: summary of results," IEEE Trans. Microw. Theory Tech., **23**, 421–429, 1975.

[46] E. Moreno, F. J. Garcia-Vidal, S. G. Rodrigo, L. Martin-Moreno, and S. I. Bozhevolnyi, "Channel plasmon-polaritons: modal shape, dispersion, and losses," Opt. Lett., **31**, 3447–3449, 2006.

[47] For example, the Corning SMF-28e+ optical fiber has dispersion $\leq$18ps/nm/km at 1550nm wavelength, http://www.corning.com.

[48] S. S. Xiao, L. Liu, and M. Qiu, "Resonator channel drop filters in a plasmon-polaritons metal," Opt. Exp., **14**, 2932–2937, 2006.

[49] C. Manolatou, M. J. Khan, S. Fan, P. R. Villenueve, H. A. Haus, and J. D. Joannopoulos, "Coupling of modes analysis of resonant channel add-drop filters," J. Quantum Electron. **35**, 1322–1331, 1999.

[50] T. W. Ebbesen, C. Genet, and S. I. Bozhevolnyi, "Surface plasmon circuitry," Phys. Today, **61**, 44–50, 2008.

[51] Z. Salamon, H. A. Macleod, and G. Tollin, "Surface plasmon resonance spectroscopy as a tool for investigating to biochemical and biophysical properties of membrane protein systems," Biochim. Biophys. Acta., **1331**, 117–129, 1997.

[52] W. P. Chen and J. M. Chen, "Use of surface plasma waves for determination of the thickness and optical constants of thin metallic films," J. Opt. Soc. Am., **71**, 189–191, 1981.

[53] S. Scheerlinck, J. Schrauwen, F. V. Laere, D. Taillaert, D. V. Thourhout, and R. Baets, "Efficient, broadband and compact metal grating couplers for silicon-on-insulator waveguides," Opt. Exp., **15**, 9625–9630, 2007.

[54] L. Eurenius, C. Hgglund, E. Olsson, B. Kasemo, and D. Chakarov, "Grating formation by metal nanoparticle-mediated coupling of light into waveguided modes," Nat. Photonics, **2**, 360–364, 2008.

[55] G. Veronis and S. Fan, "Theoretical investigation of compact couplers between dielectric slab waveguides and two-dimensional metal-dielectric-metal plasmonic waveguides," Opt. Exp., **5**, 1211–1221, 2007.

[56] G. I. Stegeman, R. F. Wallis, and A. A. Maradudin, "Excitation of surface polaritons by end-fire coupling," Opt. Lett., **8**, 386–388, 1983.

[57] N. N. Feng, and L. D. Negro, "Plasmon mode transformation in modulated-index metal-dielectric slot waveguides," Opt. Lett., **32**, 3086–3088, 2007.

[58] R. Yang, M. A. G. Abushagur, and Z. Lu, "Efficiently squeezing near infrared light into a 21nm-by-24nm nanospot," Opt. Exp., **16**, 20142–20148, 2008.

[59] L. Chen, J. Shakya, and M. Lipson, "Subwavelength confinement in an integrated metal slot waveguide on silicon," Opt. Lett., **31**, 2133–2135, 2006.

[60] J. Tian, S. Q. Yu, W. Yan, and M. Qiu, "Broadband high-efficiency surface-plasmon-polariton coupler with silicon-metal interface," Appl. Phys. Lett., **95**, 013504, 2009.

[61] A. E. Craig, G. A. Olson, and D. Sarid, "Experimental observation of the long-range surface-plasmon polariton," Opt. Lett., **8**, 380–382, 1983.

[62] Y. Kuwamura, M. Fukui, and O. Tada, "Experimental observation of long-range surface plasmon polaritons," J. Phys. Soc. Jpn., **52**, 2350–2355, 1983.

[63] B. Prade, J. Y. Vinet, and A. Mysyrowicz, "Guided optical waves in planar heterostructures with negative dielectric constant," Phys. Rev. B, **44**, 13556–13572, 1991.

[64] G. Veronis and S. Fan, "Modes of subwavelength plasmonic slot waveguides," J. Lightw. Technol., **25**, 2511–2521, 2007.

[65] E. Feigenbaum and M. Orenstein, "Modeling of complementary (void) plasmon waveguiding," J. Lightwave Technol., **25**, 2547–2562, 2007.

[66] S. I. Bozhevolnyi, V. S. Volkov, E. Devaux, and T. W. Ebbesen, "Channel plasmon-polariton guiding by subwavelength metal grooves," Phys. Rev. Lett., **95**, 046802, 2005.

[67] M. I. Stockman, "Nanofocusing of optical energy in tapered plasmonic waveguides," Phys. Rev. Lett., **93**, 137404, 2004.

[68] M. Yan, L. Thylén, M. Qiu, and D. Parekh, "Feasibility study of nanoscaled optical waveguide based on near-resonant surface plasmon polariton," Opt. Exp., **16**, 7499–7507, 2008.

[69] A. Karalis, E. Lidorikis, M. Ibanescu, J. D. Joannopoulos, and M. Soljacic, "Surface-plasmon-assisted guiding of broadband slow and subwavelength light in air," Phys. Rev. Lett., **95**, 063901, 2005.

[70] D. Parekh, L. Thylén, and C. Chang-Hasnain, "Metal nanoparticle metamaterials for engineering dielectric constants and their applications to near resonant surface plasmon waveguides," in *Frontiers in Optics, OSA Technical Digest Series*, p. FThF6, Optical Society of America, 2007. San Jose, California.

[71] B. Wang and G. P. Wang, "Planar metal heterostructures for nanoplasmonic waveguides," Appl. Phy. Lett., **90**, 013114, 2007.

[72] P. B. Johnson and R. W. Christy, "Optical constants of the noble metals," Phys. Rev. B, **6**, 4370–4379, 1972.

[73] F. Kusunoki, T. Yotsuya, J. Takahara, and T. Kobayashi, "Propagation properties of guided waves in index-guided two-dimensional optical waveguides," Appl. Phy. Lett., **86**, 211101, 2005.

[74] J. A. Dionne, L. A. Sweatlock, H. A. Atwater, and A. Polman, "Plasmon slot waveguides: towards chip-scale propagation with subwavelength-scale localization," Phys. Rev. B, **73**, 035407, 2006.

[75] G. Agrawal, *Nonlinear Fiber Optics*, 3rd edn, Academic Press, 2001.

[76] E. N. Economou, *Green's Functions in Quantum Physics*, Springer-Verlag, Berlin, 1990.

[77] Y. Zeng, X. Chen, and W. Lu, "Modified spontaneous emission from a two-dimensional photonic crystal," Phys. Rev. E, **70**, 047601(3), 2004.

[78] N. E. Hecker, R. A. Hopfel, N. Sawaki, T. Maier, and G. Strasser, "Surface plasmon-enhanced photoluminescence from a single quantum well," Appl. Phys. Lett., **75**, 1577–1579, 1999.

[79] A. Neogi, C. Lee, H. O. Everitt, T. Kuroda, A. Tackeuchi, and E. Yablonovitch, "Enhancement of spontaneous recombination rate in a quantum well by resonant surface plasmon coupling," Phys. Rev. B, **66**, 153305(4), 2002.

[80] S. Wedge, I. R. Hooper, I. Sage, and W. L. Barnes, "Light emission through a corrugated metal film: The role of cross-coupled surface plasmon polaritons," Phys. Rev. B, **69**, 245418(8), 2004.

[81] T. D. Neal, K. Okamoto, A. Scherer, M. S. Liu, and A. K. Jen, "Time resolved photoluminescence spectroscopy of surface-plasmon-enhanced light emission from conjugate polymers," Appl. Phys. Lett., **89**, 221106(3), 2006.

[82] K. Okamoto, I. Niki, A. Shvartser, Y. Narukawa, T. Mukai, and A. Scherer, "Surface-plasmon-enhanced light emitters based on InGaN quantum wells," Nat. Mater., **3**, 601–605, 2004.

[83] C. Girard, "Near fields in nanostructures," Rep. Prog. Phys., **68**, 1883–1933, 2005.

[84] A. V. Krasavina and N. I. Zheludev, "Active plasmonics: controlling signals in Au/Ga waveguide using nanoscale structural transformations," Appl. Phys. Lett., **84**, 1416(3), 2004.

[85] A. V. Krasavin, K. F. MacDonald, N. I. Zheludev, and A. V. Zayats, "High-contrast modulation of light with light by control of surface plasmon polariton wave coupling," Appl. Phys. Lett., **85**, 3369(3), 2004.

[86] A. V. Krasavin, K. F. MacDonald, A. S. Schwanecke, and N. I. Zheludev, "Gallium/aluminum nanocomposite material for nonlinear optics and nonlinear plasmonics," Appl. Phys. Lett., **89**, 031118(3), 2006.

[87] A. V. Krasavin, A. V. Zayats, and N. I. Zheludev, "Active control of surface plasmon polariton waves," J. Opt. A: Pure Appl. Opt. **7**, S85–S89, 2005.

[88] J. A. Sanchez-Gil and J. G. Rivas, "Thermal switching of the scattering coefficients of terahertz surface plasmon polaritons impinging on a finite array of subwavelength grooves on semiconductor surfaces," Phys. Rev. B, **73**, 205410(8), 2006.

[89] A. A. Govyadinov and V. A. Podolskiy, "Gain-assisted slow to superluminal group velocity manipulation in nanowaveguides," Phys. Rev. Lett., **97**, 223902(4), 2006.

[90] R. P. van Duyne, "Molecular plasmonics," Science, **306**, 985–986, 2004.

[91] P. Andrew and W. L. Barnes, "Energy transfer across a metal film mediated by surface plasmon polaritons," Science, **306**, 1002–1005, 2004.

[92] K. L. Kliewer and R. Fuchs, "Collective electronic motion in a metallic slab," Phys. Rev., **153**, 498–153, 1967.

[93] E. N. Economou. "Surface plasmons in thin films," Phys. Rev., **182**, 539–554, 1969.

[94] D. Sarid. "Long-range surface-plasma waves on very thin metal films," Phys. Rev. Lett., **47**, 1927–1930, 1981.

[95] J. J. Burke, G. I. Stegeman, and T. Tamir, "Surface-polariton-like waves guided by thin, lossy metal films," Phys. Rev. B, **33**, 5186–5201, 1986.

[96] F. Yang, J. R. Sambles, and G. W. Bradberry, "Long-range surface modes supported by thin films," Phys. Rev. B, **44**, 5855–5872, 1991.

[97] C. Jung, S. Yee, and K. Kuhn, "Integrated optics waveguide modulator based on surface plasmon resonance," J. Light. Technol., **12**, 1802–1806, 1994.

[98] N. A. Janunts and Kh. V. Nerkararyan, "Modulation of light radiation during input into waveguide by resonance excitation of surface plasmons," Appl. Phys. Lett., **79**, 299–301, 2001.

[99] J. S. Schildkraut, "Long-range surface plasmon electrooptic modulator," Appl. Opt., **27**, 4587–4590, 1988.

[100] K. Welford, "Surface plasmon-polaritons and their uses," Optical and Quantum Electron., **23**, 1-27, 1991.

[101] J. Johnstone, G. Stewart, T. Hart, and B. Culshaw, "Surface plasmon polaritons in thin metal films and their role in fiber optic polarizing devices," J. Lightw. Technol., **8**, 538–543, 1990.

[102] J. Tominaga *et al.* "Local plasmon photonic transistor," Appl. Phys. Lett., **78**, 2417–2419, 2001.

[103] A. Tredicucci, C. Gmachl, F. Capasso, A. L. Hutchinson, D. L. Sivco, and A. Y. Cho, "Single-mode surface-plasmon laser," Appl. Phys. Lett., **76**, 2164–2166, 2000.

[104] D. J. Bergman and M. I. Stockman, "Can we make a nanoscopic laser?" Laser Phys., **14**, 409–411, 2004.

[105] J. Lambe and S. L. McCarthy, "Light emission from inelastic electron tunneling," Phys. Rev. Lett., **37**, 923–925, 1976.

[106] S. I. Bozhevolnyi, J. Erland, K. Leosson, P. M. W. Skovgaard, and J. M. Hvam, "Waveguiding in surface plasmon polariton band gap structures," Phys. Rev. Lett., **86**, 3008–3011, 2001.

[107] M. Quinten, A. Leitner, J. R. Krenn, and F. R. Aussenegg, "Electromagnetic energy transport via linear chains of silver nanoparticles," Opt. Lett., **23**, 1331–1333, 1998.

[108] I. V. Novikov and A. A. Maradudin, "Channel polaritons," Phys. Rev. B, **66**, 035403(13), 2002.

[109] J. P. Kottmann and O. J. F. Martin, "Plasmon resonant coupling in metallic nanowires," Opt. Exp., **8**, 655–663, 2001.

[110] P. Berini, "Plasmon-polariton modes guided by a metal film of finite width," Opt. Lett., **24**, 1011–1013, 1999.

[111] P. Berini, "Plasmon-polariton waves guided by thin lossy metal films of finite width: bound modes of symmetric structures," Phys. Rev. B, **61**, 10484–10503, 2000.

[112] R. Charbonneau, P. Berini, E. Berolo, and E. Lisicka-Shrzek, "Experimental observation of plasmon-polariton waves supported by a thin metal film of finite width," Opt. Lett., **25**, 844–846, 2000.

[113] C. Girard, "Near fields in nanostructures," Rep. Prog. Phys., **68**, 1883–1933, 2005.

[114] C. Girard and E. Dujardin, "Near-field optical properties of top-down and bottom-up nanostructures," J. Opt. A: Pure Appl. Opt., **8**, S73–S86, 2006.

[115] S. A. Maier and H. A. Atwater, "Plasmonics: Localization and guiding of electromagnetic energy in metal/dielectric structures," J. Appl. Phys., **98**, 011101(10), 2005.

[116] W. L. Barnes, A. Dereux, and T. W. Ebbesen, "Surface plasmon subwavelength optics," Nature, **424**, 824–830, 2003.

[117] T. W. Ebbesen, H. J. Lezec, H. F. Ghaemi, T. Thio, and P. A. Wolff, "Extraordinary optical transmission through sub-wavelength hole arrays", Nature, vol. 391, p. 667–9, 1998.

[118] H. A. Bethe, "Theory of diffraction by small holes," Phys. Rev., **66**, 163–182, 1944.

[119] L. Martin-Moreno, F. J. Garcia-Vidal, H. J. Lezec, K. M. Pellerin, T. Thio, J. B. Pendry, and T. W. Ebbesen, "Theory of extraordinary optical transmission through subwavelength hole arrays," Phys. Rev. Lett., **86**, 1114–1117, 2001.

[120] K. L. van der Molen, K. J. K. Koerkamp, S. Enoch, F. B. Segerink. N. F. van Hulst, L. Kuipers, "Role of shape and localized resonance in extraordinary transmission through periodic arrays of subwavelength holes: experiment and theory," Phys. Rev. B, **72**, 045421(9), 2005.

[121] W. L. Barnes, W. A. Murray, J. Dintinger, E. Devaux, and T. W. Ebbesen, "Surface plasmon polaritons and their role in the enhanced transmission of light through periodic arrays of subweavelength hiles in a metal film," Phys. Rev. Lett., **92**, 107401(4), 2004.

[122] E. Altewischer, M. P. van Exter, and J. P. Woerdman, "Plasmon-assisted transmission of entangled photons," Nature, **418**, 304–306, 2002.

[123] H. J. Lezec, A. Degiron, E. Devaux, R. A. Linke, L. Martin-Moreno, F. J. Garcia-Vidal, and T. W. Ebbesen, "Beaming light from a subwavelength aperture," Science, **297**, 820–822, 2002.

[124] K. J. Webb and J. Li, "Analysis of transmission through small apertures in conducting films," Phys. Rev. B, **73**, 033401(4), 2006.

[125] Z. Ruan and M. Qiu, "Enhanced transmission through periodic arrays of subwavelength holes: the role of localized waveguide resonance," Phys. Rev. Lett., **96**, 233901(4), 2006.

[126] L. Salomon, F. Grillot, A. V. Zayats, F. de Fornel, "Near-field distribution of optical transmission of periodic subwavelength holes in a metal film," Phys. Rev. Lett., **86**, 1110–1113, 2001.

[127] J. B. Pendry, L. Martin-Moreno, and F. J. Garcia-Vidal, "Mimicking surface plasmons with structured surfaces," Science, **305**, 847–848, 2004.

[128] P. Lalanne, C. Sauvan, J. P. Hugonin, J. C. Rodier, and P. Chavel, "Perturbative approach for surface plasmon effects on flat interfaces periodically corrugated by subwavelength apertures," Phys. Rev. B, **68**, 125404(11), 2003.

[129] E. Popov, M. Neviere, S. Enoch, and R. Reinisch, "Theory of light transmission through subwavelength periodic hole arrays," Phys. Rev. B, **62**, 16100–16108, 2000.

[130] S. A. Darmanyan and A. V. Zayats, "Light tunnelling via resonant surface plasmon polariton states and the enhanced transmission of periodically nanostructures metal films: an analytical study," Phys. Rev. B, **67**, 035424(7), 2003.

[131] K. J. Webb and J. Li, "Analysis of transmission through small apertures in conducting films," Phys. Rev. B, **73**, 033401(4), 2006.

[132] F. I. Baida and D. V. Labeke, "Three-dimensional structures for enhanced transmission through a metallic film: annular aperture arrays," Phys. Rev. B, **67**, 155314(7), 2003.

[133] F. J. G. Abajo, G. G.Santos, L. A. Blanco, A. G. Borisov, and S. V. Shabanov, "Tunneling mechanism of light transmission through metal films," Phys. Rev. Lett., **95**, 067403(4), 2005.

[134] http://www.thebritishmuseum.ac.uk/science/lycurguscup/sr-lycugus-p1.html

[135] L. J. Sherry, R. Jin, C. A. Mirkin, G. C. Schatz, and R. P. van Duyne, "Localized surface plasmon resonance spectroscopy of single silver triangular nanoprisms," Nano Lett., **6**, 2060–2065, 2006.

[136] L. J. Sherry, S. Chang, G. C. Schatz, R. P. Van Duyne, B. J. Wiley, and Y. Xia, "Localized surface plasmon resonance spectroscopy of single silver nanocubes," Nano Lett., **5**, 2034–2038, 2005.

[137] M. Maillard, P. Huang, and L. Brus, "Silver nanodisk growth by surface plasmon enhanced photoreduction of adsorbed Ag^{+}," Nano Lett., **3**, 1611–1615, 2003.

[138] C. Sonnichsen and A. P. Alivisatos, "Gold nanorods as novel nonbleaching plasmon-based orientation sensors for polarized single-particle microscopy," Nano Lett., **5**, 301–304, 2004.

[139] J. J. Mock, S. J. Oldenburg, D. R. Smith, D. A. Schultz, and S. Schultz, "Composite plasmon resonant nanowires," Nano Lett., **2**, 465–469, 2002.

[140] Y. Sun, B. T. Mayers, and Y. Xia, "Template-engaged replacement reaction: a one-step approach to the large-scale synthesis of metal nanostructures with hollow interiors," Nano Lett., **2**, 481–485, 2002.

[141] H. S. Zhou, I. Honma, H. Komiyama, and J. W. Haus, "Controlled synthesis and quantum-size effect in gold-coated nanoparticles," Phys. Rev. B, **50**, 12052–12056, 1994.

[142] L. R. Hirsch, A. M. Gobin, A. R. Lowery, F. Tam, R. A. Drezek, N. J. Halas, and J. L. West, "Metal nanoshells," An. of Biomed. Eng., **34**, 15–22, 2006.

[143] K. Hasegawa, C. Rohde, and M. Deutsch, "Enhanced surface-plasmon resonance absorption in metal-dielectric-metal layered microspheres," Opt. Lett., **31**, 1136–1138, 2006.

[144] S. J. Oldenburg, S. L. Westcott, R. D. Averitt, and N. J. Halas, "Surface enhanced Raman scattering in the near infrared using metal nanoshell substrates," J. Chem. Phys., **111**, 4729–4735, 1999.

[145] K. Zhao, H. Xu, B. Gu, and Z. Zhang, "One-dimensional arrays of nanoshell dimers for single molecule spectroscopy via surface-enhanced Raman scattering," J. Chem. Phys., **125**, 081102(6), 2006.

[146] C. A. Rohde, K. Hasegawa, and M. Deutsch, "Coherent light scattering from semi-continuous silver nanoshells near the percolation threshold," Phys. Rev. Lett., **96**, 045503(4), 2006.

[147] R. D. Averitt, S. L. Westcott, and N. J. Halas, "Linear optical properties of gold nanoshells," J. Opt. Soc. Am. B, **16**, 1824–1832, 1999.

[148] Y. Fu, Y. Zeng, and H. Ågren, "Surface-plasmon-assisted electromagnetic field enhancement in semiconductor quantum dots," Appl. Phys. A, **87**, 167–169, 2007.

[149] E. Prodan, C. Radloff, N. J. Halas, and P. Nordlander, "A hybridization model for the plasmon response of complex nanostructures," Science, **302**, 419–422, 2003.

[150] S. Nie and S. R. Emory, "Probing single molecules and single nanoparticles by surface-enhanced Raman scattering," Science, **275**, 1102–1106, 1997.

[151] C. Loo, A. Lowery, N. Halas, J. West, and R. Drezek, "Immunotargeted nanoshells for integrated cancer imaging and therapy," Nano Lett., **5**, 709–711, 2005.

[152] Y. Fu, T.-T. Han, Y. Luo, and H. Ågren, "Multi-photon excitation of quantum dots by ultra-short and ultra-intense laser pulse," Appl. Phys. Lett., **88**, 221114(3), 2006.

[153] A. Bouhelier, M. R. Beversluis, and L. Novotny, "Characterization of nanoplasmonic structures by locally excited photoluminescence," Appl. Phys. Lett., **83**, 5041(3), 2003.

[154] J. Lee, A. O. Govorov, J. Dulka, and N. A. Kotov, "Bioconjugates of CdTe nanowires and Au nanoparticles: Plasmon-exciton interactions, luminescence enhancement, and collective effects," Nano Lett., **4**, 2323–2330, 2004.

[155] A. O. Govorov, G. W. Bryant, W. Zhang, T. Skeini, J. Lee, N. A. Kotov, J. M. Slocik, and R. R. Naik, "Exciton-plasmon interaction and hybrid excitons in semiconductor-metal nanoparticle assemblies," Nano Lett., **6**, 984–994, 2006.

[156] W. Zhang, A. O. Govorov, and G. W. Bryant, "Semiconductor-metal nanoparticle molecules: hybrid excitons and the nonlinear Fano effect," Phys. Rev. Lett., **97**, 146804(4), 2006.

[157] W. L. Barnes, A. Dereux, and T. W. Ebbesen, "Surface plasmon subwavelength optics," Nature, **424**, 824–830, 2003.

[158] D. A. Schultz, "Plasmon resonant particles for biological detection," Curr. Opin. Biotechnol., **14**, 13-22, 2003.

[159] A. J. Haes, W. P. Hall, L. Chang, W. L. Klein, and R. P. van Duyne, "A localized surface plasmon resonance biosensor: First steps toward an assay for Alzheimer's disease," Nano Lett., **4**, 1029–1034, 2004.

[160] D. M. Schaadt, B. Feng, and E. T. Yub, "Enhanced semiconductor optical absorption via surface plasmon excitation in metal nanoparticles," Appl. Phys. Lett., **86**, 063106(3), 2005.

[161] K. J. Webb and M. Yang, "Subwavelength imaging with a multilayer silver film structure," Opt. Lett., **31**, 2130–2132, 2006

[162] S. I. Bozhevolnyi, V. S. Volkov, K. Leosson and A. Boltasseva, "Bend loss in surface plasmon polariton band-gap structures," Appl. Phys. Lett., **79**, 1076(3), 2001.

Index

Color Insert

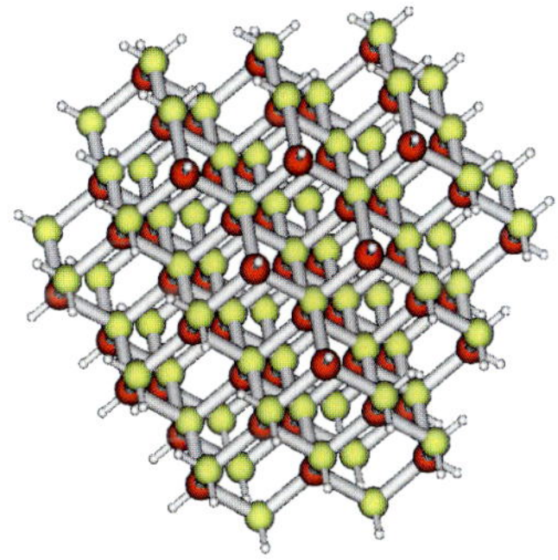

Fig. 3.1 Lattice structures of a GaAs nano cluster.

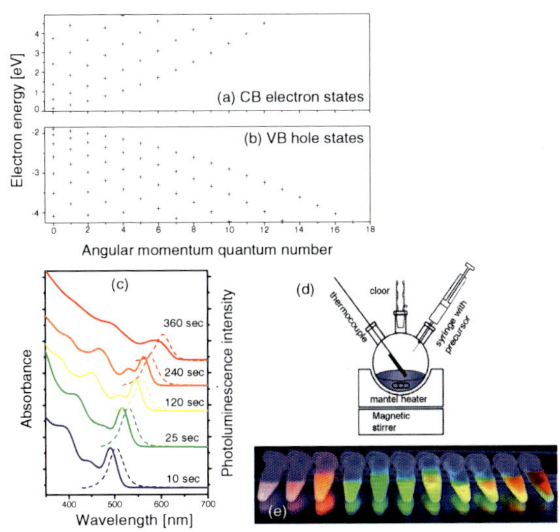

Fig. 3.4 (a) Conduction-band (CB) electron states and (b) valence-band (VB) hole states confined in a spherical CdSe QD with a radius of 4 nm as functions of angular momentum quantum number ℓ. (c) Absorbance (solid lines) and fluorescence (dashed lines), (d) basic chemical synthesis, and (e) fluorescence under UV radiation.

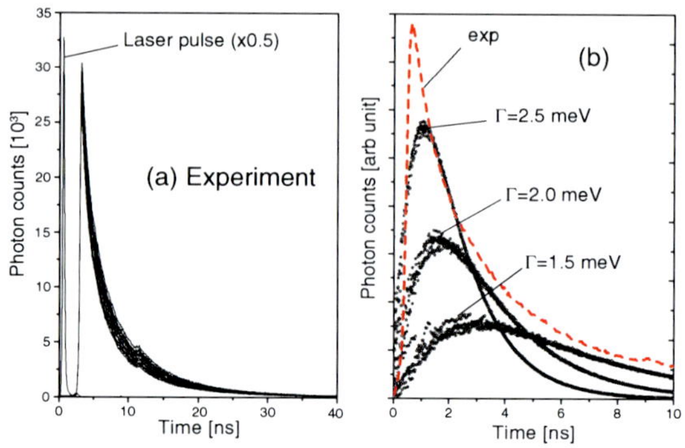

Fig. 3.7 (a) Ten repetitive measurement results of the temporal development of luminescence from II-VI CdSe core ZnS/CdS multi-shell QDs. The small bump at about 12 ns in the fluorescence decay was caused by a ringing in the laser pulse. (c) Monte Carlo simulations of spontaneous luminescence of an assembly of 5 million QDs for three different electron-phonon interactions Γ. Y. Fu, H. Ågren, J. M. Kowalewski, H. Brismar, J. Wu, Y. Yue, N. Dai, and L. Thylén, Radiative and nonradiative recombination of photoexcited excitons in multi-shell-coated CdSe/CdS/ZnS quantum dots, EuroPhysics Lett. vol. 86, p. 37003(6), 2009. Copyright (2009) by the European Physical Society.

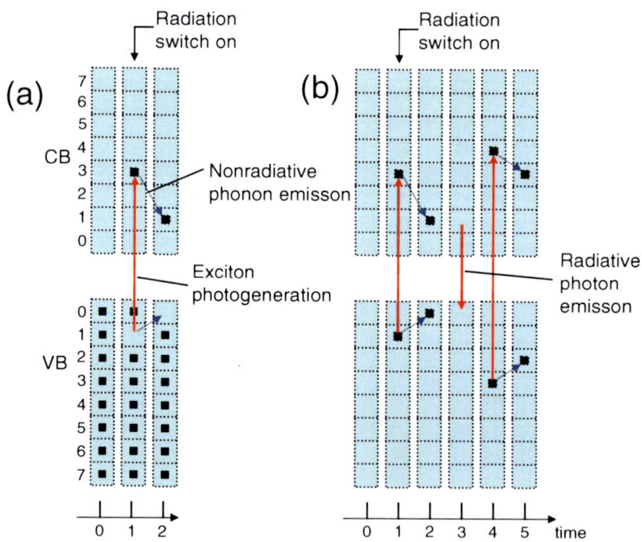

Fig. 3.8 (a) Schematic diagram of transition processes. (a) Electron picture. Black dots are electrons. (b) Electron-hole picture. Black dots in the conduction band (CB) are electrons. They are holes in the valence band (VB). T.-T. Han, Y. Fu, and H. Ågren, Dynamic photon emission from multiphoton-excited semiconductor quantum dot, J. Appl. Phys. vol. 103, p. 93703(6), 2008. Copyright (2008) by The American Institute of Physics.

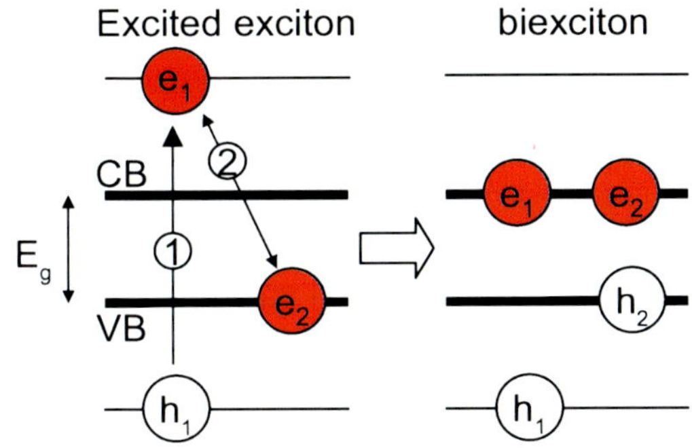

Fig. 3.14 Schematic depiction of impact ionization of a high-energy electron-hole pair. (1) Electron-hole pair (e_1 and h_1) is photogenerated. (2) e_2 gets excited from a valence-band state to a conduction-band state via Coulomb interaction with e_1, leaving hole h_2 behind. CB/VB=conduction/valence band. The reverse process is referred to be Auger recombination. Y. Fu, Y.-H. Zhou, H. Su, F. Y. C. Boey, and H. Ågren, Impact ionization and Auger recombination rates in semiconductor quantum dots, J. Physical Chemistry C, vol. 114, p. 3743–7, 2010. Copyright (2010) by American Chemical Society.

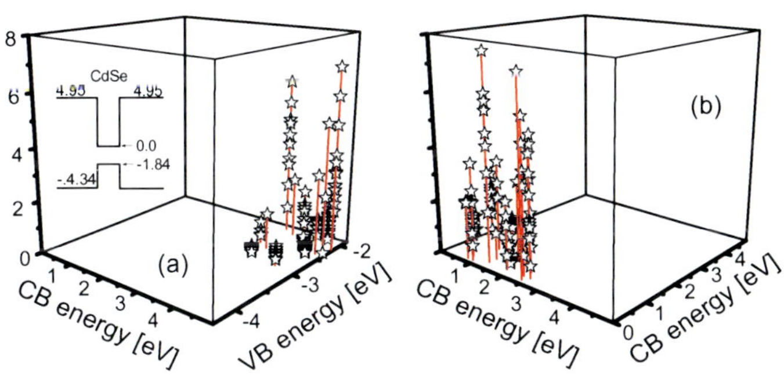

Fig. 3.15 (a) Impact ionization energy between initial CB state(s) and initial VB state(s). Inset shows the band structure of the CdSe QD in vacuum. (b) Auger recombination. The QD radius $a = 4$ nm. Y. Fu, Y.-H. Zhou, H. Su, F. Y. C. Boey, and H. Ågren, Impact ionization and Auger recombination rates in semiconductor quantum dots, J. Physical Chemistry C, vol. 114, p. 3743–7, 2010. Copyright (2010) by American Chemical Society.

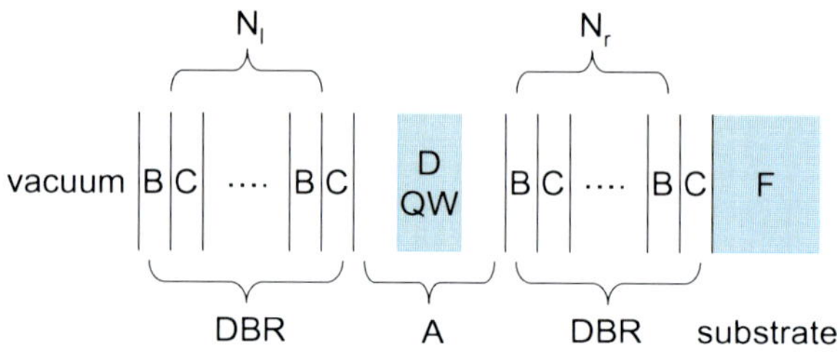

Fig. 4.1 Schematic structure of the QW cavity.

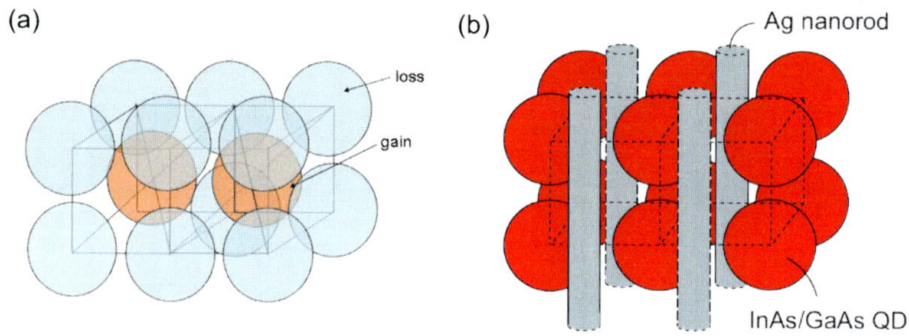

Fig. 4.10 (a) A periodic binary structure comprising gain and lossy QDs (pumped and lossy QDs, respectively) in a primitive cubic lattice. (b) A primitive cubic QD lattice interspersed with silver nanorods.

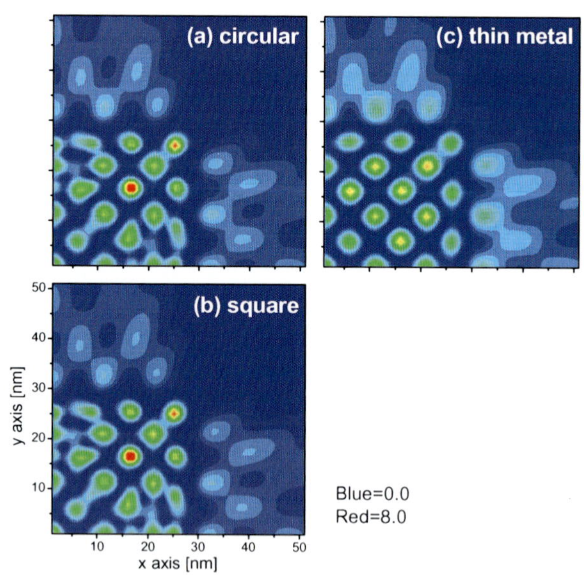

Fig. 5.18 Intensity distributions of the optical field for $1/k = 10\,\mu m$ with (a) circular diffraction apertures; (b) square diffraction apertures, and (c) thin metal reflection film ($h = 0$).

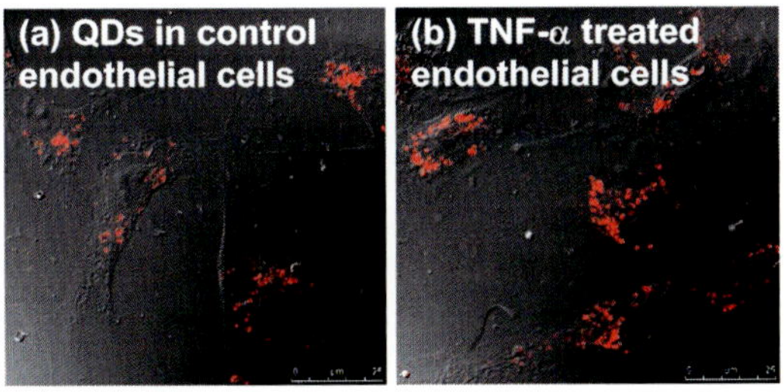

Fig. 5.22 Fluorescence of CdSe QDs (a) in control endothelial cells, and (b) in TNF-α-treated endothelial cells.

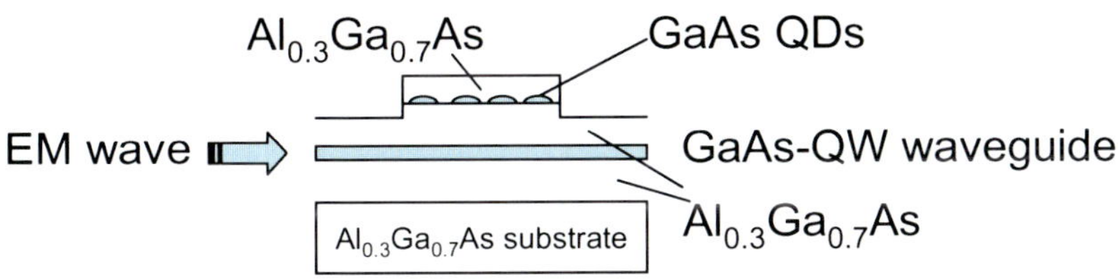

Fig. 5.24 Propagation of exciton-polariton with coupling to a QD structure generated by an incident electromagnetic (EM) wave.

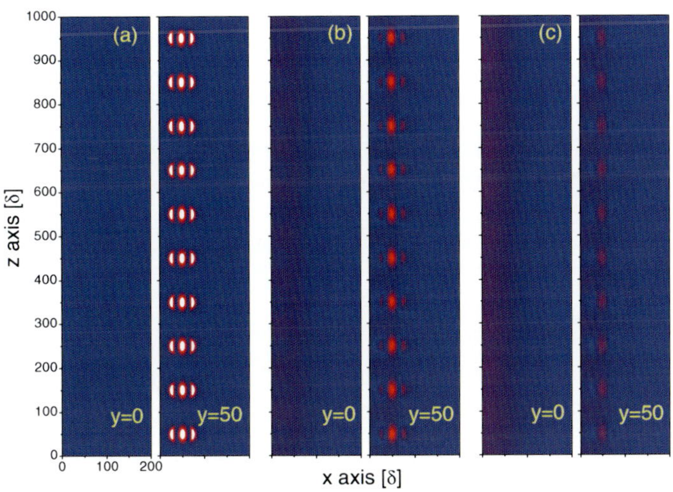

Fig. 5.27 Spatial distributions of the E_x-field amplitude. (a) $\omega = \omega_a + 0.1\omega_{\rm LT}$. (b) $\omega = \omega_a + 0.28\omega_{\rm LT}$. (c) $\omega = \omega_a + 0.5\omega_{\rm LT}$. $\hbar\omega = 1.0\,\rm eV$, $\delta = 2.2\,\rm nm$. GaAs material parameters are used for the numerical simulation, which corresponds to GaAs QDs embedded in AlGaAs barriers.

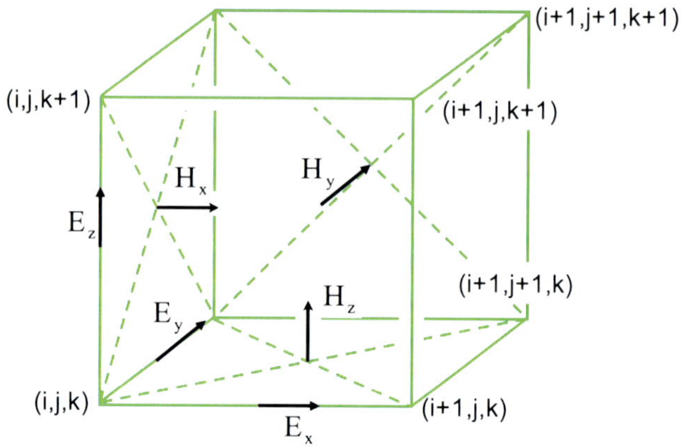

Fig. 6.3 Yee space lattice. $\boldsymbol{E}$-components are in the middle of the edges and the $\boldsymbol{H}$-components are in the center of the faces, in a rectangular coordinate system.

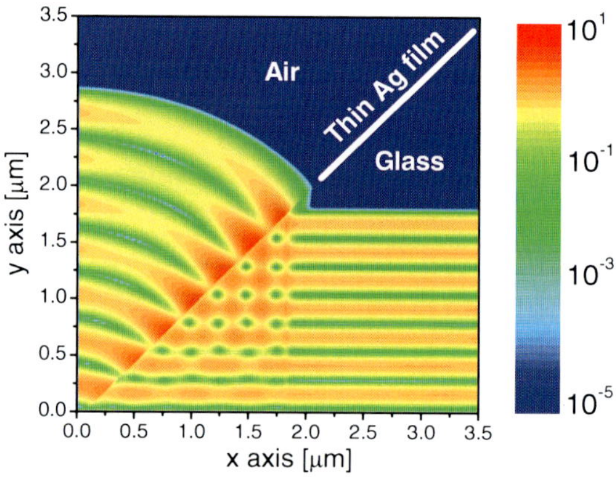

Fig. 6.4 Excitation of the SPP in the Kretschmann configuration of Fig. 6.1a. The vacuum wavelength of the incident EM plane wave is 800 nm. The plot shows the spatial distribution of the amplitude of the electric field at a time of 9.63 fs after the injection of the electromagnetic plane wave at $y = 0$ along the y direction.

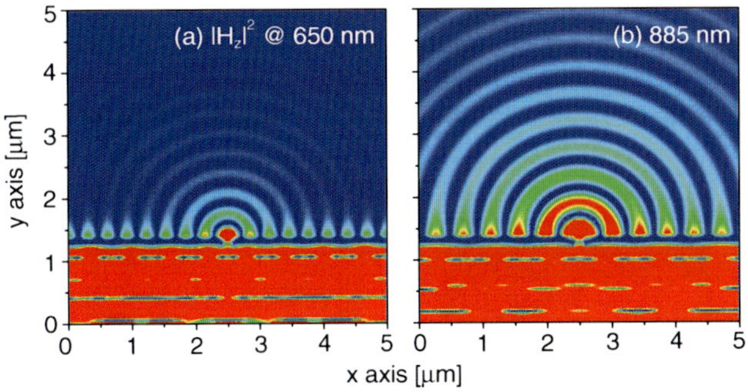

Fig. 7.1 A TM polarized plane wave, (a) $\lambda = 650$ nm, (b) $\lambda = 885$ nm, is incident to the slit (width 80 nm) on a metal film of 200 nm thickness from the bottom. A SPP is generated at the bottom of the dark-blue metal. It travels through the slit and reconverts into a radial light wave after the slit.[7]

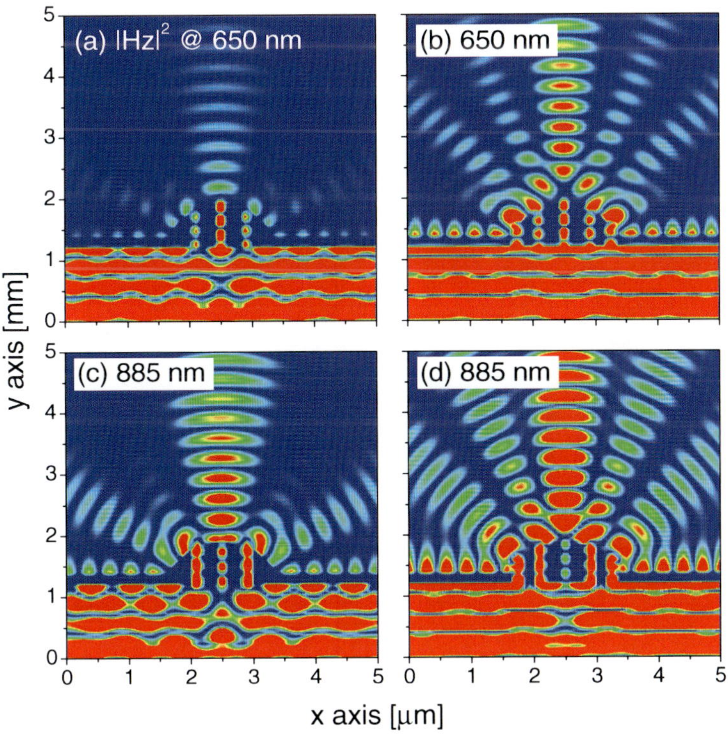

Fig. 7.2 A 2 μm wide convex bump and a planar wave ($\lambda = 650$ and 885 nm, respectively) hitting the metal film from the bottom (a,c) 80 nm-wide slits with 400 nm slit spacing and 700, 750, and 700 nm depth. (b,d) 80 nm-wide slits with 400 nm slit spacing and 450, 700, 750, 700, and 450 nm depth.[7]

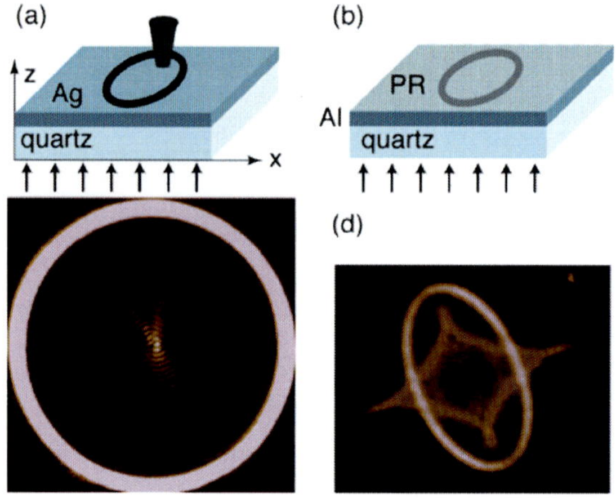

Fig. 7.3 Experimental setup for (a) near-field scanning microscopy (NSOM) and (b) plasmonic lithography measurements for recording the near-field pattern for plasmonic lenses. (c) Near-field pattern for a 14 μm-diameter circle cut into a 150 nm-thick silver film recorded with NSOM. Polarization of incident light is indicated with an arrow. (d) Near-field pattern for an ellipse with a long axis of approximately 4 μm and a short axis of 2.5 μm cut into a 70 nm-thick aluminum film recorded with plasmonic lithography. Z. Liu, J. M. Steele, W. Srituravanich, Y. Pikus, C. Sun, and X. Zhang, Focusing surface plasmons with a plasmonic lens, Nano Lett. vol. 5, p. 1726–9, 2005. Copyright (2005) by American Chemical Society.

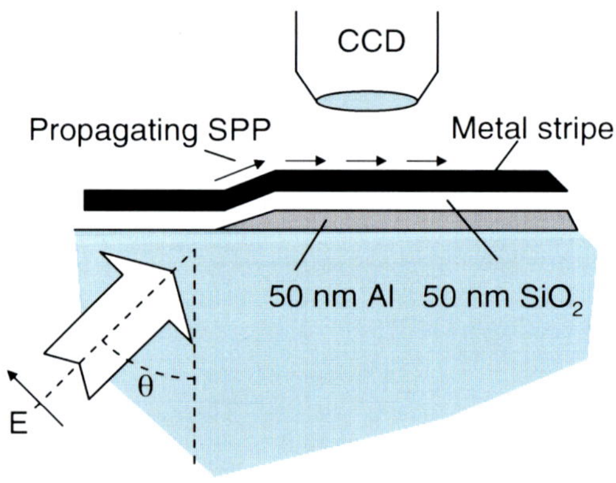

Fig. 7.5 Spatially confined SP excitation in the prism-coupling arrangement is provided by an opaque Al screen. Scattered light from the propagating SP is observed with a CCD camera.

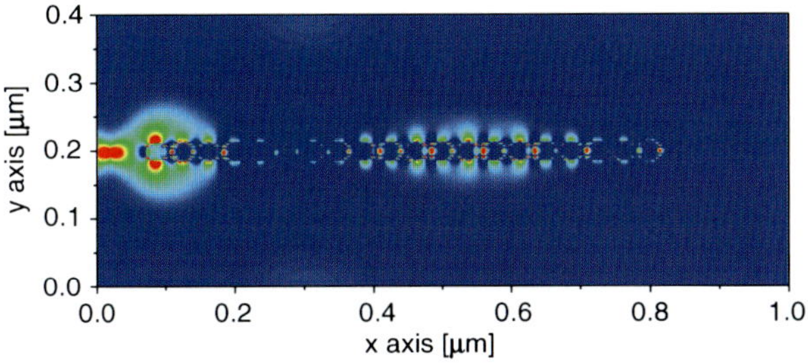

Fig. 7.7 $|H_y|$ for a 20 Ag nanoparticle chain waveguide. Excitation is at the nanoparticle on the left.

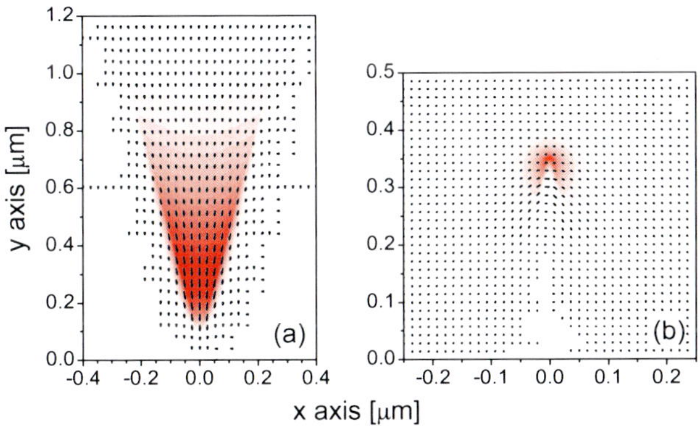

Fig. 7.10 Fundamental mode guided by two types of corner waveguides. (a) V-channel waveguide; (b) Λ-wedge waveguide. Color map is for $|\boldsymbol{H}_t|^2$, and quiver is for $\boldsymbol{H}_t$. Wavelength is at 0.633 μm. Corner angle is 30°. Corner tip is rounded with an arc with 10 nm in radius.

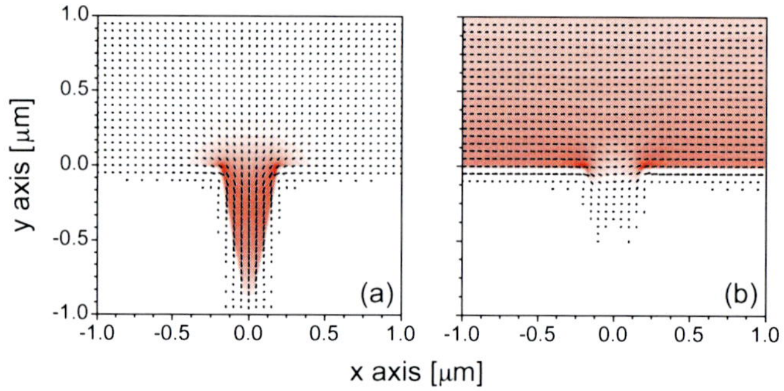

Fig. 7.15 $|\boldsymbol{H}_t|^2$ (colormap) and $\boldsymbol{H}_t$ (quiver) plots of two supermodes guided by a realistic V-channel waveguide at $\lambda = 1.4$ μm. (a) Mode with PEC symmetry; (b) Mode with PMC symmetry.

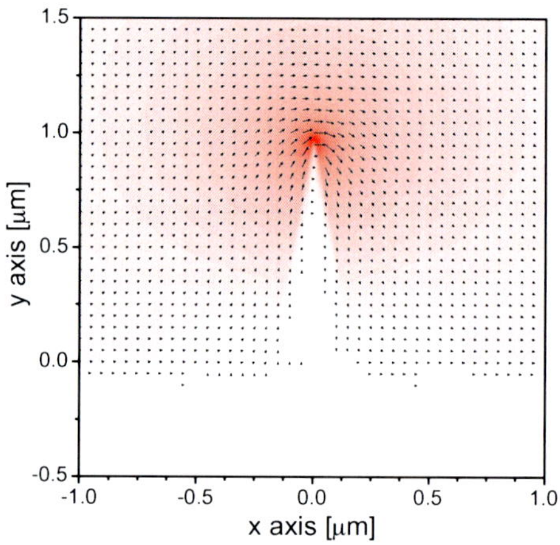

Fig. 7.17 $|\boldsymbol{H}_t|$ (colormap) and $\boldsymbol{H}_t$ (quiver) of the mode guided by a realistic Λ-wedge waveguide at $\lambda = 2$ μm. Axis unit: μm.

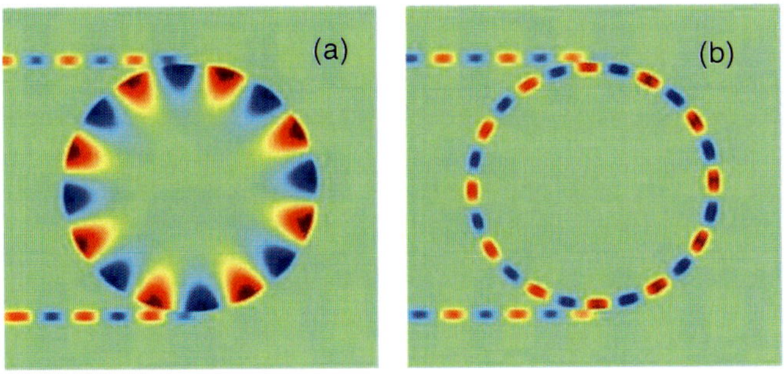

Fig. 7.20 Stead-state H_y field oscillation for a resonant frequency of (a) $f = 421.2\,\text{THz}$; (b) $f = 427.8\,\text{THz}$. Corresponding structures are shown in Fig. 7.20.

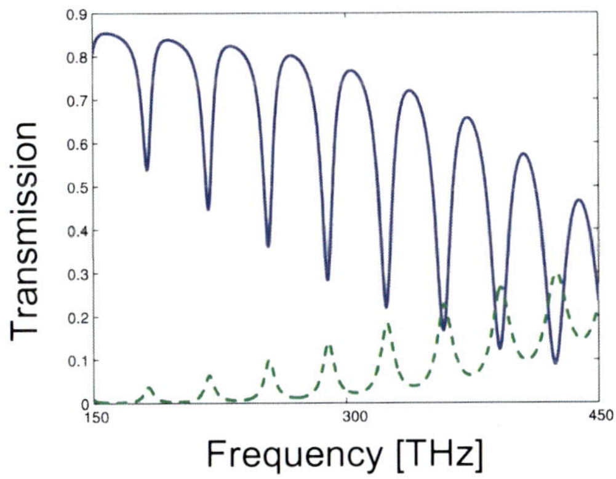

Fig. 7.21 Spectral responses of the channel drop filter (CDF) at the forward bus (solid line) and backward drop waveguide (dashed line) in a loss metal.

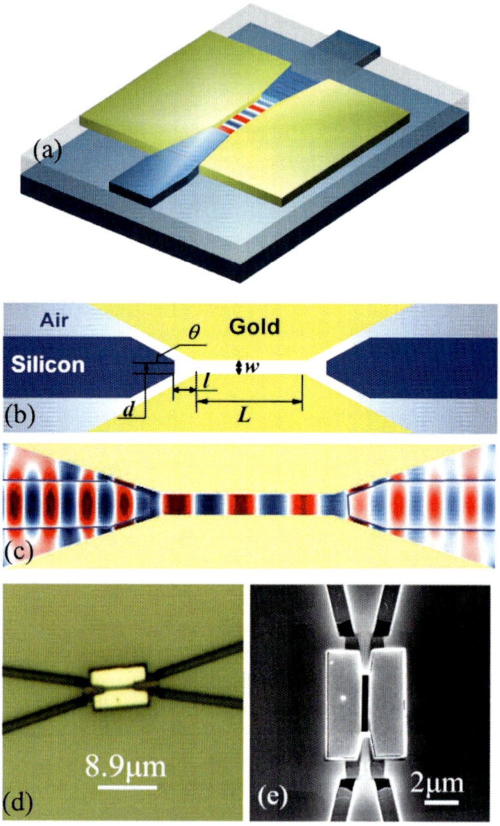

Fig. 7.22 Two tapered silicon strip waveguides coupled with a plasmonic gap waveguide. (a) The schematic picture of the silicon-metal plasmonic coupler. (b) The top-view with the taper parameters: θ is the angle between the taper and the center axis of silicon waveguide (the plasmonic taper edge is set to be parallel to the silicon taper), d is the width of the silicon taper end, l is the distance between the end of silicon waveguide and the start of the plasmonic waveguide. w and L are the width and length of the waveguide gap, respectively. (c) The H_z field distribution at the wavelength 1.5 μm simulated by 3D FDTD. (d) Optical microscope image of a fabricated silicon-gold plasmonic coupler ($\theta = 10^\circ$, $d = 200$ nm, $l = 0.25$ μm, $w = 200$ nm, $L = 3$ μm). (d) The scanning electronic micrograph of the fabricated taper. J. Tian, S. Q. Yu, W. Yan, and M. Qiu, Broadband high-efficiency surface-plasmon-polariton coupler with silicon-metal interface, Appl. Phys. Lett. vol. 95, p. 013504, 2009. Copyright (2009) by The American Institute of Physics.

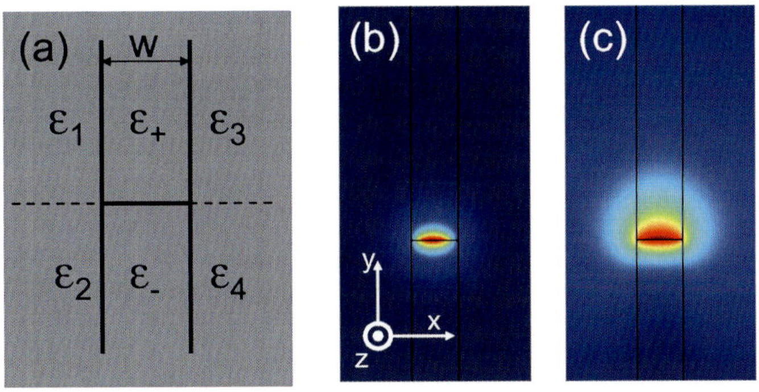

Fig. 7.25 (a) Schematic diagram of a general near-resonant SPP waveguide. (b) Mode field (H_x) supported by a sample near-resonant SPP waveguide ($\epsilon_+ = 2.1$, $\epsilon_- = -2.3$, $\epsilon_{1,2,3,4} = 1$, $w = 50\,\text{nm}$, $\lambda = 600\,\text{nm}$). (c) Field in the same waveguide but far from the singular condition ($\epsilon_- = -15$).

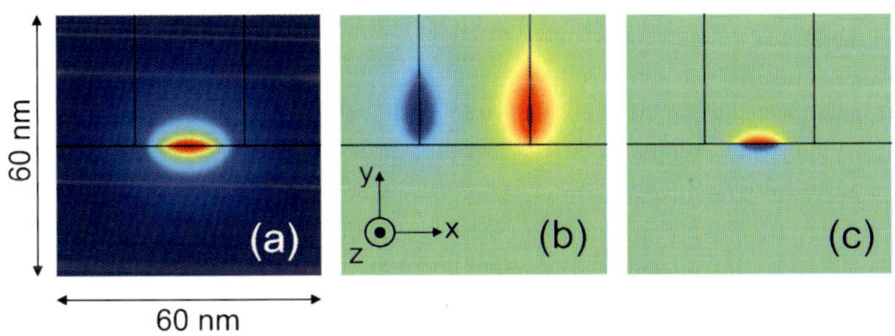

Fig. 7.27 Field plots of the SPP-based waveguide with a 25 nm-sized core. Three panels share the same color scale. (a) H_x field (min:0, max:1.27); (b) H_y field (min:-5.02e-2, max:5.02e-2); and (c) z-component Poynting vector S_z (min:-6.0e2, max:6.2e2). Axis unit: nm.

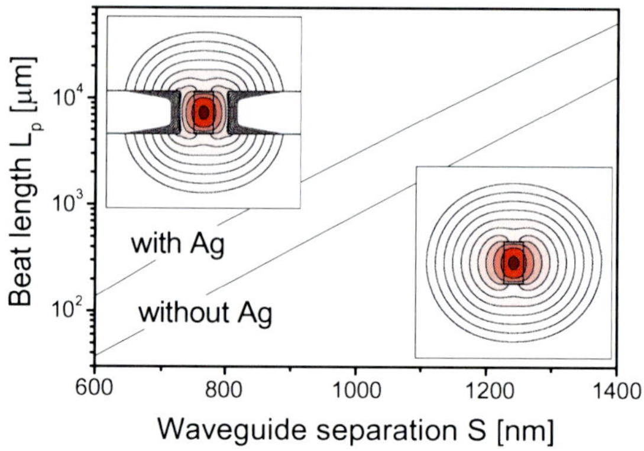

Fig. 7.34 Beat length variation as a function of waveguide separation for the quasi-TE mode at $\lambda = 1550\,\text{nm}$. The inset in the upper-left corner gives the $TE_{0,0}$ mode supported by an individual waveguide with metal cladding, while that in the bottom-right corner gives the same mode supported by a waveguide without metal cladding. The field contour lines are in 3 dB separation. The first contour line is an exception, which is at 90% of the maximum value. Domain size of the field plots: $2 \times 2\,\mu\text{m}^2$.

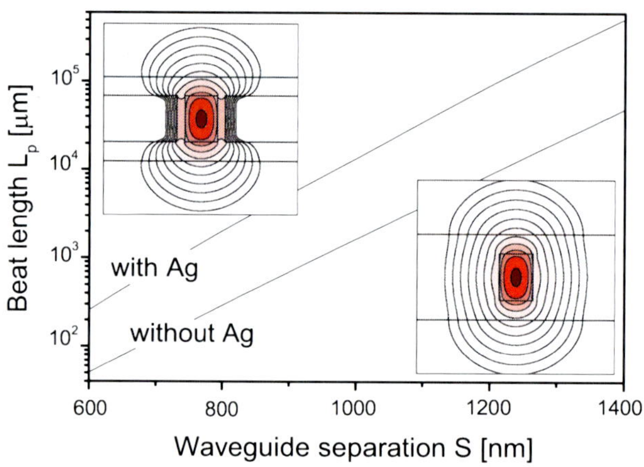

Fig. 7.36 Beat length variation as a function of waveguide separation for the quasi-TE mode at $\lambda = 1550\,\text{nm}$. The inset in the upper-left corner gives the $TM_{0,0}$ mode supported by an individual waveguide with metal cladding, while that in the bottom-right corner gives the mode supported by a waveguide without metal cladding. The field contour lines are in 3 dB separation. The first contour line is an exception, which is at 90% of the maximum value. Domain size of the field plots is $2\times2\,\mu\text{m}^2$.

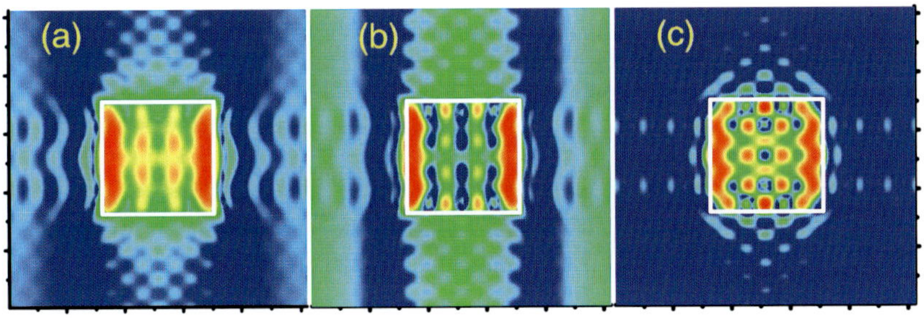

Fig. 7.42 EM field intensity distributions across a unit cell of the periodic hole array at a wavelength of (a) 802 nm at the transmission peak of $h = 600$ nm, (b) 767 nm when $h = 100$ nm, and (c) 853 nm when $h = 100$ nm. The observation plane is at a distance of 1.0 nm from the metal–air surface. The incident light is polarized along the horizontal axis of the images. The square white frames mark the spatial positions of the hole.

Fig. 7.43 The Lycurgus glass cup shows the bright red color of gold nanoparticles via transmitted light (the British museum).[134]

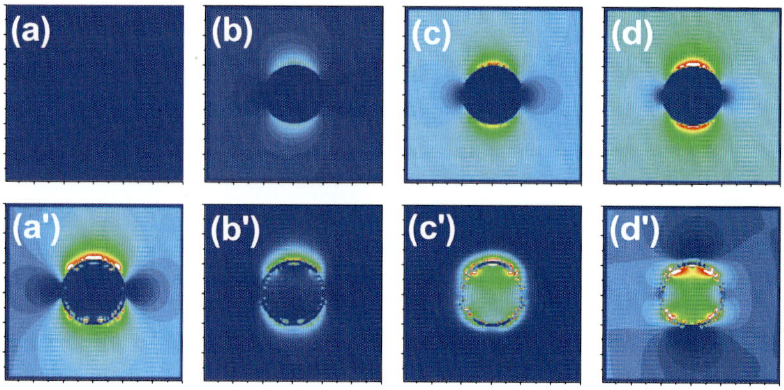

Fig. 7.47 Spatial distribution of the optical power around the CdS QD. The QD has an energy bandgap of 2.67 eV, and the photon energy of the two-photon excitation electromagnetic field is 1.34 eV. (a)+(b)+(c)+(d): bare CdS QD. (a')+(b')+(c')+(d'): 0.37 nm metal-coated CdS QD. The times are marked by arrows in Fig. 7.46.